# Utilization of By-Products and Treatment of Waste in the Food Industry

# ISEKI-FOOD SERIES

Series Editor: Kristberg Kristbergsson, *University of Iceland Reykjavík, Iceland*

---

Volume 1 FOOD SAFETY: A Practical and Case Study Approach
Edited by Anna McElhatton and Richard Marshall

Volume 2 ODORS IN THE FOOD INDUSTRY
Edited by Xavier Nicolay

Volume 3 UTILIZATION OF BY-PRODUCTS AND TREATMENT OF WASTE IN THE FOOD INDUSTRY
Edited by Vasso Oreopoulou and Winfried Russ

Volume 4 PREDICTIVE MODELING AND RISK ASSESSMENT
Edited by Rui Costa and Kristberg Kristbergsson

Volume 5 EXPERIMENTS IN UNIT OPERATIONS AND PROCESSING OF FOODS
Edited by Maria Margarida Cortez Vieira and Peter Ho

Volume 6 CASE STUDIES IN FOOD SAFETY AND ENVIRONMENTAL HEALTH
Edited by Maria Margarida Cortez Vieira and Peter Ho

# Utilization of By-Products and Treatment of Waste in the Food Industry

Edited by

Vasso Oreopoulou
*National Technical University of Athens*
*Athens, Greece*

Winfried Russ
*Technical University Munich*
*Freising, Germany*

Springer

Vasso Oreopoulou
National Technical University of Athens
School of Chemical Engineering
GR 15780 Athens
Greece
vasor@chemeng.ntua.gr

Winfried Russ
Technical University Munich
Center of Food and Life Sciences
D 85350 Freising
Germany
russ@wzw.tum.de

*Series Editor*

Kristberg Kristbergsson
Professor of Food Science
Dept. Food Science and Human Nutrition
University of Iceland
Hjardarhaga 2-6
107, Reykjavik
Iceland

Library of Congress Control Number: 2006928132

ISBN-10: 0-387-33511-0
ISBN-13: 978-0387-33511-7
eISBN-13: 978-0-387-35766-9



9 8 7 6 5 4 3 2 1

springer.com

# SERIES ACKNOWLEDGMENTS

ISEKI-Food is a thematic network on food studies, funded by the European Union as project N° 55792-CP-3-00-1-FR-ERASMUS-ETN. It is a part of the EU program in the field of higher education called ERASMUS, which is the higher education action of SOCRATES II program of the EU.

# SERIES PREFACE

The single most important task of food scientists and the food industry as a whole is to ensure the safety of foods supplied to consumers. Recent trends in global food production, distribution, and preparation call for increased emphasis on hygienic practices at all levels and for increased research in food safety in order to ensure a safer global food supply. The ISEKI-Food book series is a collection of books where various aspects of food safety and environmental issues are introduced and reviewed by scientists specializing in the field. In all of the books a special emphasis was placed on including case studies applicable to each specific topic. The books are intended for graduate students and senior-level undergraduate students as well as professionals and researchers interested in food safety and environmental issues applicable to food safety.

The idea and planning of the books originates from two working groups in the European thematic network ''ISEKI-Food'' is an acronym for ''Integrating Safety and Environmental Knowledge Into Food Studies.'' Participants in the ISEKI-Food network come from 29 countries in Europe and most of the institutes and universities involved with food science education at the university level are represented. Some international companies and nonteaching institutions also have participated in the program. The ISEKI-Food network is coordinated by Professor Cristina Silva at the Catholic University of Portugal, College of Biotechnology (Escola) in Porto. The program has a website at: http://www.esb.ucp.pt/iseki/. The main objectives of ISEKI-Food have been to improve the harmonization of studies in food science and engineering in Europe and to develop and adapt food science curricula, emphasizing the inclusion of safety and environmental topics. The ISEKI-Food network started on October 1st in 2002, and recently has been approved for funding by the EU for renewal as ISEKI-Food 2 for another three years. ISEKI has its roots in an EU-funded network formed in 1998 called ''Food Net'' where the emphasis was on casting light on the different food science programs available at the various universities and technical institutions throughout Europe. The work of the ISEKI-Food network was organized into five different working groups with specific task, all aiming to fulfill the main objectives of the network.

The first four volumes in the ISEKI-Food book series come from WG2 coordinated by Gerhard Schleining at Boku University in Austria and the undersigned. The main task of the WG2 was to develop and collect materials and methods for teaching safety and environmental topics in the food science and engineering curricula. The first volume is devoted to food safety in general with a practical and a case study approach. The book is composed of 14 chapters which were organized into three sections on preservation and protection, benefits and risk of microorganisms, and process safety. All these issues have received high public interest in recent years and will continue to be in the focus of consumers and regulatory personnel for years to come. The second volume in the series is devoted to the control of air pollution and treatment of odors in the food industry. The book is divided into eight chapters devoted to defining the problem, recent advances in analysis, and methods for prevention and treatment of odors. The topic should be of special interest to industry personnel and researchers due to recent and upcoming regulations by the European Union on air pollution from food processes. Other countries will likely follow suit with stricter regulations on the level of odors permitted to enter the environment from food processing operations. The third volume in the series is devoted to utilization and treatment of waste in the food industry. Emphasis is placed on sustainability of food sources and how waste can be turned into by-products rather than pollution or landfills. The book is composed of 15 chapters starting off with an introduction of the problems related to the treatment of waste and an introduction to the ISO 14001 standard used for improving and maintaining environmental management systems. The book then describes the treatment and utilization of both liquid and solid waste with case studies from many different food processes. The last book from WG2 is on predictive modeling and risk assessment in food products and processes. Mathematical modeling of heat and mass transfer as well as reaction kinetics is introduced. This is followed by a discussion of the stoichiometry of migration in food packaging, as well as the fate of antibiotics and environmental pollutants in the food chain using mathematical modeling and case study samples for clarification.

Volumes five and six come from work in WG5 coordinated by Margarida Vieira at the University of Algarve in Portugal and Roland Verhé at Gent University in Belgium. The main objective of the group was to collect and develop materials for teaching food safety-related topics at the laboratory and pilot plant level using practical experimentation. Volume five is a practical guide to experiments in unit operations and processing of foods. It is composed of 20 concise chapters each describing different food processing experiments outlining theory, equipment, procedures, applicable calculations, and questions for the students or trainee followed by references. The book is intended to be a practical guide for the teaching of food processing and engineering principles. The final volume in the ISEKI-Food book series is a collection of case studies in food safety and environmental health. It is intended to be a reference for introducing case studies into traditional lecture-based safety courses as well as being a basis for problem-based learning. The book consists of 13 chapters containing case studies that may be used, individually or in a series, to discuss a range of food safety issues. For convenience the book was divided into three main sections on microbial food safety, chemical residues and contaminants, and a final section on risk assessment and food legislation.

The ISEKI-Food books series draws on expertise from close to a 100 universities and research institutions all over Europe. It is the hope of the authors, editors, coordinators, and participants in the ISEKI network that the books will be useful to students and colleagues to further their understanding of food safety and environmental issues.

March, 2006 Kristberg Kristbergsson

# PREFACE

Food wastes are organic residues from the processing of agricultural raw materials to food, which arise as liquid (wastewater) and solid wastes. The wastewater results from the cleaning processes or in the form of excessive or polluted process water. Its dry material content is typically less than 5% by mass. It possibly also contains organic or inorganic cleaning agents or disinfectants. Solid food wastes with an organic origin have remarkably high water content (mostly about 80% by mass). They are usually characterized by a constant quality and purity due to the forgone processes.

The fact that these substances are removed from the production process as undesirable ingredients makes them, by definition of most European legislations, wastes. The term "by-product," which is common in industry, points up that these are mostly ulterior usable substances, often with a market value.

Waste disposal is one of the major problems facing most food processing plants. Agriculture as the traditional way of waste utilization—a consequential outcome because most raw materials are also from agricultural origin—is no longer available due to major changes in law and technology. Furthermore, new kinds of process engineering and resultant new products and markets make the utilization of waste increasingly interesting.

This book covers the main aspects of utilization of the food industry waste (defined thereby as by-product) and the treatments necessary to discard waste to environmental acceptors. It cannot cover the entire spectrum of utilization of solid and liquid wastes of the food industry. The multiplicity of possible utilizable ingredients and technologies alone would exceed such an undertaking. For this reason many utilization possibilities are briefly and exemplarily mentioned.

The first chapter shows the exigency for utilization of food wastes and gives an overview about ways of utilization. The next chapter introduces the main ideas on treatment of food waste according to the ISO 14001 standards and the EU directives concerning the environmental performance of the food industry. The following chapters cover processes for wastewater treatment in general and applications of treatment of specific wastewater from different branches. The technology of anaerobic fermentation, thereby used among others for biogas production, is described in Chapter 9 as a method for specific degradation of solid wastes. The energy generation with biogas production is economically interesting, but direct substantial utilization of food waste is also efficient. Different examples with direct practical applicability demonstrate this substantial utilization. The range of possible usable wastes is among fish, fruit, fats and spent grains (Chapters 10–13). They can be used for the production of food ingredients, e.g., polyphenols, protein isolates, and dietary fibers, but also for nonfood products like bricks or fuel. Despite the generally high water content incineration is technically feasible as exemplified by spent grains in Chapter 14. Finally, composting of agricultural and food waste is covered in Chapter 15.

Athens and Weihenstephan in February 2006
Vasso Oreopoulou
Winfried Russ

# CONTENTS

# CONTRIBUTORS

Anna Aladjadjiyan
Department of Mathematics and Physics, Agricultural University of Plovdiv 12, Mendeleev Str., 4000-Plovdiv, Bulgaria

Alexandros Aivasidis
Democritus University of Thrace, Department of Environmental Engineering, Vas. Sofias 12, GR-67100, Xanthi, Greece

Sigurjon Arason
Icelandic Fisheries Laboratories, Skulagotu 4, 101 Reykjavík, Iceland

Antonio G. Brito
University of Minho, School of Engineering – Center of Biological Engineering, Campus de Gualtar, 4710-057, Braga, Portugal

Rudolf Braun
BOKU University of Natural Resources and Applied Life Sciences, Institute for Environmental Biotechnology, 3430 Tulln, Austria

Cristina Costa
Unicer SA, Leça do Balio, Matosinhos, 4466-955 S. Mame de de Infesta, Portugal

Vasileios I. Diamantis
Democritus University of Thrace, Department of Environmental Engineering, Vas. Sofias 12, GR-67100, Xanthi, Greece

Martin Faulstich
ATZ Entwicklungszentrum, 92237 Sulzbach-Rosenberg, Germany

Vassilis Gekas
Department of Environmental Engineering, Technical University of Crete, Polytechnioupolis, GR-73100 Chania-Crete, Greece

Anselm Gleixner
INNOVAS Innovative und Umwelttecnik, Margot-Kalinke-Straße 9, D-80939 München, Germany

Werner Kepplinger
University of Leoben, A-8700 Leoben, Austria

Roland Kirchmayr
BOKU University of Natural Resources and Applied Life Sciences, Institute for Environmental Biotechnology, 3430 Tulln, Austria

Kristberg Kristbergsson
Department of Food Science and Human Nutrition, University of Iceland, Hjardarhaga 2-6, Reykjavík, Iceland

Harris N. Lazarides
Aristotle University of Thessaloniki, Department of Food Science and Technology, 54006, Thessaloniki, Greece

Dionissios Mantzavinos
Department of Environmental Engineering, Technical University of Crete, GR-73100 Chania, Greece

Martin Mayer
BOKU University of Natural Resources and Applied Life Sciences, Institute for Environmental Biotechnology, 3430 Tulln, Austria

Roland Meyer-Pittroff
Technische Universität München, Chair for Energy and Environmental Technologies of the Food Industry, Weihenstephaner Steig 22, 85350 Freising-Weihenstephan, Germany

Mladen Naydenov
Department of Microbiology and Environmental Biotechnologies, Agricultural University–Plovdiv, 12 Mendeleev Str., 4000-Plovdiv, Bulgaria

Maria Nikolopoulou
Department of Environmental Engineering, Laboratory of Transport Phenomena and Applied Thermodynamics, Campus, Technical University of Crete, Chania, GR 731 00 Greece

Regina Nogueira
University of Minho, School of Engineering – Center of Biological Engineering, Campus de Gualtar, 4710-057 Braga, Portugal.

José A. Oliveira
Adega Cooperativa de Ponte da Barca, Lugar de Agrelos, 4980-601 Ponte da Barca, Portugal

José M. Oliveira
University of Minho, School of Engineering – Center of Biological Engineering, Campus de Gualtar, 4710-057 Braga, Portugal

Vasso Oreopoulou
National Technical University of Athens, School of Chemical Engineering, Lab. of Food Chemistry and Technology, Athens, GR 15780, Greece

Christian Pecher
Flottweg GmbH & Co. KgaA, Industriestrasse 6–8, D-84137 Vilsbiburg, Germany

João Peixoto
University of Minho, School of Engineering – Center of Biological Engineering, Campus de Gualtar, 4710-057 Braga, Portugal

Gunther Pesta
Technische Universität München, Chair for Energy and Environmental Technologies of the Food Industry, Weihenstephaner Steig 22, 85350 Freising-Weihenstephan, Germany

Konstantinos B. Petrotos
Aristotle University of Thessaloniki, Department of Food Science and Technology, 54006, Thessaloniki, Greece

Stephan Prechtl
ATZ Entwicklungszentrum, 92237 Sulzbach-Rosenberg, Germany

Christoph Resch
BOKU University of Natural Resources and Applied Life Sciences, Institute for Environmental Biotechnology, 3430 Tulln, Greece

Ana Rodrigues
University of Minho, School of Engineering – Center of Biological Engineering, Campus de Gualtar, 4710-057 Braga, Portugal

Winfried Russ
Technische Universität München, Chair for Energy and Environmental Technologies of the Food Industry, Weihenstephaner Steig 22, 85350 Freising-Weihenstephan, Germany

Markus Schnappinger
Technische Universität München, Chair for Energy and Environmental Technologies of the Food Industry, Weihenstephaner Steig 22, 85350 Freising-Weihanstephan, Germany

Stefan Shilev
Department of Microbiology and Environmental Biotechnologies, Agricultural University – Plovdiv, 12 Mendeleev Str., 4000-Plovdiv, Bulgaria

Evangelia Tsagaraki
Aristotle University of Thessaloniki, Department of Food Science and Technology, 54006, Thessaloniki, Greece

Constantina Tzia
National Technical University of Athens, School of Chemical Engineering, Lab. of Food Chemistry and Technology, Athens, GR 15780, Greece

Eleni Vaiopoulou
Democritus University of Thrace, Department of Environmental Engineering, Vas. Sofias 12, GR-67100, Xanthi, Greece

Ventsislava Vancheva
Department of Microbiology and Environmental Biotechnologies, Agricultural University – Plovdiv, 12 Mendeleev Str., 4000-Plovdiv, Bulgaria

Johann Wimmer
Federal Government of Upper Austria, 4012 Linz, Austria

Gerald Zanker
Brau Union AG, A-8055 Graz, Austria

# 1

# Waste Related to the Food Industry: A Challenge in Material Loops

Winfried Russ and Markus Schnappinger

## 1.1. FOOD WASTE IN THE EC: STATUS, DEFINITION AND HANDLING

According to the data in the Eurostat database of the European Commission (EC) the waste generated in 2002 from the manufacture of food products, beverages, and tobacco products was on average about 154 kg per habitant and year. The database does not contain information about this kind of waste accumulation for all countries and years. There are no complete data obtainable for countries like Belgium, Denmark, Germany, or Luxembourg. The available data ranged from 22 kg to more than 600 kg per habitant and year depending on country (see Figure 1.1). This extremely wide range is not sufficiently explainable by the influence of different production methods or products consumed on the waste amount of a country, but it shows that more harmonized standard guidelines for definition and reckoning of the different kinds of waste are needed. An accurate data recording is important not only in order to know which amount of a certain kind of waste was produced, but even more to project how this different waste can be utilized.

There are several ways of defining waste. The EU Commission, the Secretariat of the Basel Convention, and OECD each has formally its own definitions. The EU Commission in the Waste Framework Directive 75/442/EEC provides the legal definition of waste: " 'Waste' shall mean any substance or object [. . . ] which the holder discards or intends or is required to discard" (EC, 1975). According to the EU definition the term "waste" is described similarly in different European state laws such as in Germany the law "Kreislaufwirtschafts- und Abfallgesetz (KrW-/AbfG)," which governs also recycling and waste management (Bilitewski et al.,

WINFRIED RUSS AND MARKUS SCHNAPPINGER • Technische Universität München, Chair for Energy and Environmental Technologies of the Food Industry, Weihenstephaner Steig 22, 85350 Freising-Weihenstephan, Germany.

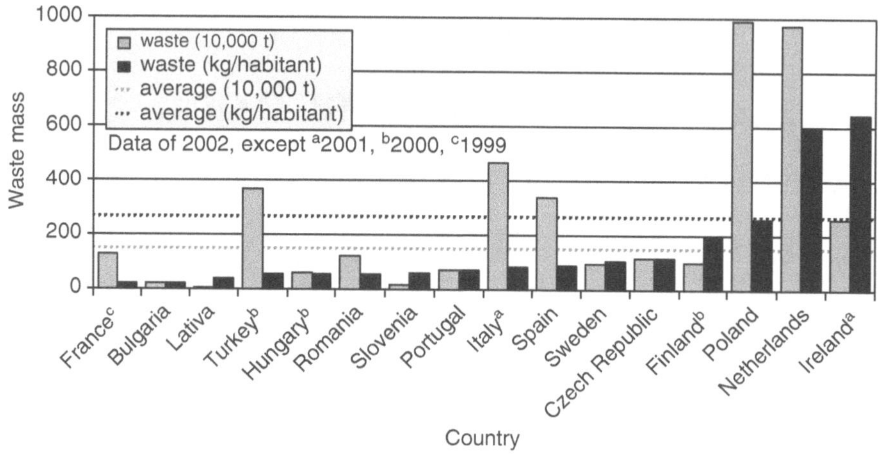


**Figure 1.1.** Waste generated in 2002 from the manufacture of food products, beverages, and tobacco products (Eurostat, 2005).

2000). The former differentiation of waste as either waste that incurs disposal costs or that which through disposal results in a monetary gain no longer exists. Up to now, the typical forms of waste created through the processing of food have never been viewed by those who produce it as waste, because it has been traditionally sold. Furthermore, this practice was considered safe from the perspective of environmental protection, because the waste was composed of food.

Working in close cooperation with the food industry, the agricultural industry has been the prime means of disposal for food waste. Because these methods have hardly been questioned, as it has not been established whether they truly constitute recycling or merely disposal and because only isolated problems have arisen, alternatives have not been actively sought (Franzen et al., 1999; van Wyk, 2001; Zanker and Kepplinger, 2002). In 2002, the food industry in Germany was spending about 300 million € for environmental protection in the area of waste management and invested about 120 million € for environmentally friendly production (Statistisches Bundesamt, 2005a,b).

Generally speaking, raw and auxiliary materials as well as processing aids enter the production process and exit as one of the following: as a desired product, non-product-specific waste, or as product-specific waste. The amount and quality of the product can be influenced almost completely independently of the type and quality of non-product-specific waste. A typical example of non-product-specific waste would be containers for chemicals involved in cleaning and disinfection. The number and type of containers to be disposed do not influence the quality of the product (Kobald and Holley, 1990; Russ, 1994).

It is characteristic of product-specific waste that the generated mass of waste relative to production levels can only be altered through technical means, which often unavoidably leads to a change in product quality. Typical examples of product-specific waste are spent grains from beer production, slaughterhouse waste from meat production, tea leaves, orange and potato peels, or old bread.

Product-specific waste unavoidably accumulates as a result of processing raw materials. It is produced during the various steps of production, in which the desired components are extracted from the raw materials. After extraction there are still often other potentially useful components present in the remaining materials.

Waste in the food industry is characterized by a high ratio of product-specific waste. This not only means that the generation of this waste is unavoidable, but also that the amount and kind of waste produced, which consists primarily of the organic residue of processed raw materials, can scarcely be altered if the quality of the finished product is to remain consistent. The utilization and disposal of product-specific waste is difficult, due to its inadequate biological stability, its potentially pathogenic nature, its high water content, its potential for rapid autoxidation, as well as its high level of enzymatic activity. The diverse types of waste generated by various branches of the food industry can be quantified based upon each branches' respective level of production.

## 1.2. FOOD WASTE: CLASSIFICATION AND PROPERTIES

The food and beverage industry is the largest manufacturing sector in the EU. With 13.6% of total turnover in the EU-15 manufacturing sector in 2002, it is larger than the automobile, chemical, machinery, and equipment sectors (CIAA, 2004). Waste disposal and by-product management in the food processing industry pose problems within the areas of environmental protection and sustainability.

The current methods for further utilization of product-specific waste have developed along traditional lines and have been closely bound to the agricultural origins of raw materials themselves. The two general methods of traditional waste utilization have been to use the waste as either animal feed (e.g., spent grains, distiller's wash) or fertilizer (filtration sludge, carbonation sludge). Many of the existing agricultural solutions of waste disposal are a balancing acts between legal regulations and the best ecological and economical solutions.

Disposal of this waste can be difficult for the following reasons (Werschnitzky et al., 1985; Russ et al., 1997; Russ and Meyer-Pittroff, 2004):

- Biological stability and the potential growth of pathogens:
  Many types of waste material either already contain large numbers of microbes and/or will be altered quickly through microbial activity. If regulations concerning infectious disease are not properly observed, then hygienically unacceptable conditions can arise, e.g., through maggots or molds. The breakdown of protein is always characterized by the generation of strong odors.

- High water content:
  The water content of meat and vegetable waste lies between 70 and 95% by mass. A high water content increases transport costs of the waste. Mechanically removing the water through use of a press can lead to further problems with waste water disposal, due to the high level of organic material in the water.
- Rapid autoxidation:
  Waste with a high fat content is susceptible to oxidation, which leads to the release of foul-smelling fatty acids.
- Changes due to enzymatic activity:
  In many types of waste arising from vegetables and fruits, enzymes are still active, which accelerate or intensify the reactions involved in spoilage.

## 1.3. DISPOSAL AND RECYCLING IN THE CONTEXT OF CHANGE IN THE AGRICULTURAL INDUSTRY AND EMERGING NUTRITIONAL TRENDS

The primary objective in waste management is to completely prevent the production of waste all together, if possible (Faulstich and Schenkel, 1994; EC, 1975). Second, if waste must be produced, then it is to be recycled. Entirely avoiding waste production or recycling food waste in the food industry is extremely difficult.

Food waste results from processing organic raw materials to foodstuffs, which is typically effected by the extraction or separation of the nutritionally valuable portion of the raw materials. The unused remains primarily consist of organic material, but further utilizing it as a food source is limited, because it possesses little nutritional value or contains inedible components. Efforts to prevent food waste from accumulating while insuring constant product quality can be realized by increasing production efficiency; however, the potential for waste reduction is very limited. Hence, a reduction in the mass of accumulated waste does not represent true waste prevention but merely a charge in product quality. Recycling standard types of food waste is, as with all raw materials of biological origin, not possible. However, an exception to this description of typical food waste are production aids. They are added during food processing but are subsequently removed, thereby either losing their functionality or being charged with the accumulation of waste material (e.g., immobilized yeast, filter aids, adsorbents).

Since these forms of waste from the processing of food are almost entirely of biological origin, further utilization as animal feed or on agricultural land have been the most convenient and obvious means of disposal available. Few problems have arisen as a result; therefore these traditional means of disposal have been neither scientifically tested nor ecologically reviewed. Studies have only focused upon increasing the efficiency of their utilization. However, problems in disposal are becoming more frequent. There are numerous reasons for this, including:

- crisis in the agricultural industry and restructuring of the food industry,
- concentration of companies in the food industry, and
- escalating waste production in the food industry.

In striving to maintain the lowest possible prices for agricultural products, the agricultural industry began approaching the limits of finding sources for animal feed. The consequences are exemplified in the outbreak of bovine spongiform encephalopathy (BSE) as well as the scandal of dioxins in animal feed that contained used oils in Belgium (van Larebeke et al., 2001). The subsequent regulations have made it difficult to use waste from the food industry containing fat and protein. Increasing automation in meat production presents a further problem, which limits the usage of this waste, because the requirements for both quality and freshness of the waste are difficult to meet due to its water content which is often high.

Through consolidation of the agricultural industry, the farms where food waste is utilized have become larger, which meets the concentration of companies in the food industry. A problem arising from the consolidation of both industries is the increase of distance between the locations where the waste is produced and where the waste is utilized, which has led to an increase in costs of logistics, making this traditionally inexpensive source for animal feed in some cases as costly as conventional animal feed, which also has a number of advantages over food waste. These advantages include the fact that conventional animal feed shows long-term stability of storage and can be produced easily.

Not only is the consolidation of waste producers in the food industry taking place, but the amount of food waste being produced also is constantly rising, because in recent years processed and semiprocessed products have jointly accounted for two thirds of global food trade. Worldwide the processed products, which are mainly convenience foods, increased from 18% in 1980 to about 34% in 1998 (Regmi and Gehlhar, 2001). Convenience foods are foods that can be consumed immediately after purchase or after being heated for a short time. Through production of such foods, waste that would normally be widely distributed across a broad range of households as a result of preparing meals accumulates at a small number of locations where these convenience foods are produced. Fewer waste collection sites and increasing waste production are two factors that have led to growing pressure on the food industry to dispose of its waste; however, the market for food waste removal in the agricultural industry is currently stagnating, which has resulted in an increase in the cost of waste disposal. Knirsch et al. (1999) described this trend using the example of wastes in the brewing industry.

At the same time, the food industry is also experiencing a structural crisis. Due to a general economic downturn, most businesses are experiencing greater pressure to lower production costs in the face of increasing disposal costs (Eberle et al., 2004).

Issues concerning the protection of consumers and of the environment have moved to the forefront in politics and public awareness. Following the series of food scandals in the 1990s the European Food Safety Authority (EFSA) was established in 2002 by the EU to ensure a higher level of consumer health protection in the area of food safety and to restore and maintain consumer confidence (EFSA, 2004). From the standpoint of consumer protection, the utilization of food waste in the production of new foods is being questioned.

With the directive 1999/31/EC the European Union sets the boundary points for waste disposal in Europe (EC, 1999a). Since June 1, 2005, in Germany disposal of untreated food waste at disposal sites is not allowed in accordance with the directive ''Ablagerungsverordnung (AbfAblV)'' of 2001. Sustainability in the cultivation of raw materials is another demand that has been placed on the food industry, which will require further consideration, planning, and rethinking (EC, 1999b). Food waste is composed of highly complex molecules, which have had a portion of their contents extracted but should still be viewed as raw materials that are still available for use.

## 1.4. ASSESSMENT OF THE ORIGIN AND THE GENERATED MASS OF THE MOST IMPORTANT TYPES OF PRODUCT-SPECIFIC FOOD WASTE

By evaluating the different branches of the food industry it was possible to determine the actual amounts of waste generated in each type of operation. Additional employee questionnaires and measurements were also carried out. In order to be able to differentiate the types of operations within individual branches of the industry, production processes were documented concurrently (Russ and Meyer-Pittroff, 2002). The specific amount of waste production is defined as the mass of accumulated waste divided by the mass of the saleable product, hereafter referred to as the ''specific waste index''.

For the branches of the food industry discussed below, enough data were gathered to allow conclusions to be reached regarding their specific amount of waste production. Table 1.1 presents an overview of the included branches, waste types, and their specific waste index. For other branches of the industry, however, insufficient data and lack of measurements did not make this possible.

### 1.4.1. Grain Products

The most important grains for human consumption are wheat, rice, corn, millet, oats, rye, and barley. They are cultivated on approximately 80% of the world's arable land and supply around 50% of the world's population with food. Wheat and rye are ground to flour and grits in flour mills, oats are made into rolled oats in oat mills, and rice is hulled in rice mills. Corn is primarily processed to starch and oil, whereas almost all barley is turned into malt (Pomeranz, 1973).

*Table 1.1.* Different branches with its types of waste and its according waste indexes (Russ and Meyer-Pittroff, 2004)

| Branch | Type of waste | Specific waste index |
|---|---|---|
| Grain products | Bran | 0.11–0.18 |
| | Middlings | 0.06–0.11 |
| | Broken grains, seeds, shells, husks | < 0.01 |
| | Fine dust, chaff, straw | < 0.01 |
| | Ergot | < 0.01 |
| | Refuse from oats containing bran and husks | 0.39 |
| | Brown rice waste | 0.11 |
| | Rice bran | 0.11–0.18 |
| | Rice flour | < 0.01 |
| | Seedlings from malt | 0.038 |
| | Malt dust | < 0.01 |
| | Grain separator waste | 0.01–0.04 |
| Noodles | Dough waste | 0.0012–0.0014 |
| | Eggshells | 0.02–0.08 |
| Potato processing industry | Potatoes peels | 0.3–0.5 |
| Coffee roasters | Silverskins | 0.02–0.04 |
| Sugar | Molasses | 0.191 |
| | Beet pulp (postextraction) | 0.517 |
| | Beet leaves and stems | 0.136 |
| | Carbonation sludge | 0.427 |
| Diary processing industry | Whey | 4.0–11.3 |
| | Cheese residue | 0.01–0.04 |
| | Waste from milk production | 0.04 |
| Meat industry | Slaughterhouse waste | 0.1–0.87 |
| Egg products | Eggshells | 0.03–0.12 |
| Beer production | Malt dust | < 0.001 |
| | Spent grains | 0.192 |
| | Break material | 0.024 |
| | Yeast | 0.024 |
| | Kieselguhr sludge | 0.006 |
| Wine cellars | Pomace | 0.136–0.145 |
| | Clarification sediment | 0.015–0.050 |
| | Yeast sediment | 0.03–0.045 |

### 1.4.2. Pasta and Noodles

The primary form of waste from this kind of production is unused dough with a specific waste index of 0.0012 to 0.0014 in relation to the mass of the finished product (Senser and Scherz, 1991). Pasta and noodles usually consist of only wheat and water, but occasionally eggs and spices are added to supplement the flavor. If fresh eggs are added to the dough, then the eggshells raise the specific waste index by 0.02 to 0.08 per mass of the finished product. The considerable fluctuation in the resultant eggshell waste is due to the varying amounts of eggs in the recipes as well as the number of products, which contain or do not contain eggs. Smaller producers create less waste per mass of finished product than larger ones.

### 1.4.3. Coffee Roasters

Silverskins, the external skin which covers coffee beans, is the resultant waste from roasting coffee. According to Kessler (1996) the specific waste index of silverskins is 0.02 to 0.04.

### 1.4.4. Sugar

In European countries sugar is generally produced by extracting sugar beets. There are several kinds of product-specific waste some of which can still be sold as by-products (molasses, beet pulp).

### 1.4.5. Dairy Processing Industry

The first waste of all milk products is sludge from centrifuging the fresh, untreated milk. Per ton of centrifuged fresh milk 275 g of sediment is generated. This waste must be pasteurized because of its high microbial content. Milk that has been rinsed from the tanks, lines, etc., as well as unusable milk due to processing error or returned products are handled as waste. This type of waste is created through milk production and the production of fluid milk products, such as yogurt, pudding, or buttermilk, at the rate of 0.04 per mass of finished product. The waste resulting from cheese production consists of whey and cheese residue. Whey arises as a result of the protein coagulation in milk. The generated amount of whey and cheese residue is dependent upon the type of cheese produced (Töpel, 1981; Sienkiewicz, 1986; Fischbach, 1987).

### 1.4.6. Meat Industry

The specific amounts of generated waste vary for each type of animal (Sielaff, 1996; Grosse, 1984) and the specific waste index ranges from 0.1 (sheep) to 0.87 (calf). The majority of the waste in the meat industry is produced during slaughtering. Slaughterhouse waste consists of the portion of a slaughtered animal that cannot be sold as meat or used in meat products. Such waste includes bones, tendons, skin, the contents of the gastrointestinal tract, blood, and internal organs.

### 1.4.7. Egg Products

In egg processing the originating waste consists almost only of eggshells, regardless of which egg products, including deep-frozen, spray-dried or chemically preserved eggs, egg yolks, egg-derived lecithin or egg oil, are produced. The high variation in the level of accumulated waste can be attributed to the variation in the size of the eggs and the thickness of their shells. The specific waste index of eggshells lies between 0.03 and 0.12, and is calculated using the mass of egg products obtained from the eggs.

### 1.4.8. Beer Production

Most kinds of wastes in the beer production process are produced by filtration and separation and consist of the filtration or separation residues or of sludge, which contains the filtration aid and residues, mostly kieselguhr and organic material (Russ and Meyer-Pittroff, 1992; Russ, 1994). Water is used to extract desired substances from coarsely grinded malt. The wort—the liquid extract—is separated from the spent grains which stay behind as waste. A protein–tannin complex known as hot trub and cool trub material also must be separated from the wort further along in the process. After fermentation is complete, the yeast that is utilized during the fermentation process also must be separated from the beer.

### 1.4.9. Wine Cellars

There are three categories to divide waste from wine cellars into: pomace, clarification sediment, and yeast sediment. According to the kind of squeezer system used the generated amount of pomace varies. As the must ferments, it clears, thus producing sediment that must be removed. The amount depends upon what processing methods are used, as well as the condition of the grapes at the time of harvest. This can result in levels of waste equal to 20% of the harvested mass of the grapes in extreme cases. Each time that sediment is removed from the must the total amount of waste increases. In some cases the amount of sediment that accumulates during fermentation can be reduced by a separation processes. Yeast sediment consists primarily of yeast cells and tartar. Their amounts are dependent upon the type of wine being produced.

## 1.5. CONTENTS OF FOOD WASTES

Up to the present the most important means of disposal has been to use waste as animal feed. The most important parameters and other information for using many types of food waste as animal feed are summarized in Table 1.2.

## 1.6. TYPICAL WAYS OF REUSE OF FOOD WASTES

### 1.6.1. General Methods

Three general methods of waste disposal not associated with agricultural practices are:

- incineration,
- anaerobic fermentation, and
- composting.

*Table 1.2.* Food waste and its contents (Russ and Meyer-Pittroff, 2002, 2004). All values are given as % by mass; the contents of the waste are given as % by dry mass

| Waste | Water content | Protein | Fat | Fibrous material[a] | Minerals | Other[b] |
|---|---|---|---|---|---|---|
| Protein-rich types of waste | | | | | | |
| Blood | 86 | 85.7 | 2.1 | | 7.1 | 5.1 |
| Seedlings from malt | 8.8 | 33.0 | 2.2 | 9.4 | 6.6 | 48.8 |
| Hot and cool trub | | 17.0–51.7 | 0.2–3.1 | 0.7–17.5 | 1.4–2.5 | 25.2–80.7 |
| Yeast | | 51.0 | 2.4 | 1.9 | 6.9 | 37.8 |
| Waste with a high cellulose/hemicellulose content | | | | | | |
| Oat husks | 7.1 | 3.4 | 1.4 | 33.5 | | 61.7 |
| Pulp from sugar beets (postextraction) | 9.4 | 10.0 | 0.9 | 20.6 | | 68.5 |
| Spent grains | 80–83 | 19–23 | 9–12 | 53–58 | 4–5.5 | 1.5–15 |
| Broken grains, seeds, peels, husks | 10–15 | 15 | 4–5 | 20–30 | 1–3 | 53–60 |
| Waste with a high carbohydrate content | | | | | | |
| Dough waste | 10.7 | 12.3 | 2.8 | 3.4 | 0.9 | 69.9 |
| Noodles | | | | | | |
| Bran | 10.9–12.2 | 14.7–18.0 | 3.5–5.2 | 5.8–15.8 | | 51.2–76 |
| Middlings | 10–15 | 13–15 | 3–4 | | 1 | 80–83 |
| Waste from oats (flour, flakes) | 8.4–9.1 | 14.3–14.9 | 7.4 | 2.2–5.3 | | 72.4–76.1 |
| Oat bran | 9.2 | 8.9 | 3.5 | 2.8 | | 84.8 |
| Brown rice waste | 13.1 | 8.3 | 2.5 | | 1.4 | 87.8 |
| Rice bran | 9.0 | 13.0 | 14.0 | 15.2 | | 57.8 |
| Rice flour | 9.7–10.8 | 14.1–14.5 | 14.0–16.7 | 8.8 | | 50.8–53.4 |
| Silverskins | 75.0 | 17.5 | 1.0 | 9.4 | 1.2 | 70.9 |
| Molasses | 33.0 | 12.9 | 0.2 | 0.5 | | 86.4 |
| Small beets, roots and other parts | 83.0 | 11.0 | 2.0 | 14.0 | | 73 |
| Whey | 93.4–94.4 | 12.1–17.9 | | | 7.6–14.3 | 76.8–80.3 |
| Potato peels | 77.8 | 2.0 | 0.1 | 2.5 | 1.0 | |
| Waste rich in minerals | | | | | | |
| Eggshells | | 5.0 | | | 95 ($CaCO_3$) | 0 |
| Carbonation sludge | 72–75 | 1.9 | | | 56.3–88.8 | 9.3 |
| Kieselguhr sludge | 55–95 | 8–15 | | | 85–88.5 | 0–7 |
| Waste with a high fat content | | | | | | |
| Slaughterhouse waste | 74 | 34.6 | 53.8 | | 7.7 | 3.9 |

[a] Cellulose and hemicellulose.
[b] This column represents the remainder of the contents not included in one of the categories in each row, and brings the total percentage of the mass up to 100%. This consists primarily of carbohydrate substances.

Because very little of value (e.g., energy, fertilizer) can be recovered using these processes, it is often necessary to pay for these kinds of disposal.

If the water content of the waste is relatively low (< 50% by mass), from a technical standpoint incineration is only a viable option. In addition, emissions play a major role when considering incineration as a form of waste disposal, since, for example, sulfur is oxidized to $SO_2$ and nitrogen to various oxides.

For food waste with a higher water content (> 50% by mass) anaerobic fermentation to methane and predominantly $CO_2$ (biogas) is most applicable. If cellulose and hemicellulose can be broken down at all, they can only be broken down very slowly in biogas fermentation systems. How quickly these compounds can be broken down defines the limits of the biogas fermentation.

Composting as a means of disposal has been developed to the point that cellulose and hemicellulose can be broken down without a problem. A high water content is not preferable, but can be tolerated given that proper composting methods are used. Since the compost market has been saturated many times over, sales of even the highest quality compost are rarely profitable.

### 1.6.2. Agricultural Methods

By dividing waste into categories of contents, can immediately be seen the logic in using the different traditional agricultural methods for reusing waste. Omnivores (e.g., pigs) can easily and very efficiently digest protein and fat. So food waste with high amounts of fat and protein is well-suited as omnivore animal feed. Because of its high protein content, this type of food waste is especially susceptible to spoilage or contamination by pathogenic microbes. Therefore, this kind of waste is generally sterilized to ensure that it is safe.

Only ruminants can be fed with waste with a high cellulose and hemicellulose content because they possess the enzymes necessary to break these substances down. For other types of animals, however, too much cellulose and hemicellulose would be physiologically harmful. Carbohydrates in the form of simple sugars should be fed to pigs.

Food waste with a high mineral content can be used as fertilizer. Organic material with certain contents that make it difficult or impossible to use this waste as animal feed is also used as fertilizer. Examples of this would be pomace and hop waste: they cannot be used as animal feed because they contain phenolic compounds and pesticide residue (Kirchgessner, 1997; Westendorf, 2000).

### 1.6.3. New Methods

New methods of disposal focus on certain contents of the food waste. The content of fibrous material (soluble and insoluble) in food is gaining importance in human nutrition. Spent grains and waste from carrots can be used for this purpose to produce fiber for human consumption.

Soluble and insoluble fiber from apple, tomato, and carrot waste, as well as peels from citrus fruits, can be extracted in various classes. In food production industry these fibers serve to bind water in foods due to their absorptive properties and ability to form gels. To produce building materials, fibrous materials from spent grains can be employed as well, where they are used as filler and structural material in fiber board. The fibers of the spent grains of brewery waste increase the strength of bricks before they are kilned. Because they burn away during the kilning process and form pores in the bricks, they increase the bricks' ability to thermally insulate as well (Russ et al., 1998; Knirsch et al., 1998; Mayer, 1998).

Pectin, a soluble fiber, can be extracted from apples, citrus fruit, and beet waste through another extraction process. However, a portion of the fibrous material remains as waste.

Fat is partially removed from slaughterhouse waste, and is then used as a basis for many products in the chemical and cosmetic industries.

Producing grappa is the classic method for reusing pomace. Since it is known that the oil in the grape seeds is nutritionally very valuable, the seeds are separated from the pomace and the oil is removed through pressing. Also, the pomace can be used as fertilizer after it has been subjected to a special fermentation process.

Processing milk generates a large amount of whey. Alternative methods for its utilization are being sought. It is possible to use whey in food (such as in beverages containing whey), but unfortunately this only accounts for a portion of the total whey produced. Methods in which whey is fermented to produce alcohol, or in which lactose is extracted from whey are technically refined and are employed in the industry in some cases; however, these methods are not always economical (Sienkiewicz, 1986; Fischbach, 1987).

## 1.7. REFERENCES


Ablagerungsverordnung, 2001, Verordnung über die umweltverträgliche Ablagerung von Siedlungsabfällen und über biologische Abfallbehandlungsanlagen – AbfAblV (February 20, 2001).

Bilitewski, B., Härdtle, G., and Marek, K., 2000, *Abfallwirtschaft – Handbuch für Praxis und Lehre*, Springer, Berlin.

CIAA, 2004, Data and trends of the EU food and drink industry 2004, Confederation of the food and drink industries of the EU, Brussels.

Eberle, U., Fritsche, U., Hayn, D., Empacher, C., Simshäuser, U., Rehaag, R., and Waskow, F., 2004, Umwelt-Ernährung-Gesundheit – Beschreibungen der Dynamiken eines gesellschaftlichen Handlungsfeldes, discussion paper, Institute for Applied Ecology, Berlin.

EC, 1975, Council Directive 75/442/EEC of July 15, 1975, on waste Regulation (EC) No. 1882/2003 of the European Parliament and of the Council (last amended September 29, 2003).

EC, 1999a, Council Directive 1999/31/EC of April 26, 1999, on the landfill of waste.

EC, 1999b, Communication from the Commission to the Council, the European Parliament, the Economic and Social Committee and the Committee of the Regions – Directions towards sustainable agriculture, COM/99/0022 final.

EFSA, 2004, Moving towards full strength, press release (February 19, 2004), European Food Safety Authority, Parma.

Eurostat, 2005, WAQ1 = Generation of waste by economic sector and households, Statistical Office of the European Communities.

Faulstich, M., and Schenkel, W., 1994, Abfallvermeidung – Naturgesetzliche, technische und geistige Möglichkeiten und Grenzen, in: *Recht auf Abfall – Vermeidung als Pflicht?*, P. Wilderer, ed., BayFORREST, Garching.

Fischbach, G., 1987, Molke als Rohstoff, dissertation, Christian-Albrechts-Universität, Kiel.

Franzen, J., Erich, E., Haep, S., Knoch, A., and Steinkamp, H., 1999, Grundlegende Untersuchungen zur Kreislaufwirtschaft in Sennereien und Molkereien durch Aufbereitung von Molke mit Hilfe der Vakuumdestillation, final report of the AiF research project no. 11452.

Grosse, C., 1984, Absatz- und Vermarktungsmöglichkeiten für Schlachtnebenprodukte und Schlachtabfälle in der Bundesrepublik Deutschland, dissertation, Universität Bonn, Bonn.

Kessler, H. G., 1996, *Lebensmittel- und Bioverfahrenstechnik,* Kessler, Munich.

Kirchgessner, M., 1997, *Tierernährung,* BLV, Munich.

Knirsch, M., Penschke, A., Krebs, S., Russ, W., Mörtel, H., Mayer, W. A., and Meyer-Pittroff, R., 1998, Application of brewery wastes in the production of bricks, *Tile & Brick Int* **14**(2).

Knirsch, M., Penschke, A., and Meyer-Pittroff, R., 1999, Disposal situation for brewery wastes in Germany, *Brauwelt Int* **VI**:477–481.

Kobald, M., and Holley, W., 1990, Emissionssituation in der Nahrungsmittelindustrie, Fraunhofer-Institut für Lebensmitteltechnologie und Verpackung, Munich.

Mayer, W. A., 1998, Verwertung von Trebern zur Herstellung von Isolier- und Dämmstoffen, dissertation, Technische Universität München, Freising.

Pomeranz, Y., 1973, *Industrial uses of cereals,* Amer. Assoc. Cereal Chem., St. Paul.

Regmi, A., and Gehlhar, M., 2001, Consumer preferences and concerns shape global food trade, *FoodReview* **24**(3):2–8.

Russ, W., 1994, Introduction of a waste management plan for breweries, *Brauwelt Int* **III**:235–239.

Russ, W., Behmel, U., Knirsch, M., Penschke, A., Schöberl, H., and Herdegen, V., 1997, Aus Abfall Wert schöpfen – Organischer Müll als Produktionsgrundlage, *Lebensmitteltechnik* **4**:46–50.

Russ, W., Behmel, U., Höhn, G., Meyer-Pittroff, R., and Penschke, A., 1998, Waste disposal – constructive options, in: *Controlling utility costs in the brewing industry,* proceedings, Birmingham, (May 13, 1998).

Russ, W., and Meyer-Pittroff, R., 1992, Solid waste handling, in: *EBC-Monograph XIX (EBC-Symposium 1992),* Hans Carl, Nürnberg.

Russ, W., and Meyer-Pittroff, R., 2002, The wastes of the food industry, in: *Reprints – The future of Waste Management in Europe,* Fuck, Koblenz, pp. 341–344.

Russ, W., and Meyer-Pittroff, R., 2004, Utilizing waste products from the food production and processing industries, *Crit Rev Food Sci* **44**(1):57–62.

Senser, F., and Scherz, H., 1991, *Lebensmitteltabelle für die Praxis,* Wissenschaftliche Verlagsgesellschaft, Stuttgart.

Sielaff, H., 1996, *Fleischtechnologie,* Behr, Hamburg.

Sienkiewicz, T., 1986, *Molke und Molkeverwertung,* Fachbuchverlag, Leipzig.

Statistisches Bundesamt Deutschland, 2005a, Investitionen für den Umweltschutz in Unternehmen des Produzierenden Gewerbes (ohne Baugewerbe) 2002.

Statistisches Bundesamt Deutschland, 2005b, Laufende Aufwendungen für den Umweltschutz im Produzierenden Gewerbe (ohne Baugewerbe) 2002.

Töpel, A., 1981, *Chemie und Physik der Milch,* Fachbuchverlag, Leipzig.

van Larebeke, N., Hens, L., Schepens, P., Covaci, A., Baeyens, J., Everaert, K., Bernheim, J. L., Vlietinck, R., and De Poorter, G., 2001, The Belgian PCB and dioxin incident of January–June 1999: exposure data and potential impact on health, *Environ Health Persp* **109**(3):265–273.

van Wyk, J. P., 2001, Biotechnology and the utilization of biowaste as a resource for bioproduct development, *Trends Biotechno,* **5**(19):172–177.

Werschnitzky, U. et al., 1985, Umwelteinwirkungen der Ernährungswirtschaft in der Bundesrepublik Deutschland, Bundesamt für Ernährung und Forstwirtschaft, Frankfurt/Main.

Westendorf, M., 2000, *Food waste to animal feed,* Iowa State Press, Iowa.

Zanker, G., and Kepplinger, W. L., 2002, Die Verwertung von Biertrebern im Brauereiverbund; *Brauwelt* **142**(46/47).

# 2

# Introduction to Food Waste Treatment: The 14001 Standards

Vassilis Gekas and Maria Nikolopoulou

## 2.1. INTRODUCTION

The aim of this chapter is to introduce the main ideas concerning treatment of food waste in the spirit of the standards and directives of the EU and to examine the environmental performance of the food industry as it seeks to comply with the ISO 14001 standards. How is the food scientist and/or engineer going to comply with these requirements? Basic tools in the hands of the food scientist and/or engineer is the knowledge of physical and physicochemical properties of the foods, the mass and energy balances, the application of these principles to the unit operations, and unit processes taken place in the industry or activity of concern. This enables the food scientist and/or engineer to predict the quantities and compositions of the various streams, especially those of the effluents. Then the environmental behavior of the organization could be obtained. From this point and on the remediation tactics and strategies could be followed. Life cycle analysis is important because not only the inputs and outputs of a given food industry or in general food activity is of importance for the environment but also the pre- and posthistory. Usually the input to an organization is the output of another organization and vice versa. Environmental aspects of a food endeavor is not a onetime concern. In the beginning, we ensure compliance with the environmental laws and regulations. Follow up is also a requirement set by the EU directives.

## 2.2. WHAT IS ISO 14001?

These standards are a part of the more general series of environmental standards ISO 14000. ISO 14001 is the most well-known standard of the series. It was first

---

VASSILIS GEKAS • Department of Environmental Engineering, Technical University of Crete, Polytechnioupolis, GR-73100 Chania-Crete, Greece. MARIA NIKOLOPOULOU • Department of Environmental Engineering, Laboratory of Transport Phenomena and Applied Thermodynamics, Campus, Technical University of Crete, Chania, GR 73100 Greece

published in 1996, and is still the only standard of the 14000 series with respect to which it is currently possible to be certified by an external certification authority. This standard can be applied by any organization that wishes to implement, maintain, and improve an environmental management system, and also to:

- assure itself of its conformance with its own stated environmental policy (those policy commitments of course must be made),
- demonstrate conformance,
- ensure compliance with environmental laws and regulations,
- seek certification of its environmental management system by an external third party, and
- make a self-determination of conformance.

In 1996, the EC directive 96/61/EC of 24 September 1996 concerning integrated pollution prevention and control also was issued. The aim is to map any organization's (not only food of course) environmental performance. It is therefore the aim of this chapter to introduce the food scientist and/or engineer to those regulations and to show a way to comply with them. The organization will order the ISO 14001 toolkit and also the above directive. The contents of the ISO 14001 are the following:

**Part 1 - Introduction and Background**

| | |
|---|---|
| Chapter 1 | Introduction to Integrated Management Systems |
| Chapter 2 | ISO 14000 and Environmental Management Systems |
| Chapter 3 | Third-Party Registration |
| Chapter 4 | Introducing ISO 14001 |

**Part 2 - Policy and Planning**

| | |
|---|---|
| Chapter 5 | Environmental Policy |
| Chapter 6 | Environmental Aspects |
| Chapter 7 | Legal and Other Requirements |
| Chapter 8 | Objectives and Targets |
| Chapter 9 | Environmental Management Program |

**Part 3 - Implementation and Operation**

| | |
|---|---|
| Chapter 10 | Structure and Responsibility |
| Chapter 11 | Training, Awareness, and Competence |
| Chapter 12 | Communication |
| Chapter 13 | Environmental Management System Documentation |
| Chapter 14 | Document Control |
| Chapter 15 | Operational Control |
| Chapter 16 | Emergency Preparedness and Response |

**Part 4 - Checking and Corrective Action**

| | |
|---|---|
| Chapter 17 | Monitoring and Measurement |
| Chapter 18 | Non-conformance and Corrective and Preventive Action |

The directive mentioned above consists of 22 articles and 4 annexes. Examples of organizations and the pollutants to examine are also given (www.homestead.com, 2005).

The above is just an introductory step toward the actions, which have to be taken by the organization. This is not the objective of this chapter. The main aim is to prepare the future engineer manager to face the environmental issues of the organization for which he/she is going to work for.

## 2.3. THE FOOD SECTOR AS AN ORGANIZATION

### 2.3.1. Inputs

Figure 2.1 shows a typical food organization with its interactions with the environment. The whole agrofood sector could be thought of as an organization or only the food packaging unit or a certain food industry or even a part of an industry. The entire agrofood sector is considered as the organization (Gekas and Balta, 2005).

It consists of:

- primary production,
- postharvesting,
- production, and
- packaging.

In the input side we have:

- raw materials,
- water, and
- energy.

Water is considered separately because of its importance not only in the manufacturing but also for cleaning and hygiene. Attention has to be paid to the life cycle analysis of the inputs. In the raw materials it could be a possible

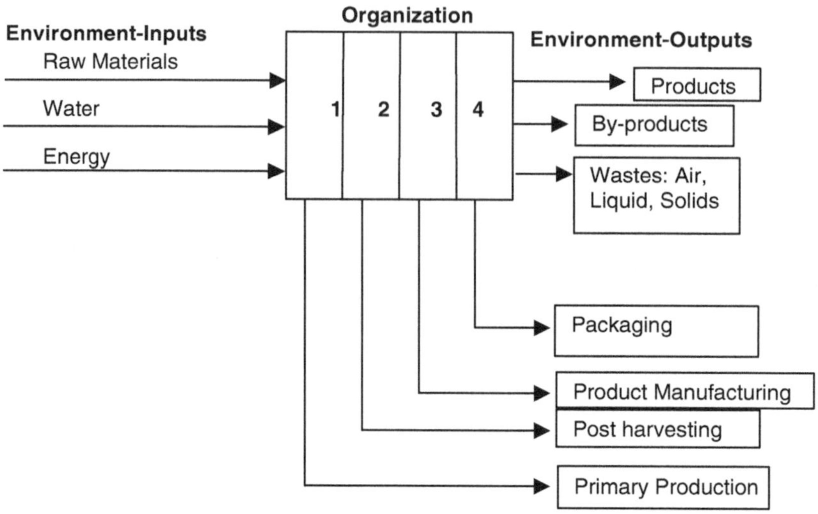


**Figure 2.1.** Structure of a food factory.

contamination by pesticides. Water also is important. Where is it taken from? What is its analysis? Is water quality conforming to the specifications in the various uses of the water inside the organization?

Where is the energy coming from? Is the energy coming from environmentally friendly sources? From renewable sources? From conventional sources? The latter question is a crucial one. If a food activity is planned for many decades to come, then it should be taken into consideration that the conventional resources, namely oil, natural gas, and hydrocarbons, are rather limited and are more or less environmentally unfriendly. Also, are the outputs of the organization recycled so that water and energy could be used as inputs?

These questions and trying to find the right answers will enable one to tackle the environmental problems in a creative way. An attractive solution is to use a part of the wastes (after having recovered valuable substances) as biomass to produce biofuels for the energy requirements of the unit or the organization. Also take into account that the conventional fuel deposits are going to be exhausted in a few decades. According to information sources from oil concerns and statistics from the energy authorities of United States and Canada the deposits of oil, natural gas, and hydrocarbons will last 26, 34, and 86 years, respectively, with the assumption that no new deposits are going to be found and that the demand increase per year will be at maximum 10%.

Foods are necessary for the survival of *Homo sapiens*. It has been found that foods are *sine qua non* for the maintenance of human health. Foodaceuticals is a modern concept and challenge.

Will nuclear energy be the future energy source? But this is environmentally the most unfriendly energy form, both short term and even more risky long term.

Will it be hydrogen? Yes, but from which source? If the answer is water splitting then again by which energy are we going to split the water molecule? Nuclear energy? The same problem arises again.

Sun and wind, the solar and eolian forms of energy are quite attractive but depend on the weather facilities at a place. Geothermal field is another example. In Crete, it is thought that such energy sources be applied in connection with food manufacturing. But biomass is international. This is a smart choice to make. Solve the negative consequences problem with a positive supply of energy solution. Incineration of municipal wastes after a proper predrying in Italy or Greece could be possible, for example, to produce energy, solving at once two serious problems. There are areas with shortages of both energy and water. There are technological solutions to solve the problem of both input demands by the food organization. Sea or brackish water desalination plants can be driven by sun or by the wind. This is really a compliance of the input side to the organization in an environmental protection spirit. The engineer will face three big questions:

- How to avoid the contamination of raw materials from pesticides and other pollutants?
- How to provide ''environmentally friendly'' produced water to the organization?
- How to provide ''environmentally friendly'' produced energy to the organization?

### 2.3.2 Outputs

As shown in Figure 2.1 there are three kinds of outputs from the food, and not only from the food organization.

- The products. Special care would be taken in order to avoid
  - raw material contamination, and
  - contamination from the production line itself. Therefore, hygienic design is needed.
- The by-products. The same is valid as for the products.
- The waste effluents
  - to the air,
  - to the surface waters, and
  - to the soil or the groundwater.
- The wastes could also be
  - gas emissions,
  - liquid phases, aqueous or of different solvents, and
  - solids.

A considerable part of the latter are the used packages after food consumption. In the Royal Kingdom according to a statistic in the year 1996, one third of all solid wastes are of the food packages origin (Gekas and Balta, 2001). What will be the strategy of the environmentally responsible scientist concerning those outputs, wastes effluents in particular?

The strategies are discussed in Section 2.7. In Sections 2.8 and 2.9, the focus is on the strategies of recovery and achieving in an environmentally friendly way the production of high-value products from wastes using the membrane bioreactor concept. Before that the hygienic and risk analysis is discussed in Section 2.6. Basic knowledge of physical properties and mass and energy balances are given in Sections 2.4 and 2.5, respectively.

## 2.4. PHYSICAL PROPERTIES

The following properties of the foods and food constituents are necessary for the prediction of the environmental performance:

1. Thermodynamic properties: partition coefficients, activities, and chemical potentials.

Through the knowledge of these properties we can predict the distribution of a given component between two or more phases, for example, hexane in the kernel oil.

2. Kinetic data K, $K_m$ for enzyme catalyzed reactions.
3. Physical properties: viscosity, density, and diffusivity.
4. Biological oxygen demand (BOD) Chemical oxygen demand (COD) Suspended solids (SS)
5. Toxic and hazardous: A difference in the values of COD and BOD indicates probably a toxic or hazardous character of the effluents because in this case nonbiodegradability is occurring.

A good software, such as the Super Pro by the Intelligen, Inc., company includes a database of the above properties. Also the database produced as the outcome of the DOPPOF (the initials are ''database of physical properties of foods'') European project contains the physical properties of a great number of foodstuffs (www.nelfood.com). Other sources of physical property data are in Gekas (1992) and Rahman (1995).

## 2.5. MASS AND ENERGY BALANCES

A typical case is the combination of two unit procedures, a synthetic process, A, followed by a separation unit operation, B (Figure 2.2a). In the food industry process A is a chemical reaction, an enzyme catalyzed process, or in general biocatalyzed process.

A could be a:

- feed tank
- mixing tank
- a reactor or bioreactor

and B could be a separation module, such as a:

- decanter
- centrifuge
- membrane unit
- activated carbon fixed or fluidized bed

According to the case in the system of Figure 2.2a more pretreatment or posttreatment steps could be added. Those steps could be unit operations or unit

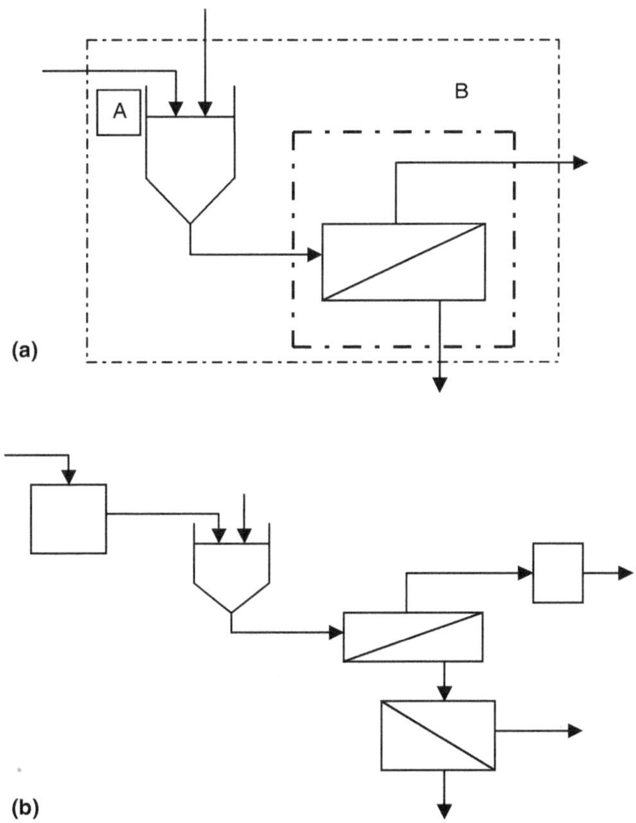


**Figure 2.2.** (**a**). A combination between a synthetic and a separation process. (**b**). More pre- and posttreatment steps added.

processes or unit bio-processes. A conserved property such as mass (total or the mass of a given component) or energy fulfill the balance equations. A basic requirement for any balanced equation to apply is to define very carefully the so-called control volume. For example, in Figure 2.2b the control volume could be the unit A or B or both procedures A+B. In the former case we have a specific balance. A balance applied to the overall process is an overall or general balance. Concerning mass balances a total balance concerns the conservation of the total mass, whereas a certain component follows a partial mass balance.

### 2.5.1. Kinds of Balances

It is understood that there are four kinds of mass balances:

1. total and overall (general)
2. partial and overall (general)
3. total and specific
4. partial and specific

The balance of any conserved (extensive) property, F, can be written as:

(1) $$\sum F_{in} = \sum F_{out} - \int dF/dt$$

For large-scale times or when generally there are no source or accumulation terms:

(2) $$\sum F_{in} = \sum F_{out}$$

In the partial mass balances, sources usually appear. Then the inputs equal the outputs ± the sources.

### 2.5.2. Example of Balances

2.5.2.1. Evaporation of Tomato Paste

For 200,000 kg/day of tomato paste at 35°C, 5% dry solids (DS) to be converted to 30% DS in a one-stage evaporator (Figure 2.3) operating at 93°C (77 kPa, $H = 2664$ kJ/kg, $h = 387$ kJ/kg), where $H$ and $h$ are enthalpies of the water in the steam and liquid form, respectively. Saturated steam was fed to the heat exchanger 1250 kPa ($T = 190$°C, $H = 2786$ kJ/kg, $h = 808$ kJ/kg). The Cp of the tomato paste was 4.01 kJ/kg°C.

a) What is the quantity of steam removed?
   *Solution:* Considering 24 hours of operation and neglecting the solids removed by the steam, the partial balance of solids is:

$$F x_F = L x_L.$$

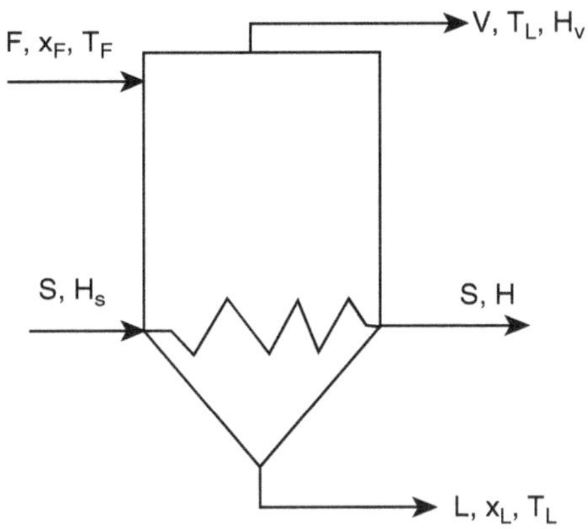


**Figure 2.3.** Single-stage evaporator.

Substituting $F = 200{,}000$ kg, $x_F = 0.05$, $x_L = 0.30$ gives $L = 33{,}333$ kg. The total mass balance is:

$$F = V + L,$$

therefore $V = 166{,}667$ kg.

b) What is the steam recovery?
*Solution*: 93°C is taken as the reference temperature. The total energy balance is

$$F\,c_{p,\,\text{pulp}}\,(T_F - T_{ref}) + S\,(H_s - h_{\text{ref}}) = Lc_L\,(T_L - T_{\text{ref}}) + V(H_V - h_{\text{ref}}) + S\,(h_s - h_{\text{ref}})$$

Substituting $F = 200{,}000$ kg, $c_{p,\,\text{pulp}} = 4.01$ kJ/kg°C, $T_F = 35$°C, $T_{\text{ref}} = T_L =$ 93°C, $H_s = 2{,}786$ kJ/kg, $h_{\text{ref}} = 387$ kJ/kg, $V = 166{,}667$ kg, $H_v = 2{,}664$ kJ/kg, $h_s = 808$ kJ/kg, and finally $S = 215{,}378$ kg. Recovery of steam:

$$V/S = 77.4\%.$$

c) Which is the contact heat area, if U the overall heat transfer coefficient is 3,100 W/m$^2$K?
*Solution*: If U the overall heat transfer coefficient is 3,100 W/m$^2$K

$$q = U\,A\,\Delta T = S(H_s - h_s).$$

Substituting $S = 2.5$ kg/s, $H_s = 2{,}786$ kJ/kg, $h_s = 808$ kJ/kg, $U = 3{,}100$ W/m$^2$K and $\Delta T = 190\text{-}93 = 97$ K, A is calculated to be 16.4 m$^2$.

2.5.2.2. Two-stage Evaporation

For the same operation as previously described but with a two-stage evaporation carrying out (Figure 2.4), the previously calculated heat transfer contact area

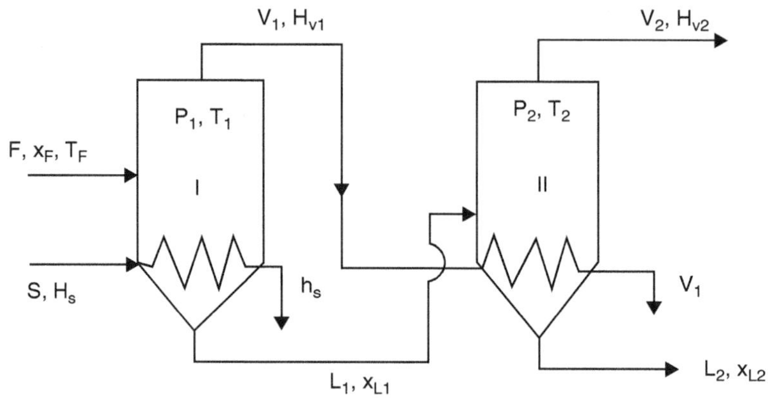


**Figure 2.4.** Two-stage evaporator.

16.4 m$^2$ was taken as the invariant. First evaporator is at 150°C (476 kPa, $H$ = 2,747 kJ/kg, $h$ = 632 kJ/kg), and the second at 93°C (77 kPa, $H$ = 2,664 kJ/kg, $h$ = 387 kJ/kg). $U_1 = 3{,}750$ W/m$^2$K and $U_2 = 3{,}300$ W/m$^2$K.

a) What is the heat balance of the second evaporator
*Solution*:

$$q_2 = U_2 A_2 \Delta T_2 = V_1 (H_{v1} - h_{v1}).$$

Substituting $H_{v1} = 2{,}747$ kJ/kg, $h_{v1} = 632$ kJ/kg, $U_2 = 3{,}300$ W/m$^2$K, $A_2 = 16.4$ m$^2$, $\Delta T = 150$–$93 = 57$ K, the quantity of steam removed from the first evaporator is $V_1 = 1.28$ kg/s $= 110{,}744$ kg/day.

b) What is the solid content (DS) of the concentrated effluent of the first evaporator?
*Solution*: Total mass balance specific in evaporator 1:

$$F = V_1 + L_1.$$

$F = 200{,}000$ kg/day and $V_1 = 110{,}744$ kg/day so that $L_1 = 89{,}256$ kg/day. Neglecting solids removed by the steam effluent $V_1$, the partial specific solids balance is:

$$F x_F = L_1 x_{L1}.$$

$F = 200{,}000$ kg, $L_1 = 89{,}256$ kg and $x_F = 0.05$ gives us the fraction $x_{L1} = 0.11$.

c) What is the steam quantity required in the first evaporator?
*Solution*: 150°C is taken as the reference temperature and the specific total energy balance is:

$$F\,c_{p,\text{pulp}}\,(T_F - T_{\text{ref}}) + S(H_s - h_{\text{ref}}) = L_1\,c_{L1}\,(T_{L1} - T_{\text{ref}}) + V_1(H_{v1} - h_{\text{ref}}) + S(h_s - h_{\text{ref}})$$

Substituting $F = 200{,}000$ kg, $c_{\text{p, pulp}} = 4.01$ kJ/kg°C, $T_F = 35$°C, $T_{\text{ref}} = T_L = 150$°C, $H_s = 2{,}786$ kJ/kg, $h_{\text{ref}} = 632$ kJ/kg, $V_1 = 110{,}744$ kg, $H_V = 2{,}747$ kJ/kg, $h_s = 808$ kJ/kg we find $S = 106{,}786$ kg.

d) What is the new steam economy?
*Solution*: The total overall balance is:

$$F = V_1 + V_2 + L_2,$$

Substituting $F = 200{,}000$ kg, $V_1 = 110{,}744$ kg, και $L_2 = 33{,}333$ kg (from the previous problem), we find $V_2 = 55{,}923$ kg. Steam recovery $= (V_1 + V_2)/S = 1.56 = 156\%$. The economy (recovery) of the steam is very much improved with the new concept of the two-stage evaporation unit (Geankoplis, 2003).

## 2.6. INSIDE THE ORGANIZATION: HYGIENIC DESIGN, HACCP

A very important coordinate of the ultimate goal for food quality is the microbiological safety (prevention of the development of microorganisms) and the elimination of foreign objects that possibly could contaminate our products and by-products. The hygienic design of any production, as well as rules to avoid risks are required (HACCP: Hazard Analysis of Critical Control Point) (www.ehedg.org, 2005).

The hygienic design concerns the machines, tanks, pipes, pumps, valves, instrumentation, the locals (roofs, corners, walls, grounds, storage areas, etc.), and the personnel. Guidelines for the hygienic design have been edited by the European organization EHEDG—the European Hygienic Engineering and Design Group. The EHEDG provides guidance on the hygienic engineering aspects of manufacturing of safe and wholesome food. This is achieved through production, publication, and updating of guidelines available in several languages. To bridge the gap between theory and practice, training modules will be created based on the guidelines equipment approval through certification to assist equipment suppliers and food manufacturers. The EHEDG has EC support through the thematic network, HYFOMA, which is the European network for hygienic manufacturing of food. Its goal is guideline development and dissemination of information.

Concerning the HACCP there are seven principles:

- Analyze hazards: Potential hazards associated with a food and measures to control those hazards are identified. The hazard could be biological, such as a microbe; chemical, such as a toxin; or physical, such as ground glass or metal fragments.
- Identify critical control points: These are points in a food's production—from its raw state through processing and shipping to consumption by the

consumer—at which the potential hazard can be controlled or eliminated. Examples are cooking, cooling, packaging, and metal detection.

- Establish preventive measures with critical limits for each control point. For a cooked food, for example, this might include setting the minimum cooking temperature and time required to ensure the elimination of any harmful microbes.
- Establish procedures to monitor the critical control points. Such procedures might include determining how and by whom cooking time and temperature should be monitored.
- Establish corrective actions to be taken when monitoring shows that a critical limit has not been met, for example, reprocessing or disposing of food if the minimum cooking temperature is not met.
- Establish procedures to verify that the system is working properly, for example, testing time-and-temperature recording devices to verify that a cooking unit is working properly.
- Establish effective recordkeeping to document the HACCP system. This would include records of hazards and their control methods, the monitoring of safety requirements and action taken to correct potential problems. Each of these principles must be backed by sound scientific knowledge, for example, published microbiological studies on time and temperature factors for controlling foodborne pathogens.

## 2.7. WASTE TREATMENT STRATEGIES

In general there are three sorts of strategies:

1. The end of pipe abatement
2. The reduction at source
3. The zero-point discharge

If the first strategy is to be followed, the organization can just discharge the wastes to a nearby biological station, which will treat the industrial wastes along with the municipal ones or may have an installation of primary, secondary, and tertiary wastewater treatment. A thorough knowledge of the production process and its unit operations and processes is required in order to be able to apply the second strategy in the practice.

For example, you are the new responsible man (or woman) for the environmental management policy in your company, which is a potato fries production. One of the unit operations is blanching in hot water. Blanching in hot water has some adverse effects. You have followed the food engineer course so you know everything about blanching. You know that due to driving forces there will be a loss of, for example, $Ca^{2+}$ in the hot water. This calcium quantity will pass into the waste of the industry. The previous policy of the company was to pay a fee to discharge this waste to the nearby aeration installation for the municipal wastes. The community asked for an increase of the fee because they found that your calcium causes troubles in the tertiary treatment of the aeration installation. What are you going to do? If you follow the end of pipe strategy, probably you think to remove $Ca^{2+}$ before the discharge of the potato waste to the community's biological plant. Another alternative would be to intervene in the blanching operation. Change from hot water to steam, or even use a membrane operation already in that step and not end of pipe to avoid having large volumes at the end. These measures are reduction at source.

Finally, in the food industry zero-point discharge is desired, since it is assumed that there is no environmental contamination through the (raw materials and water) inputs, everything found in the wastes is in principle possible to retrieve and utilize. An example of zero discharge is provided by the dairy industry, which is the only food industry so far advanced in what is regarded as the environmental performance. Zero discharge means not only that a waste like whey, for example, is 100% utilizable by retrieving all valuable substances from it, but also that cleaning waters containing milk are treated to milk and pure water achieve which is recycled or otherwise allowed to be discharged.

## 2.8. THE KEYWORD: RECOVERY

The magic word expressing the particularity of food wastes is "recovery." Food waste should not be considered as wastes but as raw materials to develop high additive value as new products.

Thus monosaccharides can be obtained through selective hydrolysis of lactose that has been recovered from whey. Oligopeptides can be obtained through peptic hydrolysis of whey protein concentrate (WPC) isolated from whey. Valuable phenol compounds could be recovered from olive oil mill wastes and then used as raw materials for the development of cosmetics and pharmaceuticals. Ethanol could be produced through the enzymatic conversion of cellulose-rich wastes. Pectin could be produced from fruit juice effluents; the list is long indeed.

An example of using appropriate technology is the production of WPC, which can be obtained from whey following the steps below:

- Membrane separation, reverse osmosis or ultrafiltration (RO and UF)
- Evaporation
- Drying

Protein content in the whey is 1.5%. The membrane concentration factor could be as high as 18. This means that the effluent leaving the membrane step could have a protein concentration of 27%. This could be brought up to a value of 54% through a two-stage evaporation unit and the final protein concentration could be 80% after the drier.

## 2.9. APPLICATION OF MEMBRANE REACTORS TO BY-PRODUCTS TREATMENT

Enzymes are usually involved in the conversions mentioned above. Beta galactosidase is the enzyme hydrolyzing lactose to glucose and galactose. The same enzyme under appropriate conditions of the reaction also has synthetic properties, i.e., galactose molecules get attached to a lactose molecule giving oligosaccharides. Peptidases convert the proteins to peptides. Alpha amylase converts the starch into less viscous intermediate hydrolysates, which upon further hydrolysis by glycol amylase yield glucose.

Beta amylase converts the starch to maltose. Maltose is hydrolyzed by alpha glycosidase to glucose. Those are a few examples. The device to carry out such a reaction with a subsequent separation is known as a membrane reactor or bioreactor. A frequent application is the CSTR bioreactor, which is a continuously stirred tank reactor with a membrane unit, usually ultrafiltration, downstream. This principle is shown in Figure 2.5. One of the advantages with this kind of reactor is when the substrate is macromolecular and the product micromolecular. Then as the product is removed in the permeate and the substrate with the enzyme recirculates in the retentate and to the reactor tank, possible product inhibition effects are avoided.

Another kind of reactor uses the immobilized enzyme form, the enzyme being immobilized on the surface of or entrapped inside the membrane matrix. Then the membrane unit is both the place for the reaction and the separation. Other concepts used are loading the biocatalyst enzymes or whole cell in the shell side of hollow fibers, passing the substrate through the hollow fiber's lumen. The module follows a passive mode of operation, with no pressure exerted. The substrate being permeable to the membrane diffuses from the lumen (the hole) of the fiber to the shell, it reacts there with the biocatalyst and the products diffuse back to the lumen and are taken away (Figure 2.6) (Gekas, 1986).

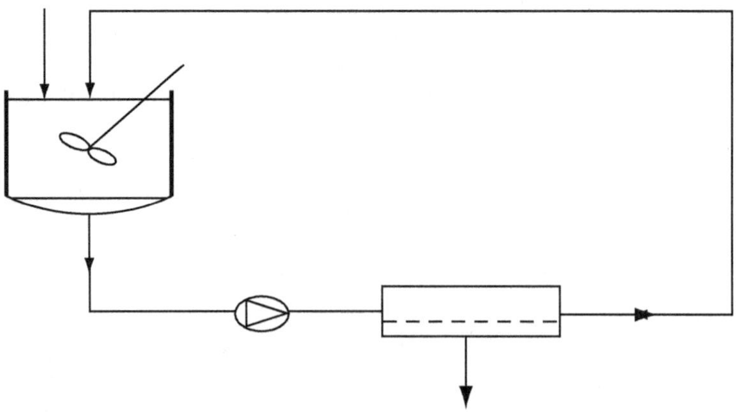

**Figure 2.5.** CSTR-UF reactor.

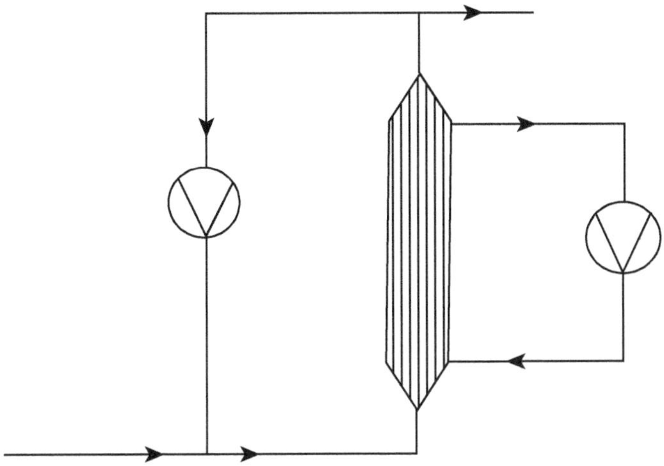

**Figure 2.6.** Hollow fiber reactor.

## 2.10. REFERENCES

Geankoplis, C., 2003, *Transport Processes and Separation Process Principles*, 4th edition, Prentice Hall, New York, USA.

Gekas, V. C., 1986, Artificial membranes as carriers for the immobilization of biocatalysts, *Enzyme Microb. Technol.* **8**(88)**:450.**

Gekas, V., 1992, *Transport Phenomena of Foods and biological materials*, C.R. C Press, Florida.

Gekas, V., and Balta, K., 2005, *Food and Environment*, Tziolas Publications, Thessaloniki, Greece.

Rahman, S., 1995, *Physical Properties of Foods*, C.R. C Press, Florida.

www.homestead.com/iso14001, 2005.

www.ehedg.org, 2005.

www.nelfood.com, 2005.

www.intelligen.com, 2005.

# 3

# Basic Unit Operations in Wastewater Treatment

Dionissios Mantzavinos

## 3.1. INTRODUCTION

Municipal wastewater treatment typically comprises a string of physical, chemical, and biological processes aimed at the removal of the polluting load and the production of a final product that can safely be disposed of in watercourses and/or reused. Of the various processes involved, physical processes, which are also commonly referred to as unit operations, play a major role throughout the various treatment stages.

In a conventional wastewater treatment plant (WWTP), which is schematically illustrated in Figure 3.1, one can easily identify several such stages and typical efficiencies in terms of biochemical oxygen demand (BOD) and suspended solids (SS) removal.

1. Preliminary treatment (also referred to as pretreatment): it aims to remove bulky and large solids, thus preconditioning the effluent prior to the remainder of the treatment. This stage involves several unit operations such as screening, comminution, sedimentation for grit removal, skimming, and flow equalization.
2. Primary treatment: unit operations such as sedimentation and flotation are employed to remove the suspended and colloid fractions of the effluent.
3. Secondary treatment: it aims to remove organic material through biological processes.
4. Tertiary or advanced treatment: it aims to remove nutrients such as nitrogen and phosphorous, residual suspended solids, inorganics and toxic and refractory organics that have escaped from previous stages. This can be done through physical (e.g., filtration, screening, air stripping, adsorption, ion exchange), chemical (precipitation, oxidation), or biological processes. However, with the exception of very few countries that have nutrient

DIONISSIOS MANTZAVINOS • Department of Environmental Engineering, Technical University of Crete, GR-73100 Chania, Greece

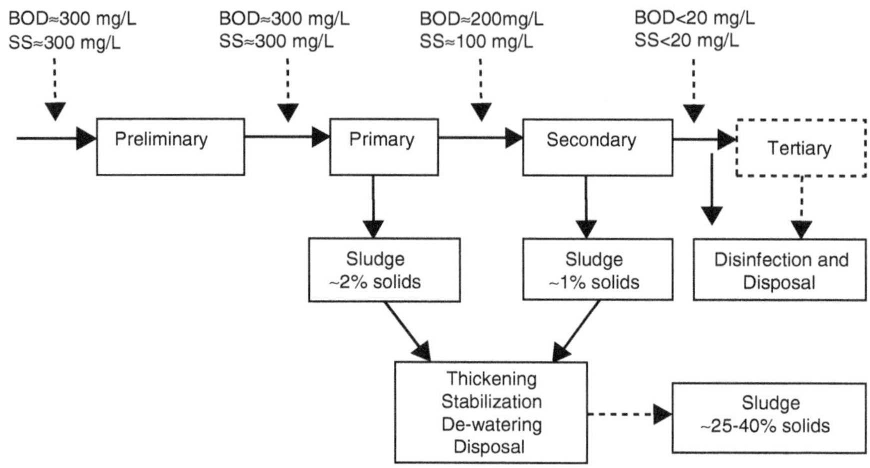


**Figure 3.1.** Simplified flowsheet of WWTP.

removal regulations for ecological reasons, typical WWTPs do not go beyond secondary treatment.

5. Disinfection removes water pathogens by chemical (e.g., chlorination or ozonation) or physical processes (e.g., UV irradiation). The treated effluent can then be safely discharged in natural receivers or partially reused (e.g., for irrigation).
6. Sludge treatment includes processes such as thickening, dewatering, drying, and digestion and aims to reduce the volume of the sludge to be handled as well as to stabilize biologically the final product, which is usually then sent to landfills (although it also may be used as fertilizer or fuel).

The aim of this chapter is to describe the principles of the main unit operations typically employed in WWTPs such as sedimentation, flotation, screening, skimming, and flow equalization. It should be pointed out that particular emphasis is given on sedimentation, since it is found in nearly every single treatment stage. On the other hand, unit operations associated with tertiary treatment are not discussed here, since they lie outside the scope of this book.

## 3.2. UNIT OPERATIONS IN PRELIMINARY TREATMENT

### 3.2.1. Screening

Screening aims to remove bulky materials that may block and consequently damage pumps, pipelines, and other mechanical equipment. This is done through the use of screens (or racks) that retain the oversized materials of the effluent. According to the size of their openings screens are classified as follows:

1. Bar or coarse screens with openings between 20–150 mm; these are commonly used in preliminary treatment to remove very large solids.
2. Fine screens with openings between 3–6 mm; they are also used in preliminary treatment to remove medium-sized solids as well as in primary treatment to remove some suspended solids.
3. Microscreens with openings between 0.02–0.3 mm; they are used in tertiary treatment for suspended solids removal.

#### 3.2.1.1. Bar Screens

They comprise vertical or inclined steel bars placed at equal intervals across a channel through which the effluent flows and may be either manually or mechanically cleaned. The former may be used in plants with flow rates up to $1000\,m^3/d$, while the latter are suitable for larger capacities.

Manually cleaned screens have openings between 25–50 mm. The flow channel, which is usually rectangular and horizontal, is 1.5–4 m deep and 0.6–2 m wide with the bars set 30–80° from the vertical to facilitate solids removal. Mechanically cleaned screens have openings between 10–100 mm with the flow channel being 1.5–8 m deep and 0.8–4 m wide. Screenings have the following characteristics: density ranging from 600 to $1000\,kg/m^3$, humidity from 75–90%, volatile solids content from 80–90%, and a fuel value from 13–18 MJ/kg. Nevertheless, screenings are extremely malodorous and should be removed as soon as possible. In light of this, they are either disposed of immediately through land filling (in small plants) or incineration (in large plants) or, following shredding, are reintroduced in the effluent to undergo further treatment in the following stages.

#### 3.2.1.2. Basic Calculations

The width of channel is computed as follows:

$$W = \frac{\rho + \delta}{\delta} \frac{Q_{max}}{V_{max} h} + W_O \tag{1}$$

where $W$: channel width, m
$\rho$: maximum width of the bar facing the flow, m
$\delta$: minimum clear spacing of bars, m
$Q_{max}$: maximum flow rate, $m^3/s$
$V_{max}$: maximum velocity of flow through the openings of the screen, m/s
$h$: flow depth, m
$W_O$: correction factor for side support, m

To safeguard proper operation, the velocity of flow approaching the screen as well as that of flow through the openings should be chosen in such a way that (1) solids are not deposited in the channel and (2) screenings already captured by the screen are not dislocated and consequently reenter the flow. In this respect, the approach velocity always should be greater than 0.4 m/s, while the velocity through the openings should take values between 0.6 and 1.2 m/s (these values correspond to the maximum flow rate).

Hydraulic losses through screens are given as follows:

$$h_L = \beta\left(\frac{\rho}{\delta}\right)^{4/3}\frac{V^2}{2g}sin\theta \tag{2}$$

where $h_L$: hydraulic loss, m

$\beta$: bar shape factor (typical values are 1.79 for circular bars, 2.42 for sharp-edged rectangular bars, and 1.67 for rectangular bars with semicircular faces)

$V$: velocity of flow approaching the screen, m/s

$\theta$: angle of screen with horizontal

It should be noted that the hydraulic losses given by Eq. (2) correspond to clean screens and are in the order of few millimeters. In practice, losses of about 0.15–0.2 m are expected for clogged screens.

### 3.2.2. Sedimentation

Sedimentation (or settling) refers to the separation of solid particles from suspensions due to gravity. In preliminary treatment, sedimentation aims to remove grid, sand, and other inorganic materials with diameter greater than about 0.15–0.20 mm. Depending on the concentration of solids and the possible interactions between the particles, four types of settling can occur:

1. Discrete particles settling (type I). It occurs in highly dilute suspensions where the particles settle without interacting with each other. Grit removal belongs to type I sedimentation.
2. Flocculent settling (type II). It occurs in relatively dilute suspensions where the particles tend to agglomerate resulting in the formation of larger particles, which settle faster. Type II sedimentation is a typical primary treatment process.
3. Hindered or zone settling (type III). It occurs in suspensions with moderate to high concentrations of solids. Particles interact in such a way that hinders settling of neighboring solids.
4. Compression (type IV). It occurs in highly concentrated suspensions, where the solids are brought into physical contact with each other. Both type III and type IV settling occurs in secondary sedimentation tanks.

#### 3.2.2.1. Analysis of Grit Removal

On the assumption that type I sedimentation involves perfectly spherical particles of common size, the following forces apply to each particle:

(3) $$F_1 = (\rho_s - \rho)g\frac{\pi d^3}{6} = \text{gravity} - \text{buoyancy}$$

(4) $$F_2 = C_D\frac{\pi d^2}{4}\rho\frac{V^2}{2} = \text{frictional drag force}$$

where $\rho_s$: density of solids, $kg/m^3$
$\rho$: density of liquid, $kg/m^3$
$C_D$: drag coefficient
$d$: diameter of spherical particle, m
$\pi d^3/6$: volume of sphere, $m^3$
$\pi d^2/4$: cross sectional area of sphere, $m^2$
$V$: particle velocity, m/s.

The terminal particle velocity ($V_s$) can then be derived equating Eqs. (3)–(4):

(5) $$V_s = \left[\frac{4g(\rho_s - \rho)d}{3C_D\rho}\right]^{1/2} \quad \text{(Newton's law)}$$

Drag coefficient is a function of Reynolds number (Re) (e.g., laminar or turbulent flow) and also depends on particle geometry. For spheres and Re < 500–1000, drag coefficient is given as follows:

(6) $$C_D = \frac{24}{Re} + \frac{3}{Re^{1/2}} + 0.34$$

As far as common wastewater treatment practice is concerned, Re is usually less than 0.2; in this case, the last two terms of Eq. (6) can easily be omitted and $C_D \approx 24/Re$ (Stokes' flow regime). Replacing drag coefficient in Eq. (5) and bearing in mind that $Re = (dV_s\rho)/\mu$ (where $\mu$ is the viscosity of the liquid), Newton's law becomes:

(7) $$V_s = \frac{\rho_s - \rho}{18\mu}gd^2 \quad \text{(Stokes' law)}$$

Stokes' law dictates that the terminal settling velocity is proportional to the square value of particle size and also depends on the physical properties of solids and liquid. Since the latter remain practically unchanged (although, for instance, a change of effluent temperature could change viscosity and consequently settling velocity), the design of settling tanks focuses on the selection of particles with terminal velocity $V_s$ (this corresponds to a certain diameter $d$) so that all particles with velocities greater than or marginally equal to $V_s$ settle, and therefore get removed from the stream. Figure 3.2 shows the ideal particle trajectory in a type I sedimentation tank of length ($L$), width ($W$), and depth ($H$). Particle trajectories depend on both the horizontal flow velocity ($V$) and the vertical settling velocity ($V_s$). Particles are considered to have been settled and consequently removed from the effluent after they have struck the bottom of the settling zone and gotten trapped in the settling zone where they remain. The horizontal flow velocity is given as follows:

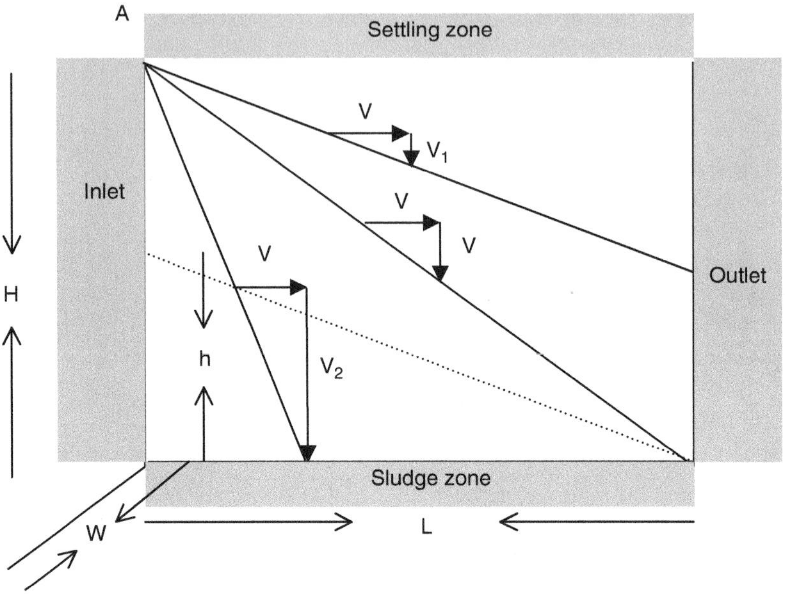


**Figure 3.2.** Ideal particle trajectory in type I sedimentation tank.

(8)
$$V = \frac{Q}{WH}$$

where $Q$: volumetric flow rate.

For particles entering the settling zone at $A$ with settling velocity $V_s$:

(9)
$$\frac{V}{V_s} = \frac{L}{H} \Leftrightarrow V = V_s \frac{L}{H}$$

From Eq. (8)–(9):

(10)
$$V_s = \frac{Q}{WL} = \frac{Q}{A}$$

where $A$: tank surface area.

It is obvious from Eq. (10) that the terminal settling velocity is independent of the tank depth. The $V_s$, which is also known as overflow rate or surface loading rate (i.e., volumetric flow rate per unit area), is the most important design parameter in settling tanks.

All particles with velocities greater than or equal to $V_s$ (e.g., $V_2$ in Figure 3.2) will definitely be removed. However, particles with velocities smaller than $V_s$ (e.g., $V_1$ in Figure 3.2) cannot be removed unless they enter at point B or lower (trajectory shown by the dotted line); in this case, only a fraction of solids equal to $V_1/V_s = h/H$, i.e., the ratio of entry depth to the total tank depth will be removed. The total fraction ($X$) of solids that have been removed is then given as follows:

$$X = (1 - X_s) + \int_0^{X_s} \frac{V_1}{V_s} dx \tag{11}$$

where $1-X_s$: fraction of solids with $V_2 \geq V_s$

$\int_0^{X_s} \frac{V_1}{V_s} dx$: fraction of solids that are eventually removed with $V_1 < V_s$.

3.2.2.2. Experimental Determination of Solids Removal

To determine the extent of solids removal as a function of settling velocity a series of experiments are carried out in a laboratory with settling column which is shown in Figure 3.3. The effluent is introduced in the column and stirred to achieve uniform solids concentration. Samples withdrawn from the column at regular time intervals and from a fixed depth (usually 1.5 m) are analyzed with respect to the concentration of suspended solids. Data are used to plot the fraction of suspended solids ($X_s$) as a function of the settling velocity.

Let us consider the measurements of Table 3.1 which are used to plot Figure 3.4. The aim is to calculate total solids removal at a given settling velocity, let us say $V_s = 8\,m/h$. The solids fraction with velocities greater than and equal to 8 m/h is $(1-X_s)=(1-0.8)=0.2$ and will be removed completely. However, solids with velocities less than 8 m/h only will be partly removed; this is given by the integral of Eq. (11) and corresponds to the shaded area of Figure 3.4.

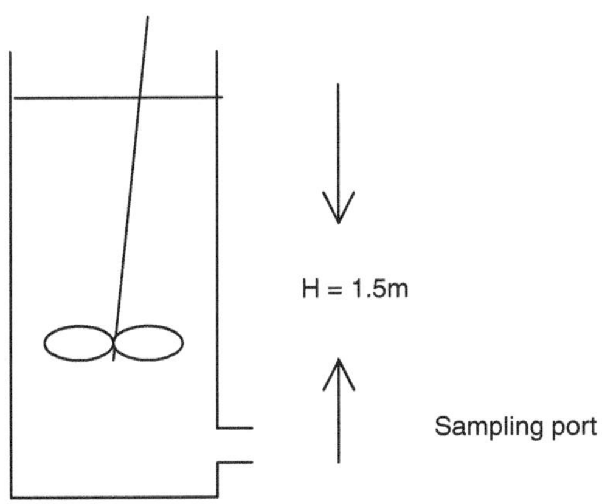


**Figure 3.3.** Sketch of settling column for type I sedimentation.

*Table 3.1.* Example of discrete solids sedimentation

| Settling time (t), min | Settling velocity (H/t), m/h | Fraction in suspension ($X_s$) |
|---|---|---|
| 5 | 18 | 0.95 |
| 10 | 9 | 0.82 |
| 15 | 6 | 0.68 |
| 20 | 4.5 | 0.50 |
| 25 | 3.6 | 0.32 |
| 30 | 3 | 0.25 |
| 50 | 1.8 | 0.08 |

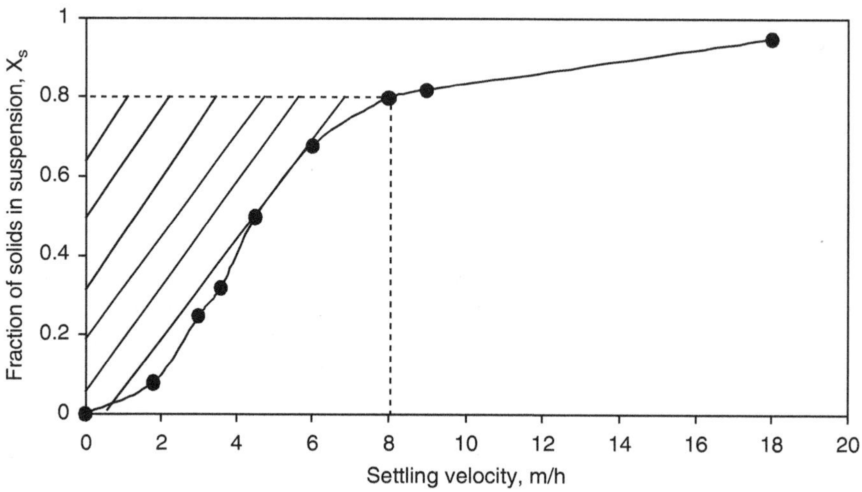


**Figure 3.4.** Fraction of suspended solids at various settling velocities.

The total fraction of removed solids is:

$$X = (1 - 0.8) + \frac{1}{8}\int_{0}^{0,8} V_1 dx = 0.2 + \frac{1}{8}\left[\frac{8+6}{2}(0.8 - 0.68) + \frac{6+4.5}{2}(0.68 - 0.5) + \frac{4.5+3.6}{2}(0.5 - 0.32) + \frac{3.6+3}{2}(0.32 - 0.25) + \frac{3+1.8}{2}(0.25 - 0.08) + \frac{1.8-0}{2}(0.8 - 0)\right]$$

$$= 0.2 + \frac{1}{8}(3.2) = 0.2 + 0.4 = 0.6 \quad (12)$$

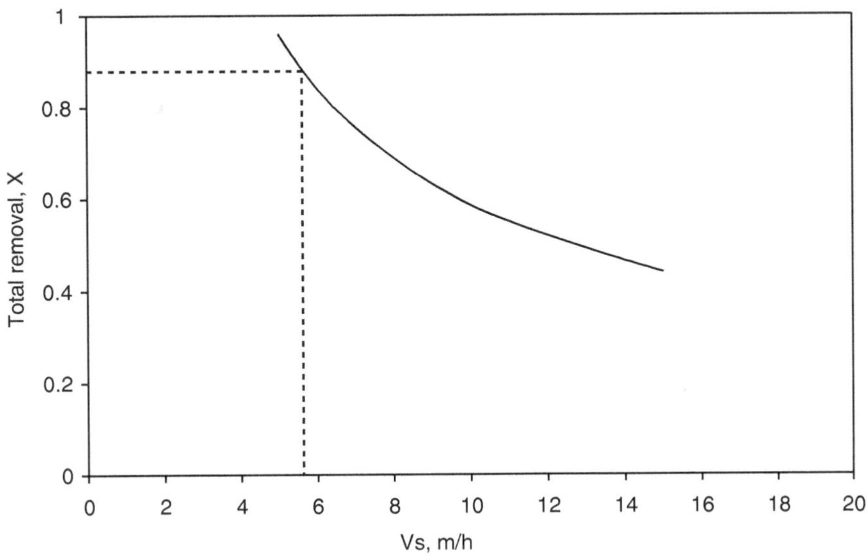


**Figure 3.5.** Total solids removal at various settling velocities.

In a similar manner, one can compute solid removals at various settling velocities, and consequently plot Figure 3.5.

Figure 3.5 is used to determine the settling velocity needed to achieve certain solids removal. This value is then divided by a factor typically between 1.2 and 1.8 to account for any differences between the laboratory, settling column, and the actual tank. The corrected settling value is then used to calculate tank size for a given flow rate.

### 3.2.2.3. Determination of Solids Removal from Particle Size Distribution

Should the particle size distribution and the specific weight be known, total solids removal can be computed according to the Bhargava–Rajagopal relationship:

$$\frac{1}{R} = \left(\frac{U}{177.88 + 44.71U}\right) + \left\{\frac{1}{\exp\left[(0.003186U + 2.036)\ln \mathrm{P}_{10} + \exp\left(\frac{lnU + 4.627}{2.809}\right)\right]}\right\} V_s \left(\frac{2.65-1}{s-1}\right)\left(\frac{\nu_T}{\nu_{30}}\right) \tag{12}$$

where $R$: solids removal, %

$U$: uniformity coefficient of particle size distribution

$P_{10}$: effective size, mm

$V_s$: settling velocity, m/s

$s$: specific gravity

$\nu_T$, $\nu_{30}$: kinematic viscosity at temperature (T) και 30°C, respectively, $m^2/s$.

This relationship was originally developed for relatively concentrated sand (s = 2.65) suspensions at 30°C. For other materials and temperatures correction according to the last two terms of Eq. (12) is needed. The major drawback of this approach is the need to determine particle size distribution, which is done by sieving a known amount of dry solids. The $P_{10}$ value corresponds to that particle size for which 10% w/w of solids are smaller. Uniformity coefficient is defined as the $P_{60}/P_{10}$ ratio, where $P_{60}$ is the particle size for which 60% w/w of solids are smaller.

### 3.2.3. Skimming

Skimming may be required for effluents containing considerable amounts of fats, grease, and oily materials (Figure 3.6). This is done in skimming tanks where these substances float, since they are lighter than water, and can then easily be overflowed. Skimming does not take longer than 3–5 min and can process up to 40 liter of fats for 60 liter/min flow rate.

### 3.2.4. Flow Equalization

Flow equalization is the final, polishing step of preliminary treatment aiming at achieving a uniform feed in terms of composition and flow rate for the following treatment processes. As shown in Figure 3.7, there are two possible configurations for flow equalization, namely: (1) in-line and (2) off-line. In-line conditioning is highly effective since all the flow gets through the equalization tank; nonetheless, this increases treatment cost as large tank volumes are required. Off-line equalization costs less since only a fraction of the flow is equalized; however, this compromises the quality of the effluent as the degree of equalization may be inadequate.

To determine the required flow equalization tank size information regarding the quantities of effluents produced daily is needed. Data are plotted in the form of cumulative effluent volume at various time intervals; Figure 3.8 shows the resulting curves for two typical flow patterns. Moreover, the line corresponding to the

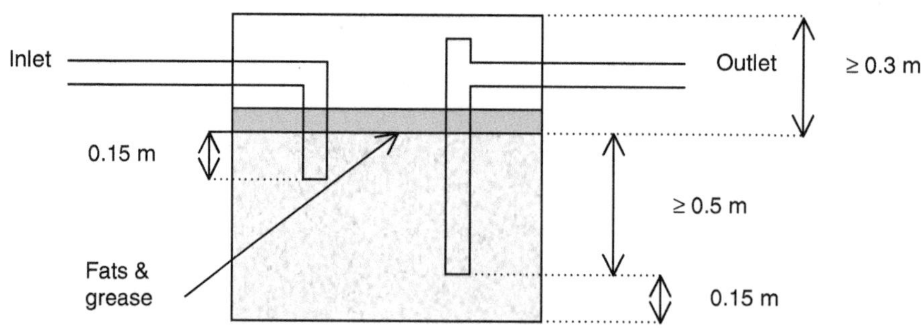


**Figure 3.6.** Sketch of skimming tank.

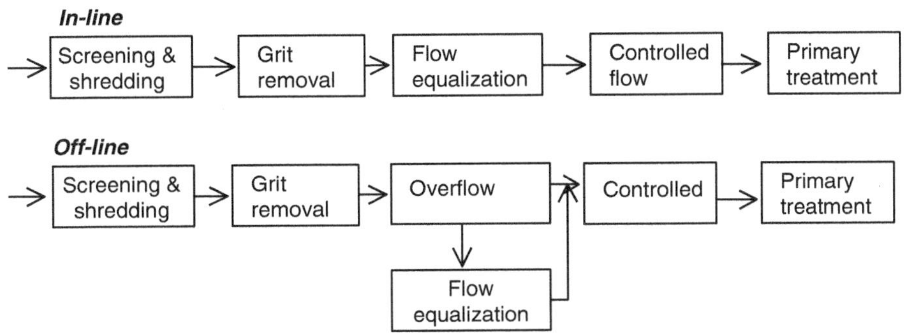


**Figure 3.7.** In-line and off-line flow equalization.

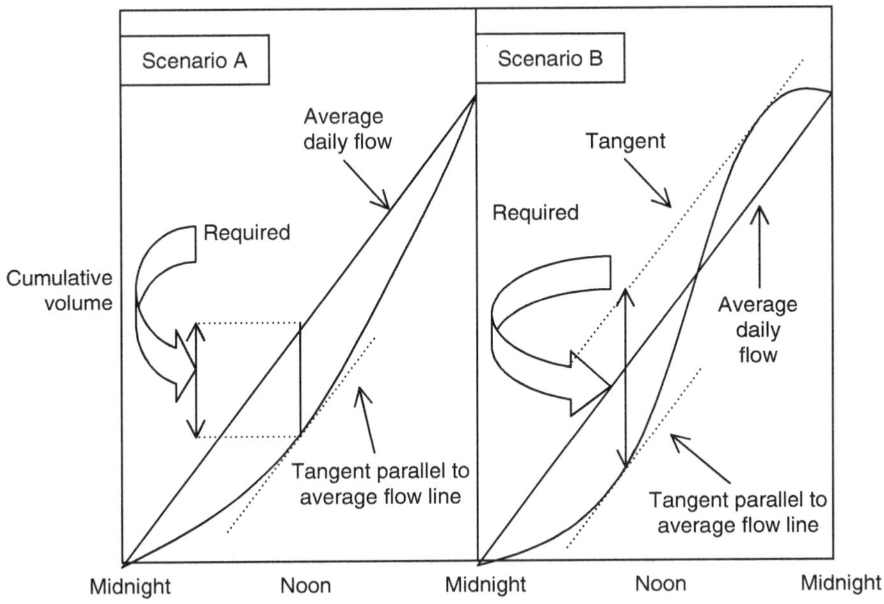


**Figure 3.8.** Graphic determination of flow equalization tank volume.

average daily flow is also drawn; this is a straight line passing through the origin and the last point of the cumulative volume curve.

In the case where the volume curve does not cross the average flow line (e.g., scenario A in Figure 3.8), the required tank volume is equal to the vertical distance between the tangent to the volume curve which is parallel to the average flow line and the average flow line itself. On the other hand, if the volume curve does cross the average flow line (e.g., scenario B in Figure 3.8) the required tank volume is equal to the vertical distance between the tangents to the upper and lower parts of the curves which are parallel to the average flow line. In practice, the tank size is always 10–20% greater than that computed according to the aforementioned procedure to account for any unexpectedly high flow fluctuations.

## 3.3. UNIT OPERATIONS IN PRIMARY TREATMENT

### 3.3.1. Primary Sedimentation

Primary sedimentation aims to remove organic and inorganic suspended solids whose size ranges from 0.001 to 0.1 mm. The process may achieve SS and BOD removal up to about 70% and 40%, respectively, and is a flocculent (type II) settling.

#### 3.3.1.1. Experimental Determination of Solids Removal

To design a primary sedimentation tank experiments are carried out in a laboratory with a settling column (Figure 3.9) whose height is equal to the depth of the sedimentation tank (e.g., 3 m). Unlike the settling column used in grit removal, this one has several sampling ports located at equal intervals, typically 0.6 m. The effluent is introduced in the column and stirred to achieve uniform solids concentration. Samples are then withdrawn from the various sampling ports at regular time intervals and are analyzed with respect to the concentration of suspended solids.

From the measured solids concentration ($c$) and the known initial concentration ($c_0$) the fraction of solids that have settled at a given time and depth is calculated as $100 \times (c_0 - c)/c_0$. These values then can be used to plot the so-called isoconcentration trajectories shown in Figure 3.10; these trajectories are essential in

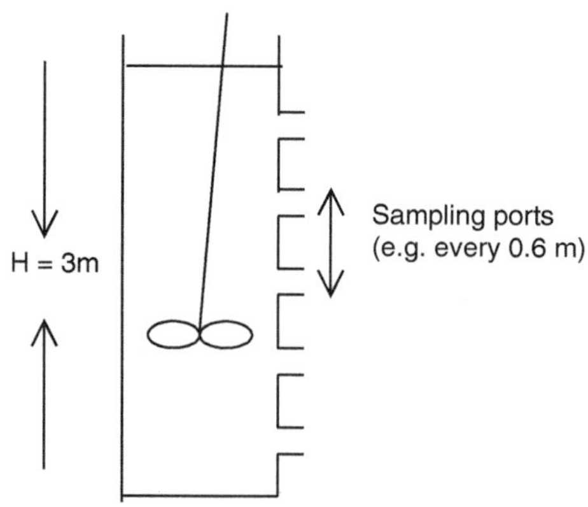


**Figure 3.9.** Sketch of settling column for type II sedimentation.

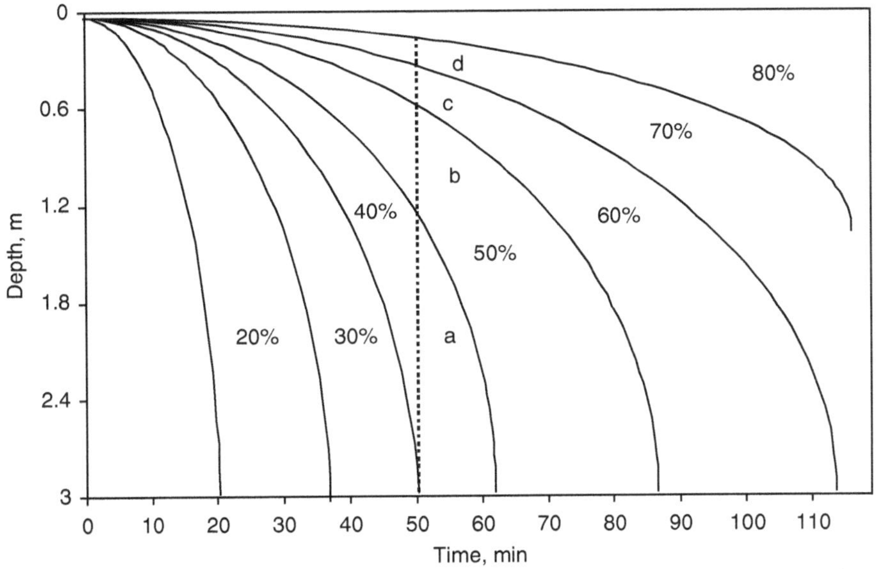


**Figure 3.10.** Isoconcentration trajectories for flocculent sedimentation.

*Table 3.2.* Example of flocculent sedimentation

| Interval | Mean depth ($h$), m | Solids removal ($h/H \times$ interval), % |
|---|---|---|
| 0–40 | 3 | $3/3 \times (40–0) = 40$ |
| 40–50 | 3-a/2 | $2.14/3 \times (50–40) = 7.1$ |
| 50–60 | 3-a-b/2 | $0.92/3 \times (60–50) = 3.1$ |
| 60–70 | 3-a-b-c/2 | $0.43/3 \times (70–60) = 1.4$ |
| 70–80 | 3-a-b-c-d/2 | $0.24/3 \times (80–70) = 0.8$ |

calculating the extent of suspended solids removal as a function of the tank residence time and the settling velocity.

Let us consider a 3 m deep tank where we want to calculate solids removal after 50 min (this corresponds to a settling velocity of $3/50 = 0.06$ m/min $= 86.4$ m/d) according to the trajectories shown in Figure 3.10. It is obvious that, at the conditions in question, 40% of solids are removed completely; for the remaining intervals (40–50%, 50–60%, and so on) a fraction of solids will be removed; this is equal to the $h/H$ ratio where ($h$) is the mean settling depth of the 40–50%, 50–60%, etc., intervals and ($H$) is the total tank depth.

For the various intervals shown in Figure 3.10, the mean settling depth and the respective removal are shown in Table 3.2. This procedure then can be repeated for several other residence times (and consequently settling velocities) and the computed overall removal values can be used to plot Figures 3.11 and 3.12.

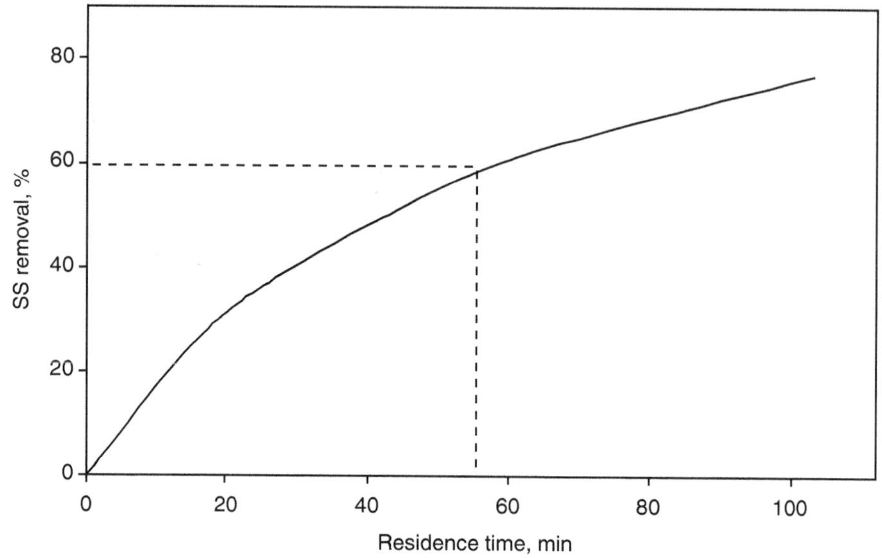


**Figure 3.11.** Solids removal for various residence times.

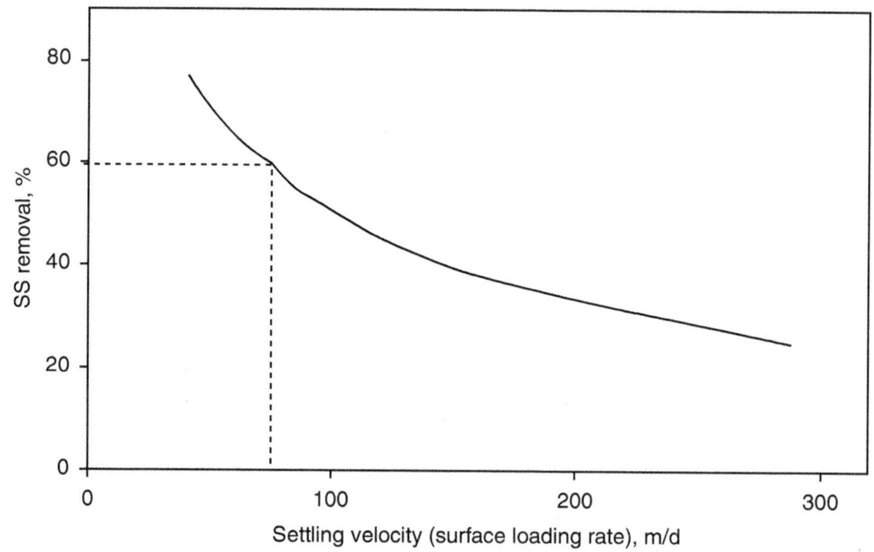


**Figure 3.12.** Solids removal for various settling velocities.

In practice, the experimentally computed residence times are multiplied by a factor of about 1.2–1.5 to account for differences between the laboratory column and the actual tank. In a similar fashion, overflow loadings are divided by the same correction factor.

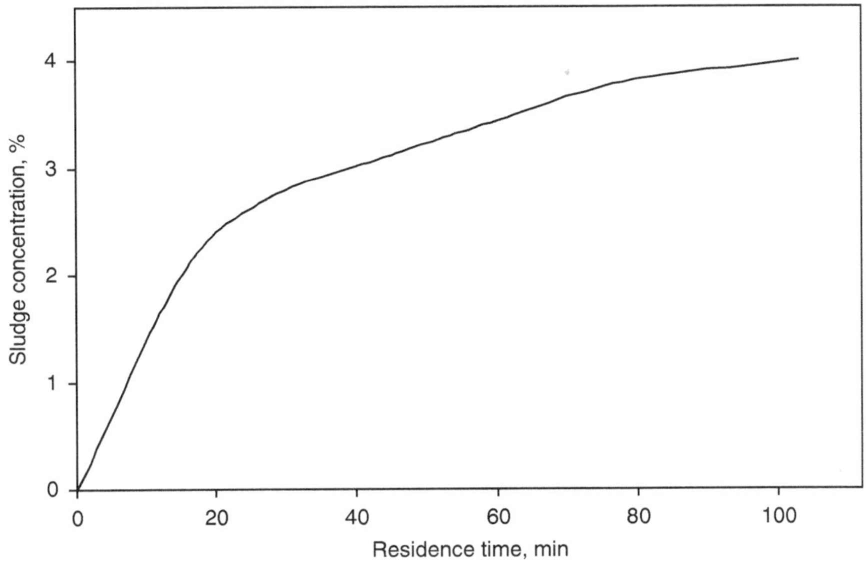


**Figure 3.13.** Sludge concentrations for various residence times.

Other than the aforementioned parameters, it is interesting to know the concentration of sludge accumulating at the bottom of the tank. Samples taken from the bottom are analyzed with respect to solids concentration and the results are plotted as in Figure 3.13.

The primary sludge (including solids and liquid) flow rate is calculated as follows:

$$Q_s = \frac{M}{1000 c_s s} \tag{13}$$

where $Q_s$: volumetric flow-rate, $m^3/d$
$M$: mass of solids that settle, kg/d
$C_s$: sludge concentration, % (typically up to about 6%)
$s$: sludge specific gravity (typically between 1.03 and 1.05).

Since solids removal is a strong function of residence time and surface loading rate (as was seen in Figures 3.11 and 3.12), various correlations have been proposed and some typical expressions are as follows:

$$R = 0.82 \exp(-0.008q) \tag{14}$$

$$R = 0.51 \exp(-0.0029q) \tag{15}$$

$$R = (0.00043 S_0 + 0.51)[1 - \exp(-0.7\theta)] \tag{16}$$

$$R = 1 - \frac{1}{1.1(\log S_0 \theta^{\log S_0/10})} \tag{17}$$

where $R$: solids removal, %
$q$: surface loading rate (settling velocity), $m^3/(m^2d)$
$S_0$: initial concentrations of suspended solids, mg/L
$\theta$: residence time, h.

It should be noticed that Eqs. (14)–(17) (as well as others that are not cited here) should be used with caution and skepticism in designing a specific process since the conditions (e.g., type of effluent, type and size of tank) from which they were originally developed may be significantly different from those of the process in question.

### 3.3.2. Flotation

Flotation aims at the removal of fine suspended particles whose separation by sedimentation is practically infeasible as it would require very long treatment times. In general, particles with specific gravity less than unity tend to float spontaneously (e.g., as in the case of skimming); on the other hand, flotation of particles that are only marginally heavier than water can be assisted by introducing fine air bubbles in the liquid. In this case, the bubbles attach to the particulate matter and the buoyant force of the combined particle and gas bubbles is great enough to bring the particle to the surface.

There are three distinct methods to introduce the gas into the liquid as follows:

1. Dissolved air flotation. Air is dissolved in the liquid under a pressure of several atmospheres which is released and the air is then liberated in the form of fine bubbles 40–70 μm in diameter. This is the most common method used in wastewater treatment.
2. Air flotation. In this case, air is introduced directly to the liquid through diffusers or revolving impellers.
3. Vacuum flotation. It consists of saturating the liquid with air followed by the application of partial vacuum which causes the dissolved air to come out of the solution as fine bubbles.
4. Electrolytic flotation. It is based on the formation of oxygen and hydrogen bubbles from water electrolysis.

Regardless of how air is dissolved in the liquid phase, removal efficiency can be enhanced with the addition of various chemicals such as aluminum and iron salts. These tend to increase and/or alter the particle surface, thus facilitating bubble attachment.

#### 3.3.2.1. Analysis of Dissolved Air Flotation

The efficiency of dissolved air flotation is a function of the ratio of the amount of air ($A$) needed to the amount of solids ($S$) to achieve a certain degree of separation, which in turn strongly depends on the type of effluent under consideration. The $(A/S)$ ratio is correlated to gas solubility, solids concentration, and the applied pressure as follows:

$$(A/S) = \frac{1.3 s_a (fP - 1)}{s_s} \tag{18}$$

where $s_a$: solubility of air, ml/L (equal to 18.7 ml/L at 20°C)
$P$: applied pressure, atm
$f$: fraction of air that is dissolved at pressure $P$ (this is a correction factor typically equal to 0.5 to take into account inefficient saturation)
$s_s$: concentration of solids, mg/L.

It should be noted that Eq. (18) assumes that all the flow is pressurized and this indeed is the case for relatively small treatment units. For units handling large effluent volumes, only a portion of the flow (about 10–30%) is pressurized to keep treatment costs low (e.g., pressurizing systems, tank construction); a stream taken from the exit of the flotation tank is pressurized to a pressure tank and finally recycled back to the flotation tank. In this case, Eq. (18) becomes:

$$(A/S) = \frac{1.3 s_a (fP - 1)}{s_s} r \tag{19}$$

where $r$: recycle ratio (i.e., pressurized flow over total flow).

Equations (18)–(19) are used to calculate the required operating pressure. In either equation, 1.3 is the mass (mg) of 1 ml of air, while the $(-1)$ term shown in brackets is due to the fact that the system, following pressure release, operates at atmospheric conditions.

The area of the flotation tank is computed from the volumetric flow rate and the rise velocity of the combined particle air bubble as follows:

$$A = \frac{Q}{V_r} \tag{20}$$

where $A$: tank area
$Q$: volumetric flow rate
$V_r$: rise velocity of the combined particle and air bubble.

Like the terminal settling velocity for discrete particles sedimentation [i.e., Eq. (7)], the terminal rise velocity of the combined particle and air bubble for spherical agglomerates and Stokes' flow regime (i.e., $Re < 0.2$) is:

$$V_r = \frac{\rho - \rho_s}{18\mu} g d^2 \tag{21}$$

where $\rho_s$, $d$: density and diameter of the agglomerate, $kg/m^3$ and m, respectively
$\rho$, $\mu$: density and viscosity of the liquid, $kg/m^3$ and kg/(m.s), respectively.

## 3.4. THE ACTIVATED SLUDGE PROCESS

Secondary treatment is the heart of a typical WWTP and aims to remove organic matter by biological processes. The activated sludge process is the most common treatment employed in this stage and basically comprises an aeration tank where microorganisms aerobically degrade dissolved organics and a secondary settling tank where the sludge is separated from the clarified water. The process is schematically shown in Figure 3.14.

Air bubbling through porous or nonporous diffusers is the preferred method of aeration. Alternatively, aeration by mechanical aerators (these may be surface or submerged) may be employed. Aeration is a gas–liquid mass transfer process with oxygen concentration changing with time as follows:

$$\frac{dC}{dt} = K_L a(C_s - C_L) \tag{22}$$

where $dC/dt$: rate of oxygen concentration change, mol/(L per s)
$K_L a$: mass transfer coefficient, 1/s
$C_s$: oxygen saturation concentration, mol/L
$C_L$: oxygen concentration in the liquid, mol/L.

To achieve increased oxygen mass transfer high values of $K_L a$ are needed. This can be done increasing the temperature and the depth from which bubbling is provided as well as stirring the effluent. Furthermore, $K_L a$ values are strongly affected by the effluent composition and concentration; for instance, the presence of surfactants and other organics in the effluent may block oxygen mass transfer from air to the liquid. Depending on the effluent characteristics, oxygen mass transfer to the liquid phase may be as much as 60% lower than that to pure water.

The purpose of secondary sedimentation is twofold: (1) to remove the suspended solids from the mixed liquor, thus clarifying the effluent of the secondary treatment, and (2) to thicken the sludge which is partly recycled to the aeration tank to maintain a constant concentration of microorganisms, while the rest is removed from the system and undergoes further treatment. Both hindered settling (type III)

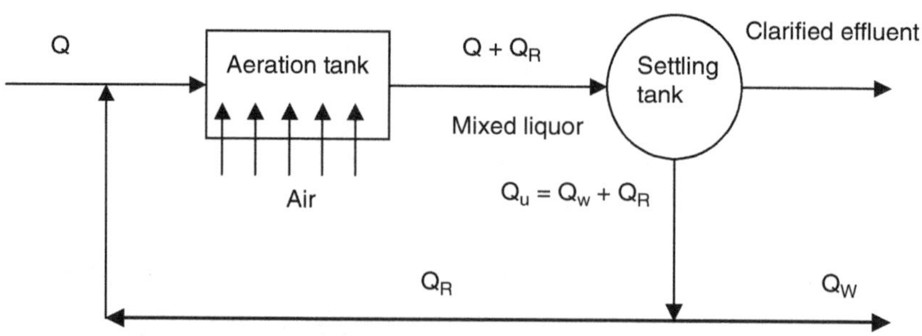


**Figure 3.14.** The activated sludge process.

and compression (type IV) occur in secondary settling tanks since the concentration of solids is considerably high. Figure 3.15 shows the various settling phases occurring in a secondary sedimentation tank as these would occur in a graduated settling column of height $H_o$.

(A) The mixed liquor is well stirred to achieve a uniform solids concentration. At $t = 0$, the initial (uniform) concentration is $C_o$.
(B) For $t < t_1$, the particles are continuously settling as a zone, while a clear layer of liquid above the particles is being formed. At the same time, a solids compression zone is formed at the bottom of the column accompanied by a second clear layer above the particles.
(C) Phase (B) is a dynamic, transitional state that ends at $t = t_1$ when the two interfaces merge and the resulting zone has a uniform solids concentration ($C_1$).
(D) Compression occurs and the solids reach their final concentration ($C_u$) at $t = t_u$.

### 3.4.1. The Initial Settling Velocity (ISV) Method

With reference to Figure 3.15, Figure 3.16 shows the change of solids–liquid interface height with time when the mixed liquor is left to settle in a graduated column. This information is used to determine the settling tank area needed for clarification and thickening as follows:

The area needed for clarification is given by:

$$A_c = \frac{Q}{V_s} = \frac{Q}{-dH/dt} \tag{23}$$

where $A_c$: clarification area, m$^2$
$Q$: volumetric flow rate to secondary settling tank
$V_s$: initial settling velocity, i.e., just before the solids concentration becomes $C_1$.

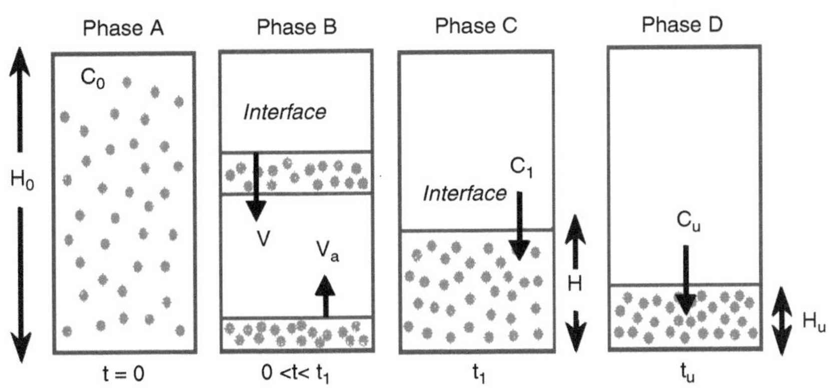


**Figure 3.15.** Processes occurring in activated sludge settling tank.

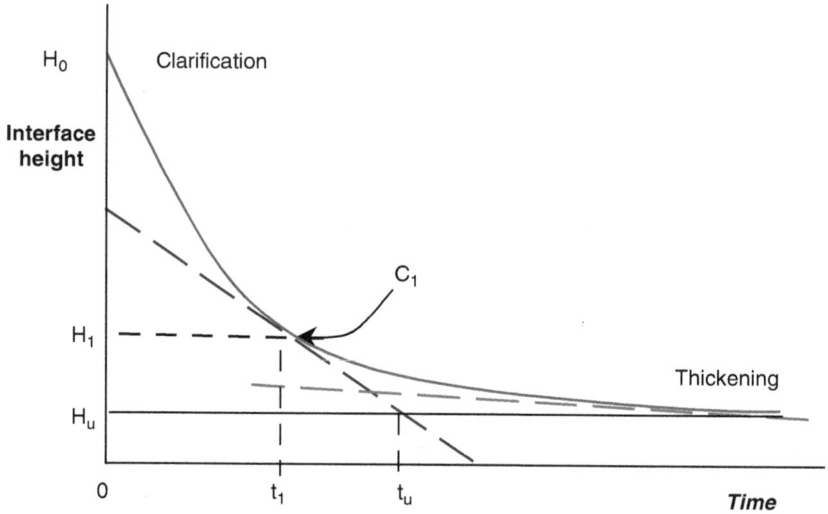


**Figure 3.16.** The ISV approach to calculate secondary settling tank areas.

The initial settling velocity is determined graphically from the tangent to the *H-t* curve in the clarification region.

The area required for thickening is given as follows:

$$A_t = \frac{Qt_u}{H_o} \tag{24}$$

where $A_t$: thickening area, m$^2$
$H_o$: initial height of interface (i.e., height of column)
$t_u$: time needed to achieve the desirable concentration of thickened sludge $C_u$.

Graphic determination of $t_u$ includes the following steps:

1. Draw the tangents to the *H-t* curve in the clarification and thickening regions. The intersection between the bisection of the resulting angle and the *H-t* curve corresponds to the critical concentration $C_1$.
2. Draw the tangent to the *H-t* curve that passes from $C_1$.
3. The height of the interface in the thickening region ($H_u$) is calculated as follows:

$$H_u = \frac{H_o C_o}{C_u} \tag{25}$$

4. From $H_u$, draw the horizontal whose intersection with the tangent passing from $C_1$ corresponds to $t_u$.

In practice, the area needed for thickening will expectedly be greater than that needed for clarification; therefore, determination of the former suffices for the rational design of the settling tank. Other than the ISV method, the area needed for thickening (but not that of clarification) may be determined experimentally according to the solids flux method. However, detailed analysis of this method lies outside the scope of this chapter.

## 3.5. RECOMMENDATIONS FOR FURTHER READING

There are several comprehensive textbooks in the literature covering various aspects of wastewater treatment including, among others, the fundamentals and the engineering of unit operations. Some the author found particularly useful are:

Droste, R. L., 1997, *Theory and Practice of Water and Wastewater Treatment*, John Wiley & Sons.

Hammer, M. J., and Hammer M. J. Jr., 1996, *Water and Wastewater Technology*, 3rd Edition, Prentice Hall.

Lin, S., 2001, *Water and Wastewater Calculations Manual*, McGraw-Hill.

Tchobanoglous, G., and Burton, F.L., 1991, *Wastewater Engineering – Treatment, Disposal and Reuse*, 3rd Edition, McGraw-Hill.

Vesilind, P. A., 2003, *Wastewater Treatment Plant Design*, IWA Publishing.

# 4

# Anaerobic Digestion of Organic Residues and Wastes

Gunther Pesta

## 4.1. INTRODUCTION

Anaerobic microbial conversion of organic matter into a renewable energy source, so-called biogas, is a well-established process and state of the art. New techniques and technologies offer possibilities to treat pasty and solid organic wastes by means of anaerobic digestion (AD) as well as liquids.

The implementation of this environmentally friendly technique depends widely on a political framework, creating and providing an economically attractive incentive for running AD plants (biogas plants). In Germany changing legislative restrictions and acts on granting priority to renewable energy sources offer thought-provoking impulses. In 2005, approximately 2,400 biogas plants in Germany with a total capacity of 450 $MW_{el}$ demonstrate the technical and economical feasibility of AD of organic wastes and waste water (Fachverband Biogas, 2005).

During the last years legislative restrictions concerning waste treatment and environmental protection were aggravated. Among others companies of the food industry were forced to rearrange their waste management. AD is one of the most interesting techniques available for organic production residues and wastes.

This chapter illustrates principles of anaerobic codigestion and typical factors for operating a biogas plant. To highlight the benefits of treating organic residues and wastes by means of AD differences between aerobic and anaerobic microbial degradation are described in principle.

## 4.2. PRINCIPLES OF ANAEROBIC DIGESTION: TERMS OF COMPREHENSION

The primary aim of AD depicts a single substrate treatment, mainly animal manures at agricultural biogas plants for producing biogas and supplying a well-balanced

GUNTHER PESTA • Technische Universität München, Chair for Energy and Environmental Technologies of the Food Industry, Weihenstephaner Steig 22, 85350 Freising-Weihenstephan, Germany

liquid fertilizer. Further traditional applications are municipal wastewater treatment plants, where surplus sewage sludge is stabilized by using an AD process as well as industrial plants, pretreating organically high-loaded wastewaters. By using only one substrate the digestion process is termed ''monodigestion,'' synonymously spoken ''monofermentation.'' The single substrate is customarily called monosubstrate.

Nowadays AD offers an alternative treatment for a wide range of substrates of both agricultural or industrial origin. Basically any liquid or solid organic wastes from, e.g., food and agricultural industry or municipal waste collections, like whey, slaughterhouse wastes, flotation fat, manure and energy crops, etc., can be utilized for AD. Table 1 exemplifies this wide range of substrates for AD, so-called cosubstrates. Besides the channeling of waste treatment the production and usage of biogas is a primary aspect using AD for industrial waste management.

Although most of substrates listed in Table 4.1 are proper for monodigestion the majority of biogas plants are operated with a mixture of various substrates. By using a homogenous mixture of two or more substrates the digestion process is termed ''co-digestion,'' as well as ''cofermentation.'' The applied substrates are so called ''co-substrates.'' The ratio of a single substrate within the applied mixture of substrates is irrelevant for the termination.

Braun and Wellinger (2002) quoted two major drivers that helped to promote co-digestion. On the one hand usually there are oversized digesters in wastewater

*Table 4.1.* Variety of substrates for anaerobic digestion (adapted from Kuhn, 1995)

| Substrate | DM[a] (%) | ODM[b] (% DM) | C/N | Biogas yield ($m^3CH_4$/kg ODM) |
|---|---|---|---|---|
| Chicken manure | 15 | 77 | 7 | 0.2–0.4 |
| Pig manure | 5–7 | 77–85 | 5–10 | 0.2–0.3 |
| Raw glycerol (biodiesel) | > 98 | 90–93 | – | 0.69–0.72 |
| Rotten potatoes | 25 | 79 | 25 | 0.5–0.6 |
| Clover | 20 | 80 | 12 | 0.4–0.5 |
| Apple slops | 25 | 86 | 30 | 0.3–0.4 |
| Spent grain | 20–22 | 87–90 | 10 | 0.6–0.7 |
| Bread (waste) | 90 | 96–98 | 42 | 0.7–0.75 |
| Cacao peels | 95 | 91 | 2.8 | 20–22 |
| Molasses | 80 | 95 | 14–27 | 0.3 |
| Whey | 95 | – | 27 | 0.5–0.6 |
| Rape seed slops | 92 | 97 | 9–12 | 0.58–0.62 |
| Green waste | 60–75 | 30–70 | 40–80 | 0.2–0.6 |
| Flotating sludges (fat) | 5–24 | 83–98 | – | 0.6–0.8 |
| Feces (intestinal) | 12–15 | 80–84 | 17–21 | 0.2–0.3 |
| Rumen (pressed) | 20–45 | 90 | 11–20 | 0.6–0.7 |
| Animal meal | 8–25 | 90 | – | 0.5–0.8 |
| Fat (separators) | 35–70 | 96 | – | 0.7 (1.0) |
| Grass | 21–23 | 76–80 | 22–24 | 0.45–0.5 |

[a] dry matter
[b] organic dry matter

treatment plants. Addition of cosubstrates helps to produce more biogas. Otherwise there are agricultural biogas plants, where biogas production from manure alone is not economically viable. Cosubstrates increase the income not only from a higher biogas yield but also through tipping fees.

According to this, most common applications of codigestion can be found in agricultural biogas plants by using a basic substrate like cow manure and adding minor amounts of additional substrates, like flotation fats, biowastes, maize or grass silage. Nearly 85% of agricultural biogas plants in Germany are running by means of codigestion. Codigestion also is a well-known practice at municipal wastewater treatment plants, when digesting aerobic sludge surplus with organic residues out of (agro)industrial processing, commercial wastes, or municipal solid organic wastes.

These two areas of AD application differ widely in the actions needed, e.g., technical equipment, economic calculation. Nevertheless the basic principles of AD and the influence on the process are the same, when adding cosubstrates to the digester.

## 4.3. PRINCIPLES OF ANAEROBIC MICROBIAL TREATMENT OF ORGANIC MATTER

### 4.3.1. Aerobic Treatment with Anaerobic Stabilization of Sludge

Biological wastewater treatment in municipal sewage plants is usually an aerobic process. To stabilize the sludge surplus AD processes are widely used. Symbiotic metabolizing bacteria convert organic carbon under anaerobic conditions predominantly into methane and carbon dioxide, so-called biogas.

Figure 4.1 shows a carbon mass balance of aerobic wastewater treatment with an anaerobic stabilization of sewage sludge surplus. According to the diagrammed carbon flow scheme about 50% organic carbon is lost by producing worthless $CO_2$ and about 50% by producing biomass and sewage sludge, respectively. After stabilizing the sludge surplus by AD or drying, the sludge residual needs to be treated adequately, causing additional costs.

The utilization of the AD by-product biogas in combined heat and power units (CHP) is technically and economically feasible. In general, electric power as well as the heat are directly used in the sewage plant, the latter to temper the digester.

### 4.3.2. Anaerobic Treatment and Biogas Production

The organic carbon freight of wastewater is a valuable energy source rather than a worthless wastewater component. Due to this fact it is obvious to obtain a benefit by cleaning the wastewater and thereafter providing an energy source. Instead of spending additional energy to reduce carbon freight by aeration it is reasonable to produce biogas, a renewable energy source.

Figure 4.2 shows a carbon mass balance of anaerobic wastewater treatment. Treating wastewater by AD converts more than 90% of its organic carbon freight

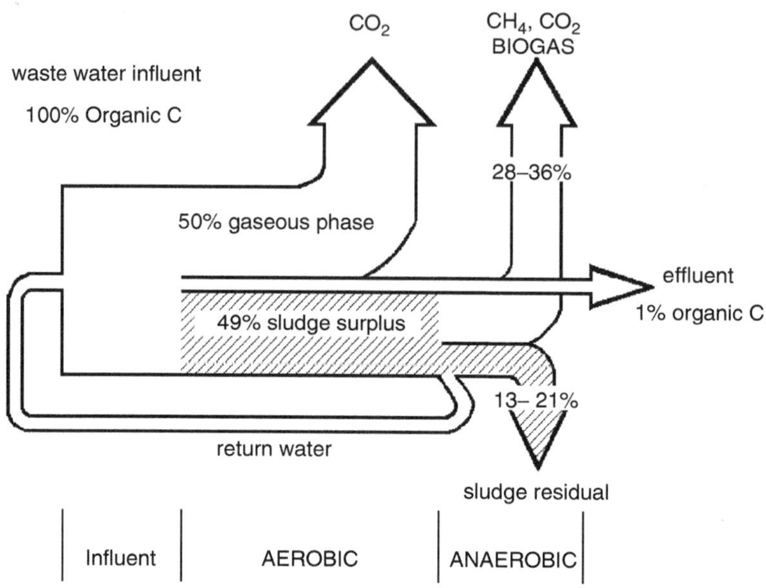


**Figure 4.1.** Carbon mass balance of aerobic waste water treatment.

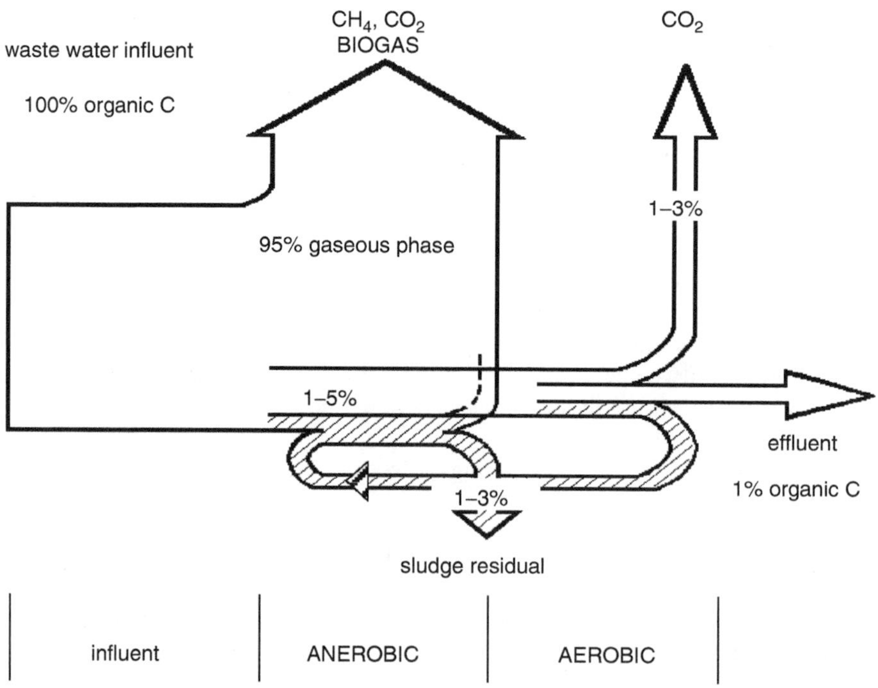


**Figure 4.2.** Carbon mass balance of anaerobic wastewater treatment.

into biogas. To reduce finally the COD of the fermentation residues to a runoff ditch level, an advanced effluent treatment by aeration is still necessary.

### 4.3.3. Anaerobic Digestion and Biogas Production

The degradation of substrate by anaerobic bacteria takes place within two main phases: hydrolysis (decomposing of organic material) and methanogenesis (production of biogas). The degree of degradation as well as biogas quality can be affected in various ways by technological and procedural measures. Figure 4.3 shows the single steps of anaerobic digestion processes within the two main phases.

- Step 1: Hydrolysis. Fermentative bacterias hydrolyze by means of their excellular enzymes within the first step of AD complex structured substrate, e.g., carbohydrates, proteins, or lipids, into simple water-soluble compounds, such as monosugars, amino acids, or fatty acids.
- Step 2: Acidification. The intermediates out of hydrolysis are further broken down into water-soluble organic end products, mainly short-chained fatty acids, alcohols, and carbon dioxide gas ($CO_2$). These conversions are undertaken by the same fermentative bacterias that are responsible for the hydrolysis, displacing the process into the cell itself. The educts are incorporated and digested by the fermentative bacterias that segregate the products mentioned above.
- Step 3: Acetogenesis. Acetogenesis is a linkage between the degradation into water-soluble compounds and methane formation. End products of microbial metabolisms are converted into short-chained volatile fatty acids (VFAs), mainly into actetate and $CO_2$. This conversion can only take place at a low concentration of hydrogen ($H_2$), which is produced during acetogenesis as a by-product. That is why the acetogenic bacterias live in symbiosis with methanogenic bacterias, which are using $H_2$ for the formation of methane. Beside this, both acetate and $CO_2$ are the main educts for the methane-forming bacterias.
- Step 4: Methanogenesis. Methanogenesis or methane formation depicts the last step of the anaerobic digestion process. Methanogenic bacteria (species: archea) belong to the oldest organisms of the world. Their metabolism is very specialized: All species are able to convert $CO_2$, most of them $H_2$, a few acetate, and only one methanol. Methanogenics are strictly anaerobic. Oxygen inhibits their metabolism or mortifies the microorganisms.

Apparently hydrolysis of polymeric substances in anaerobic environments seems to be the rate-limiting step in the overall conversion of organic matter to methane. Hydrolysis is the primary step in anaerobic digestion, but in fact is not the one that controls the process. Methanogenis rather is the rate-limiting step. Methanogenic bacteria are slow growing and sensitive to changes in their environmental parameters, like pH or temperature. Due to their strict anaerobic metabolism any

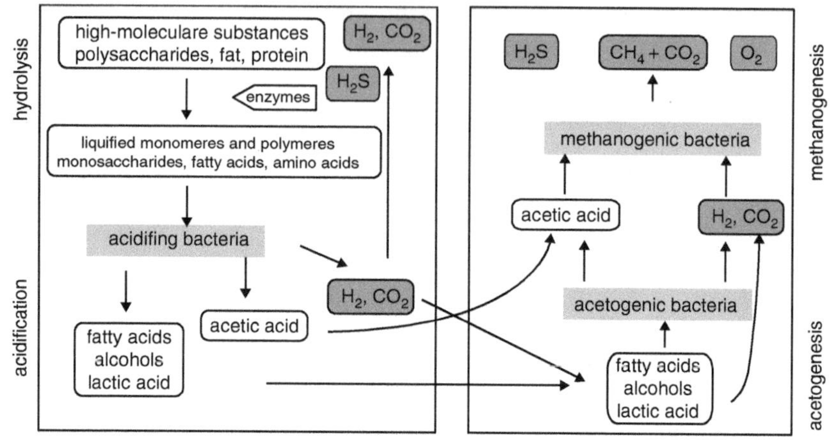


**Figure 4.3.** Degradation steps of anaerobic digestion (according to Grepmeier, in Pesta, 2004).

source of oxygen inhibits their growth. Nevertheless, the products of the hydrolytic reactions are fermented by both hydrolytic and nonhydrolytic bacteria.

### 4.3.4. Biogas: a Renewable Energy Source

As a result of the anaerobic digestion process, biogas is produced. Biogas is a mixture of methane ($CH_4$, 50–85% by volume), carbondioxide ($CO_2$, 15–50% by volume), and trace gases, e.g., water ($H_2O$), hydrogen sulfide ($H_2S$), or hydrogen ($H_2$). Although the small concentrations of these trace gases have little effect on the physical properties of biogas, they influence the selection of technologies for cleaning and utilizing biogas. At least dewatering and the removal of $H_2S$ are necessary before biogas is utilizable.

Substrate composition is a major factor affecting the methane yield and methane production rates. Table 4.2 shows the specific biogas yields and qualities of carbohydrates, lipids, and proteins.

Due to the composition of the biodegradable organic compounds such as lipids and proteins, a higher percentage of methane is produced, compared to oxidized compounds like sugars. Ultimate methane yields, however, are influenced principally by the biodegradability of the organic components. Certain natural organic compounds such as lignin are refractory to decomposition under anaerobic conditions even at long residence time. This refractory property is related to the lack of enzymes to carry out the initial hydrolytic reactions. It also has been established that certain compounds can build a complex with others rendering them resistant to anaerobic decomposition. Nevertheless, the digestion of cosubstrates results in a better quality and higher quantity of biogas yield, compared to a monodigestion of manure or sewage sludge.

*Table 4.2*. Specific biogas yields of substrates main components

| | Gas yield (L/kg ODM) | $CH_4$ (% by volume) | $CO_2$ (% by volume) | Calorific value (kWh/kg ODM) |
|---|---|---|---|---|
| Carbohydrates | 790 | 50 | 50 | 4.0 |
| Lipids | 1250 | 68 | 32 | 4.9 |
| Proteins | 700 | 71 | 29 | 8.0 |

In general, biogas is used in combined heat and power plants (CHP) to generate electric power and thermal energy. Also, it substitutes natural gas in gaseous fuel steam vessels to produce steam. Biogas as a renewable energy source helps to reduce the consumption of fossil energy carriers and also $CO_2$ emissions. In additon to the $CO_2$-trading guidelines biogas has an emission factor of 0 tons $CO_2$ per TJ compared to fuel oil with 78 t and natural gas with 56 t $CO_2$ per TJ (EC, 2004).

The calorific value (c.v.) of biogas depends on its $CH_4$ content and varies between 4 and 7.5 $kWh/m_N^3$. The average c.v. of 6.0 $kWh/m_N^3$ is equivalent to 0.62 liter fuel oil (65% by volume $CH_4$).

### 4.3.5. Two-phase Digesting Systems

Generally, all four single steps of AD shown in Figure 4.2 take place within one single digester, termed single-phase system or digestion. To improve stability and rate of degradation preferably the first two steps of hydrolysis and acidification are specially separated from the others, a so-called two-phase system is installed. A hydrolysis digester has to be advanced to a methanogenic digester. The methanogenic digester is fed with the thoroughly hydrolyzed substrate out of the hydrolysis digester, acetogenesis, and methane formation startup. Figure 4.4 shows a simplified flow scheme of a two-phase AD plant for brewery waste water. The advantage of a two-phase fermentation system is the possibility to exert influence on the main degradation processes, hydrolysis and methanogenesis.

Two-phase systems are preferably used for inhomogeneous substrates differing in their biodegradability. Microorganisms always degrade substrates first, of which assimilation or dissimilation takes less energy. At a predicted dwell time and a continuous feeding of a mixture of easy and difficult degradeable substrates the microorganisms utilize the easier degradable ones first. The difficult degradable substrate will be utilized only fragmentarily or even not at all. Two-phase fermentation systems reach a COD-reduction up to 98%, even with inhomogeneous substrates (Behmel, 1993).

The qualitiy of biogas also may be improved by leading off the exhaust biogas of the hydrolysis digester preferably by a biofilter. Biogas produced in the hydrolysis digester consits of 85–90% $CO_2$, 5–10% $CH_4$, and an exceeding amount of olfactory active substances. Biogas from the methane digester is rich in $CH_4$ and lesser efforts are necessary to remove trace gases, like $H_2S$.

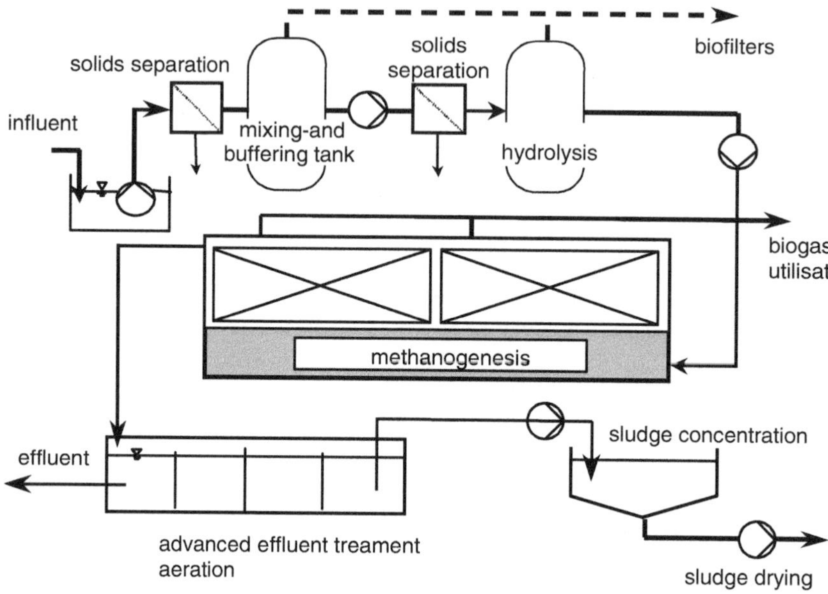


**Figure 4.4.** Scheme of a two-phase AD plant, e.g., in a brewery.

## 4.4. BENEFITS AND LIMITS OF CODIGESTION

The benefits of codigestion are obviously and mainly economical, and to a lesser extent technological and not to forget ecological advantages, such as dilution of potential toxic compounds, improved balance of nutrients, synergistic effects of microorganisms, increased load of biodegradable organic matter, and better biogas yield. There are additional advantages including hygienic stabilization, increased digestion rate, etc., when the process occurs under thermophilic conditions (Sosnowski, 2003).

It still has to be borne in mind that codigestion also requires additional equipment and investment costs. For a successful operation of a biogas plant with cosubstrates several important facts have to be considered. In the following the implementation of co-digestion is discussed in general. Details have to be checked in each single application.

### 4.4.1. Legislative Restrictions for Codigestion

First it is necessary to become acquaintanced with legislative restrictions, technical rules, and guidelines concerning the implementation of cosubstrates in biogas plants as well as handling digestate. In Europe the differences between the singular national statutes disappear due to a centralized legislative. Besides environmental legislation, specific acts on wastes need to be considered, favoring composting and AD. Most important are (Braun and Wellinger, 2002):

- landfill legislation,
- soil protection legislation,
- groundwater protection legislation,
- waste treatment legislation,
- human and animal health rules and sterilization guidelines, and
- waste recovery rules and regulations (guidelines) and the legal definition of anaerobic digestion as a waste treatment or as a recovery process.

A more detailed view on legislations in different European countries regarding implementation of AD is given in Nordberg (2004).

### 4.4.2. Composition of Cosubstrates

Although it should be taken for granted, it is necessary to mention explicitly that the origin and composition of cosubstrates have to be well-known. In general, substrates emerge from agricultural sources, agricultural-related industries such as food and beverage industry, and also municipal waste collections (refer to Table 4.1). The high variety of sources determines the variety of compositions of cosubstrates, physical and chemical. As an example the composition of municipal solid organic waste (MSOW) depends on factors like area of collection (rural, urban), seasons of the year, or peoples acting. Table 4.3 exemplifies the range of characteristics of organic fractions of municipally solid waste (OFMSW) from fruit and vegetable markets collected by Treviso municipal authority in Northern Italy.

For doubtful or missing data, additional analysis and a digestion test, respectively, must be carried out. This ensures that the substrate is appropiate for AD and does not affect the process negatively, e.g., by an exceedingly heavy metal content or detergents and disinfectant residues. A wrong addition of inhibiting substrate to the digester often leads to a rapid process breakdown, followed by a long restart phase. Interrelated to process malfunctions are financial losses, which are rarely calculated in advance by the operators.

*Table 4.3.* Characteristics of the OFMSW from Treviso, Italy (Pavan, 2002)

| Parameter | Unit | Average | Standard deviation | Max. | Min. |
|---|---|---|---|---|---|
| Dry matter (DM) | g/kg | 81.8 | 23.9 | 105.6 | 67.8 |
| Organic DM (ODM) | g/kg | 73.6 | 10.8 | 85.9 | 67.3 |
| Organic DM (ODM) | % DM | 90.0 | 9.6 | 99.5 | 81.5 |
| Total Kjeldahl nitrogen (TKN) | g/kg DM | 29.6 | 6.2 | 35.2 | 23.9 |
| Phosphorous, total ($P_{tot}$) | g/kg DM | 8.9 | 2.3 | 11.3 | 7.1 |

### 4.4.3. Energy Crops as Cosubstrates

Besides industrial or municipal cosubstrates like whey or OFMSW, energy crops are added to the digester, especially to increase biogas yields in agricultural biogas plants. Energy crops of C4 plants are cultivated primarily as feedstock for anaerobic digestion and applied directly or as silage. Despite a great choice of crops, corn has become the most popular in Germany and Sudan grass in Austria (Braun and Wellinger, 2002).

Adverse to the positive ecological balance, e.g., because of decentralized crop production, short route of transport, $CO_2$ neutrality, a merely unstable calculation base exists and has to be taken into account. Quality and quantity of crop depends on environmental conditions (crop yield) as well as on overall production costs for providing the cosubstrate (for instance, costs of seeding, cultivating, processing). In recent times great efforts were made to increase the specific crop of biomass per hectare by growing mainly biomass-forming species and investigations on the right time of harvesting (Amon et al., 2004).

Additional to increasing specific biomass yields appropriate pretreating technologies for lignocellulose-containing substrates are investigated. Among others the application of enzymes becomes increasingly interesting. Recent investigations on saccharification of biomass were carried out within a joint project (Lenz and Hölker, 2005; Pesta and Meyer-Pittroff, 2005).

Due to difficult to degrade components, like hemicellulose, cellulose, and especially lignin, biomass causes severe problems during anaerobic fermentation processes. Chemical and physical hydrolysis help to increase its biodegradability but are linked with high operating costs and demand for additional equipment. Treating biomass with enzymes for the saccharification of hemicellulose and cellulose is an economically feasible alternative. Lignin cannot be degraded under these anaerobic conditions.

Within the last years several enzyme preparations are commercially offered as fermentation boosters in biogas plants. These enzymes are added to increase the degradation of fermentable carbohydrates. However, these enzymes are too expensive for an areawide, economically feasible implementation. Concerning the structure of vegetable cell walls their extensive degradation is possible only by means of multienzyme systems.

A recently introduced technique of producing enzymes by an innovative solid-state fermentation system offers the opportunity to produce multienzyme systems with mixtures of fungi. These multienzyme systems are more reactive and cheaper than preparations made out of single enzymes.

Much higher enzyme yields can be achieved per time period in comparison to optimized submerged cultures or to other solid-state fermentations by using this new technique. Additionally these multienzyme systems act specifically on the applied substrates. During the production process the enzymatic activity is especially adapted to that kind of substrate, which is meant for the treatment with the multienzyme system.

Mixed cultures of fungi are cultivated on solid substrates and adapt their enzyme secretion. Multienzyme systems as a product are efficient in hydrolyzing and saccharificating their purpose substrates.

Highly specialized multienzyme systems are developed, produced, and implemented in AD processes to treat spent grains. Treating brewery grains for AD processing with these multienzyme systems offers an interesting alternative to reorganize the waste treatment procedures within the brewery.

### 4.4.4. Inhibitors and Toxic Substances

As mentioned above inhibitors and toxic substances affect the digestion operation negatively. In general, the influences of ammonia, sulfide, propionic acid, petrochemicals, nitrate, nitrite, tensides, cyanides, phenolic derivates, and solvents are known, as well as loads of inorganic solubles more than 30 g per liter. Additional to external sources, some of these inhibitors are created during the anaerobic degradation of organic material, e.g., ammonia, $H_2S$, and phenolic derivates, which emerges out of protein degradation.

To reduce the loading a specific selection and blend of cosubstrates should be provided. Wherever it is possible additional influence on the production process should be exerted, helping to make by-products suitable for AD.

For example, the composition of detergents and disinfectants used in the production process should be known. A simple change of the cleaning and rinsing solutions used often changes the biodegradability of wastewaters radically. Absolute acids or bases are microbially inoffensive. Excessive use of phosphoric acid may result in a phosphor elimination step before discharging the effluent. Sulfur acid increases the production of sulfuric substances and $H_2S$ under anaerobic conditions. Tenside solutions may have a negative impact on a stable process.

Nevertheless, AD plants have to be able to handle and tolerate inhibitors in a certain range, like detergents and disinfectants. Therefore, a dialogue among the biogas plant's engineer, the detergent producers, and the technical management of the company is necessary prior to the engineering of the biogas plant.

Inhibitors or toxic substances also may be brought into the system by transport facilities delivering the substrate. Faulty or even lacking cleaning of a loading area may cause severe problems, if galvanic slurries were handled before organic substrates, for instance. However, only heavy metals in solution have a toxic effect. Precipitation with sulfides reduces the concentration of soluble metals. This effect also is used for chemical desulfurization by adding liquid ferric or ferrous chloride to the digesters solution. $H_2S$ precipitates within the liquid phase instead of outgassing into the gaseous phase, deteriorating biogas quality.

### 4.4.5. Technical Adjustments

Inserting cosubstrates in an existing biogas plant sometimes demands additional technical equipment such as grinder, mixer, or sterilizer. Pretreatment systems have to be adapted to the cosubstrate. Impurities have to be removed mechanically. Glass, stones, wood, plastic electronic parts (batteries), bones, etc., may be found in biowastes. Particles have to be cut to a certain size. A thorough pretreatment not only increases the surface of substrate to support microbial degradation but also prevents pipes or pumps from blocking and sedimentation and scum layers.

Sterilization, homogenization, solid/liquid separation, or pH adjustment furthermore may be required. Because of investment costs stock grounds are usually open spaces without covers, as seen in Figure 4.5. To prevent odor nuisances from substrate stockgrounds the entire handling of substrates should be housed inside and extracted air cleaned with biofilters.

Due to inorganic and inert contents an effective removal of sand and sediments from the digester is necessary. For example, chicken manure shows an exceeding amount of sand and unsoluble solids, due to the type of feeding.

An appropriate technique and process engineering are necessary to meet the requirements of codigestion. Braun and Wellinger (2002) specify numerous organic materials in addition to their anaerobic digestibility as an evaluation of organic wastes and by-products for use in anaerobic digestion. Besides the characteristics excellent, good, and poor for digestion they discuss pretreatment or substrate characteristics. Table 4.4 lists some materials.

### 4.4.6. Adjusting the Nutrient Balance

It is necessary to maintain a well-balanced nutrition of the microorganisms by a proper composition of the daily load of substrate. For instance, microorganisms utilize carbon during AD 20 to 30 times faster than nitrogen (Schlegel, 1985). This predicts an optimal ratio from C to N of 20–30:1 within the substrate, providing mainly easily degradable carbon.

A narrow ratio of carbon to nitrogen (C/N) is indicated by the figures in Table 4.1. For an effective biogas process a ratio of at least 35–40 is required. Adding

**Figure 4.5.** Open space stockgrounds for cosubstrates.

*Table 4.4.* Evaluation of organic wastes and by-products for the use in anaerobic digestion (adapted from Braun and Wellinger, 2002)

| Material | Excellent | Good | Poor | Remarks |
|---|---|---|---|---|
| *Biogenic materials from agriculture* | | | | |
| Straw and other fibrous plant residues | | | + | Chopping or grinding required |
| Green plant material, crop, grain, silage | | + | | Chopping required; disturbing sand, stones; scum layer formation can occur |
| Silage leachate | + | | | High COD loading can result |
| Harvest residues | | + | | Chopping required; disturbing sand, stones |
| *Food industry waste* | | | | |
| Expired food | | + | | Expensive unpacking required |
| Dough, confectionary waste | | + | | Liquefaction (dilution) required |
| Whey | + | | | NPR (no pretreament required) |
| Residues from canning and frozen food | | + | | Expensive unpacking required |
| Residues from fruit juice production | | + | | Chopping advisable |

cosubstrates with a high carbon content the conditions for the biogas process can be improved. Using urine as additive is advantageous for increased biogas production (Idnani and Laura, 1971). Furthermore, trace elements are essential for microorganisms growth and metabolism: sulfide is necessary for building up cell structures and methanogenic bacteria have an essential need for nickel, cobalt, molybdenum, and wolfram (Mudrack und Kunst, 2003).

Even though the ratio of C/N is of essential importance mainly for aerobic degradation processes, the anaerobic digestion also is dependent to a certain degree. Hence, instead of analyzing and adjusting the C/N ratio currently more regard should be paid to a homogeneous and constant mixture of substrate and a quasi-continuous substrate dosage to the digester. A doubled effect is gained: a balanced and constant ratio of C/N and nutrients. Overall a better process performance and a higher biogas yield is guaranteed. The digestate also will be improved in quality as a fertilizer by an optimized nutrient balance. Inhibitory effects of nutrients and trace elements are observed over various applications and are caused by the concentrations listed in Table 4.5.

### 4.4.7. Influence of the pH

Most microorganisms grow best under neutral pH conditions and acid or alkaline media can influence their metabolism negatively. Changes in pH destroys enzymes or alters the chemical equilibrium of enzymatic reactions (Burton and Turner, 2003).

Usually cosubstrates like waste fat or distillery slops differ from manure with regard to kinetics and the available degree of degradation. Easily degradable

*Table 4.5.* Inhibitory concentrations of nutrients and trace elements (Scherer, 1989)

| Component | Inhibitory concentration |
|---|---|
| $Na^+$ (calculated NaCl) | 30 g/L |
| $K^+$ (calculated KCl) | 3 g/L |
| $Ca_2^+$ (calculated $CaCl_2$) | 2.8 g/L |
| $Mg_2^+$ (calculated $MgCl_2$) | 2.4 g/L |
| $NH_4^+$ (free $NH_3$) | pH-related (80–150 ppm) |
| Sulfide | pH-related ($< 100$ ppm) |
| $CO_2$ | $> 1$ bar and $< 0.2$ bar |
| Nitrate, nitrite | $> 50$ ppm |
| Free volatile fatty acids | pH-related |

substrates tend to acidify rapidly, lowering the pH noticeably. Process disruptions and instabilities are the consequence; in the worst case methanogenesis is severely inhibited. Simultaneously, increasing olfactory emissions caused by sulfur have been observed. Higher emissions of hydrogen sulfide are the reason for severe damages in the CHP, particularly when not noticed in time. Single-phase systems especially are affected by an uncontrolled acidification process.

Substrates containing high concentrations of sugars and carbohydrates such as distiller's grain, or glycerol show mainly lactic and propionic acid as intermediates. In this case, lactic acid is the precursor of propionic acid. If at the same time the pH gets lowered, the short-chained VFA, mainly propionic acid, become toxic for microorganisms because of its increasing cell wall permeation. This breakdown is followed by odor nuisance, poor gas quality by an increasing volume of $CO_2$ and exceeding $H_2S$ content up to 12,000 ppm, causing severe problems at the CHP. For example, biogasification process of organic-rich solid waste by two-phase fermentation was stopped at a total VFA concentration of about 13,000 mg/liter as acetic acid (Vietez and Ghosh, 1999).

The major problem is to maintain a stable pH between 7.0 and 8.0 within the methane formation reactor. This problem is associated with low pH of the substrate itself, poor buffering capacity, and the possibility of potentially high volatile fatty acid accumulation during digestion (Banks and Humphreys, 1998; Campos et al., 1999; Vietez and Ghosh, 1999; Yadvika et al., 2004).

The simplest way to level the pH value is to add concentrated pure bases, despite the disadvantages linked to it. The dosage of bases may affect the salt freight of the effluent. An increased concentration of sodium may additionally cause metabolic disruptions of the bacteria.

A well-balanced equilibrium of alkali and earthy base metals prevents these problems. If it is necessary to add base for pH-adjustment, a specified mixture of sodium (NaOH), potassium (KOH), calcium (added as $CaCO_3$), and magnesium (added as $MgCO_3$) should be taken into account.

The dosage of alkali may be supplemented or even substituted by recirculating digested effluent. This process water is saturated with $CO_2$ and its soluble salts, particularly bicarbonates. Using liquid digestate instead of alkali increases the systems buffer capacity and helps to economize the application of base. In certain cases magnesium may increase complexation and precipitation. Experiments will verify the extent.

To prevent digestion and especially methanogenesis from uncontrolled acidification a pretreatment is necessary, preferably in a hydrolysis digester of a two-phase digester system as mentioned above. Nevertheless, a pH value of hydrolysis between 4.5 and 6.3 has proved to be favorable for an optimal degradation effect. The organic load of easily degradable substrate acidifies easily and quickly, resulting in a pH value of 4 and even lower. Adjusting the pH by recirculating liquid digestate as well as a dosage of alkali is a feasible measure.

### 4.4.8. Process Monitoring and Control

Process monitoring and control is more complex and reliable than a simple measurement of pH and temperature. Legislative demands and guidelines of good manual practice ask for a consistently managed operating journal, bearing responsibilities and requirements in mind.

Process monitoring includes data on input materials (origin, type, amount) and output materials (amount of digestate sold/deployed), separated feedstocks and materials (type, amount), hygenization conditions (time, temperature), digester conditions (temperature, pH), biogas production (quality, quantity), electrical energy gained and fed to the grid (amount) or thermal energy (amount), and any remarkable incident (e.g., CHP maintainance). Appropriate inline/online analytic systems are available to simplify and automatize the daily procedure, e.g., gas analyzers for biogas composition. By calculating characteristic values such as specific biogas yields, the digestion process not only is recorded but also validated. In addition to a constant validation and process control further analysis of the digester liquid are obligatory.

During startups, changes of substrates (amount, type), or digester failures constant analysis of concentrations of VFA are essential. Occasionally measurements of nitrogen content or ammonia nitrogen content and dry matter content are indicated. Besides these external and more sophisticated measurements the quality of biogas depicts an easy to collect parameter for detecting digester failures rapidly and reliable. Therefore, automated inline/online gas analyzers should be preferred to manual measurements. In general, proper processing of substrates and monitoring of the digestion process is a precondition for a reliable codigestion with stable operations and adequate biogas yields.

### 4.4.9. Biological Cleaning of Extracted Air with Biofilters

During AD processing several flows of extracted air, loaded with vapor and various organic substances, may cause severe odor nuisances. These gases are generated from stockground, buffering and mixing tank, hydrolysis digester, and advanced effluent treatment. Some examples of organic substances are alcohols, aldehydes, esters, ketones, and dimethyl sulfide (DMS). Among others cleaning with biofilters is possible.

If the biofilter works properly, the organic load of the extracted air will be degraded under aerobic conditions into $CO_2$ and water. Sulfuric substances will be oxidized into elementary sulfur, which has to be removed regularly.

Ammonium and phosphor should be added because these nutrients are not available in the extracted air. A lack of oxygen has to be corrected by adding fresh air or discharged air from production rooms. To keep the biofilter moist sometimes extra water might be necessary.

Besides biological cleaning with biofilters olfactory loaded and extracted air might be thermically utilized. Extracted air can only substitute a certain amount of combustion air because of possible coating within the CHP or steam vessel.

### 4.4.10. Fees for Digesting Cosubstrates and Biogas Utilization

As mentioned above, by addition of cosubstrates conditions for the biogas process can be improved. A higher biogas yield is guaranteed and depending on the national or local framework the profits for the disposal of the cosubstrates improve economic feasibility tremendously. Fees for counting organic residues and wastes as cosubstrates are generally a lucrative deal for the biogas plant operating company. Additional investments like a thermal sterilization unit will be compensated sooner.

In Germany the Act on Granting Priority to Renewable Energy Sources [Renewable Energy Sources Act (EEG)] is one of the key elements to enlarge the use of renewable energy sources, following the national targets in the EU directive. The EEG regulates the feeding into the grid and compensation of electricity from renewable energy sources. The EEG obliges grid operators to purchase electricity generated from renewable energy sources and to pay minimum compensation to the generators of this electricity. The amount of compensation is based on the costs that arise from the generation of electricity to facilitate the economic viability of installations where electricity is generated from renewable energies.

Depending on the installed electric capacity of the CHP and the year of installation the generator receives between 0.0877 and 0.1133 € per kWh (installation year 2005). Additional grants are paid for using CHP or innovative energy techniques. Additional reports on providing renewable energy sources in selected EU countries are given by Kajan (2004) about Hungary, by Kovacs et al. (2004) about Czech Republic, by Navickas (2004) about Lithuania, and by Thiel (2004) about Poland, and overall information on EU standards by Nordberg (2004).

## 4.5. DIGESTATE: A SUBSTITUTE FOR MINERAL FERTILISERS

Manure has ever been utilized as a fertilizer with or without anaerobic treatment. By anaerobic treatment several characteristics of manure, such as the reduction of odor emissions, hygienic aspects, lower viscosity, and most important the fertilizing effect of nitrogen, can be improved. Codigestion increases both the quantity and the nitrogen load of digestate slurry.

Storing digestate slurry has to be undertaken in fully covered, gastight tanks. Coverless storage tanks ought to be refused due to ecological and environmental reasons. Because of seasonal restrictions for deploying the slurry adequate capacities have to be provided for storage. The annual amount of nitrogen deployed onto the soil and the loading with heavy metals is strictly limited by legislation.

High concentrations of nitrogen cause problems particularly in agricultural areas. Only 50% of the nitrogen, the ammonia, is available for plants. Most of the nitrogen is oxidized to nitrate and washed out into the ground water. Furthermore, ammonia utilization is guaranteed only during the growth of plants.

The digestate still contains nondigestible and residual substances, liquid organic and inorganic constituents of input materials, and of course bacterial biomass (Braun and Wellinger, 2002). One possibility to reduce the amount of slurry is mechanical dewatering. Surplus water may be treated aerobically in activated sludge reactors. The solid fraction is further mineralized and stabilized by aerobic composting. Selling compost as a fertilizer might achieve additional revenues.

Another worthwhile possibility to deal with high amounts of organic residues in digestate is a consideration of its influences in natural fertilizers. It becomes apparent that by using anaerobic digestion only 50% of cellulose is fermented in economically feasible hydraulic retention times and lignin not at all. For example, using brewery spent grains as a cosubstrate degradation of its structural substances is fragmentary. Solid substances of brewery spent grains are similar to those from grain slops or husks. By analyzing organic residues in digestates of brewery spent grains results in Table 4.6 are obtained. It can be seen that the targeted use of such a material can have a substantial influence upon the improvement of soil, specifically with regard to the quality of the humus.

Especially cellulose and lignin form starting substances for the production of humic substances. Humic substances are high-molecular compounds, which have the ability to store water, molecules, and ions (heavy metals, nutrients) in a reversible manner, and play a role in the capability of the soil to influence the mobility of the substances mentioned above in the ecosystem ground water, soil, and plant.

*Table 4.6.* Organic remains of brewers spent grains in digestate (Behmel, 1993)

| Component | % by mass |
|---|---|
| Lignin | 40–50 |
| Nitrogen, total | 2–2.5 |
| Cellulose, hemicellulose | 40–45 |

## 4.6. CONCLUSIONS

Anaerobic digestion (AD) is state of the art in a wide range of applications. Agricultural biogas plants as well as anaerobic wastewater treatment plants and anaerobic treatment plants for biological wastes are becoming increasingly widespread. An important part in applications plays the simultaneous digestion of a mixture of at least two substrates, or cosubtrates.

AD of cosubstrates combines two effects: On the one hand it depicts an effective and hygienic procedure for treating organic residues and wastes. On the other hand AD produces a renewable energy source, biogas, which is more than a simple by-product. Biogas fired in steam vessels or combined heat and power units, or even in combination of absorption chillers, helps to substitute fossil fuel sources.

Running an AD plant is an extension of the agricultural or industrial production process and demands engagement and knowledge of the process. To ensure a constant quality, besides technologically and procedurally excellent equipment, trained staff and accompanying documentation are necessary.

The time frame putting the biogas plant in operation and starting the process are for short too transfer to specific knowledge on the process to the operator of a biogas plant. Especially knowledge on digestion problems including accurately timed identification and troubleshooting is essential. Long-term support by the plant constructor and engineer should be given to the operator.

Proving economic feasibility is not only done by calculating biogas plant as an end-of-pipe solution; it is also necessary to elaborate a total energy consumption and energy distribution scheme. Each application needs to be treated as an unique project with an unique general framework. This ensures a customized biogas plant design and absolutely reliable processing. Additionally, new techniques and technologies, such as multienzyme systems, offer an interesting alternative to reorganize the waste treatment procedures within the companies.

Thus the necessity of developing a sensible strategy for the use of organic residues and wastes and to optimize the protection of the soils becomes highly important. This strategy must be applied to the requirements of the areas used and the plants growing there. With aerobic and anaerobic composts, as well as with liquid fertilizers produced with anaerobic methods and municipal solid wastes, one can reach a good condition of the soils and save mineral fertilizers.

The energetic infrastructure as well as the economic conditions in Germany and in wide areas of the European Union the disposal safety of wastes is the foremost issue. The energetic and ecological benefits of AD are much more noticeable under the situations prevalent in third-world countries. In many instances a shoddy infrastructure of the (decentralised) energy supply, energy shortages as well as ground erosion and desertification occur regionally at the same time.

The preconditions for agricultural production could be improved significantly through an increased and specifically targeted use of anaerobic technology. The necessity for a hygienic removal of organic wastes and wastewater on the one hand, as well as the supply of energy and the preparation of fertilizers and soil improvement materials on the other hand is obvious. These are the most plausible conditions for the use of anaerobic technology. The availability and spreading of this technology in suitable regions must certainly be one of the more sensible forms of sustainable technical cooperation programs.

## 4.7. REFERENCES

Amon, T. et al., 2004, Biogaserträge aus landwirtschaftlichen Gärgütern, in: *Proceedings of the 10th Alpenländisches Expertenforum*, Bundesanstalt für Alpenländische Landwirtschaft, Gumpenstein.

Banks, C. J., Humphreys, P. N., 1998, The anaerobic treatment of a ligno-cellulosic substrate offering little natural buffering capacity, *Water Sci Technol*, **38**(4/5):29–35.

Behmel, U., 1993, Mehrstufige Methanisierung von Brauereireststoffen, dissertation, Technische Universität München, Freising.

Behmel, U., and Meyer-Pittroff, R., 1996, Risiken bei der Cofermentation organischer Reststoffe in Biogasanlagen, *Korrespondenz Abwasser* **10**(43):2172–2178.

Braun, R., 1982, *Biogas – Methangärung organischer Abfallstoffe*, Springer-Verlag, Wien.

Braun, R., and Wellinger, A., 2002, *Potential of Co-digestion*, IEA Bioenergy TASK 37, Nova Energie GmbH, Aadorf.

Burton, C. H., and Turner, C., 2003, Anaerobic treatment options for animal manures, in: *Manure management: treatment strategies for sustainable agriculture*, Silsoe Research Institute, Bedford.

Callaghan, F. J. et al., 1999, Co-digestion of waste organic solids: batch studies. *Bioresource Technol* (67):117–122.

Campos, E., Palatsi, J., and Flotats, X., 1999, Co-digestion of pig slurry and organic wastes from food industry, in: *Proceeding of the Second International Symposium on Anaerobic Digestion of Solid Waste*, J. Mata-Alvarez Barcelona, ed., pp. 192–195.

Demirel, B., Yenigun, O., and Onay, T. T., 2005, Anaerobic treatment of dairy waste waters: a review. *Process Biochem* (40):2583–2595.

EC, 2004, Commission Decision of January 29, 2004, establishing guidelines for the monitoring and reporting of greenhouse gas emissions pursuant to Directive 2003/87/EC of the European Parliament and of the Council, C(2004)130 final.

EEG, 2000, Act on Granting Priority to Renewable Energy Sources (Renewable Energy Sources Act) of April 1, 2000 (status quo: October 1, 2005).

Fachverband Biogas e. V., 2005, *Biogas – das Multitalent für die Energiewende*. Freising.

Grepmeier, M., 2002, Mit der Gasanalyse decken Sie Störungen schneller auf. In: *top agrar Fachbuch: Biogas – Strom aus Gülle und Biomasse*, Landwirtschaftsverlag, Münster, pp. 84–85.

Idnani, M. A., and Laura, R. D., 1971, Increased production of biogas from cowdung by adding agricutural waste materials, *J Sci Food Agricult*:164–167.

Kaendler, C., Langhans, G., 2000, Plant scale co-fermentation of farm manure and industrial organic wastes, in: *Proccedings of the 14th Annual Residuals and Biosolids Management Speciality Conference*, February 27–29, 2000, WEF Biosolids, Boston..

Kajan, M., 2004, Possibilities and limits of biogas utilization in Czech Republic, in: *Proceedings of the 13th Annual Meeting of The German Biogas Association*, January 28–30, 2004.

Kaparaju, P., and Rintala, J., 2005, Anaerobic co-digestion of potato tuber and its industrial by-products with pig manure, *Resour Conserv Recy* (43):175–188.

Kovacs, K. L. et al., 2004, Possibilities, limits and new research results related to biogas production in Hungary, in: *Proceedings of the 13th Annual Meeting of The German Biogas Association*, January 28–30, 2004.

Kuhn, E., 1995, *Kofermentation, Arbeitspapier 219*, Kuratorium für Technik und Bauwesen in der Landwirtschaft e. V. (KTBL), Darmstadt.

Lenz, J., and Hölker, U., 2005, *Saccharification of Spent Grains by Means of Multi-Enzyme-System Treatment*, project documentation, Wissenschaftsförderung der deutschen Brauwirtschaft e.V., Berlin.

Mata-Alvarez, J., Mace, S., and Llabres P., 2000, Anaerobic digestion of organic solid wastes. An overview of research achievements and perspectives. *Bioresource Technol* (74):3–16.

Mudrack, K., and Kunst, S., 2003, *Biologie der Abwasserreinigung*, Spektrum Verlag, Heidelberg.

Navickas, K. 2004, Possibilities and limits of biogas utilization in Lithuania, in: *Proceedings of the 13$^{th}$ Annual Meeting of The German Biogas Association*, January 28–30, 2004.

Nordberg, A., 2004, Policies in different EU-countries influencing implementation of anaerobic digestion, in: *Proceedings of the 13$^{th}$ Annual Meeting of The German Biogas Association*, January 28–30, 2004.

Pavan, P. et al., 2002, Thermophilic anaerobic codigestion process in the AF-BNR-SCP approach: pilot scale study, in: *Proceedings of the 3$^{rd}$ International Symposium Anaerobic Digestion of Solid Wastes*, September 18–20, 2002, Munich/Garching.

Pesta, G., and Meyer-Pittroff, R., 2004, Anaerobe Reinigung von Abwässern, *Getränkeindustrie* **2**(58):31–33.

Pesta, G., and Meyer-Pittroff, R., 2005, What you should know about implementing an anaerobic digestion plant in a brewery?, in: *Proceedings of the 30$^{th}$ International European Brewery Convention (EBC) Congress*, May 14–19, 2005, Prague.

Scharf, P., 1992: Energieversorgung eines Brauereibetriebes mit Brauereisreststoffen, dissertation, Technische Universität München, Freising.

Scherer, P. A., 1989, Biogastechnik – Grundlagen und Besonderheiten der anaeroben Prozessführung, in: *Biogas Anaerobtechnik in der Abfallwirtschaft*, K. J. Thome-Kozmiensky, Verlag für Energie- und Umwelttechnik, Berlin.

Schlegel, H. G. 1985, *Allgemiene Mikrobiologie*, Georg Thieme, Stuttgart.

Sosnowski, P.; Wieczorek, A.; and Ledakowicz, S., 2003, Anaerobic digestion of sewage sludge and organic fraction of municipal solid wastes, *Adv Environ Res* (7):609–616.

Stroot, P. et al., 2001, Anaerobic codigestion of municipal solid waste and biosoilids under various mixing conditions – I. Digester Performance, *Water Res* **7**(35):1804–1816.

Thiel, R., 2004, Opportunities and limits of biogas utilization in Poland, in: *Proceedings of the 13$^{th}$ Annual Meeting of The German Biogas Association*, January 28–30, 2004.

Vietez, E. R., and Ghosh, S., 1999, Biogasification of solid wastes by two-phase anaerobic fermentation, *Biomass Bioenerg* (16):299–309.

Yadvika et al., 2004, Enhancement of biogas production from solid substrates using different techniques – a review, *Bioresource Technol* (95):1–10.

# 5

# Fundamentals and Applications of Anaerobic Digestion for Sustainable Treatment of Food Industry Wastewater

Vasileios I. Diamantis, Eleni Vaiopoulou, and Alexandros Aivasidis

## 5.1. INTRODUCTION

Developments in production technology have frequently resulted in the concentrated local accumulation of highly organic-laden wastewaters, a factor that therefore must be taken seriously into account with respect to the disposal technology. With conventional aerobic wastewater treatment (the activated-sludge process) the greater the organic load, the more problematic the process, since the required level of oxygen transfer rapidly approaches limits that are both technical and economically unattainable in practice. The anaerobic treatment of industrial wastewater has become increasingly important in recent years as a result of environmental protection legislation, increased energy costs, and problems with the disposal of excess sludge formed in aerobic treatment processes. Efforts are now being made to remove organic contaminants insofar as possible without external resources of energy, taking advantage of the biogas produced, where the desired level of purification is ultimately achieved with the aid of a subsequent aerobic biological clarification step.

The fundamental advantages of anaerobic wastewater treatment are:

1. Energy-intensive oxygen transfer is avoided.
2. The energy content of the resulting biogas can be put to further use.

---

VASILEIOS I. DIAMANTIS, ELENI VAIOPOULOU, AND ALEXANDROS AIVASIDIS • Democritus University of Thrace, Department of Environmental Engineering, Vas. Sofias 12, GR-67100, Xanthi, Greece.

3. Very little excess sludge is formed, while at the same time up to 95% of the organic contamination is converted into a combustible gas, constituting a case of true disposal.
4. Heavy metals are not subjected to oxidative mobilization, but rather to reductive precipitation (as heavy-metal sulfides), facilitating their orderly disposal rather than wider dispersal.
5. Aerosol formation accompanying oxygen transfer is avoided.
6. Space–time yields (bioreactor performances) reach a level fundamentally unattainable in the activated-sludge process because of the limitations imposed by oxygen transfer.
7. Anaerobic biomass can be preserved without feed for long periods of time without any significant deterioration of its properties.

## 5.2. BASIC PRINCIPLES

To acquire a more than empirical understanding of anaerobic wastewater treatment forces, one has to examine more closely the interactions among the various microorganisms in a food chain. The basic scheme is depicted in Figure 5.1. In simplified terms, at least three groups of microorganisms are involved in the degradation of complex organic molecules. To begin with, biopolymers that cannot be directly utilized by the anaerobic organisms are first hydrolytically degraded to soluble monomers (amino acids, sugars, and fatty acids). Acidogenic bacteria that form short-chain fatty acids and alcohols ferment the latter. This is followed by acetic

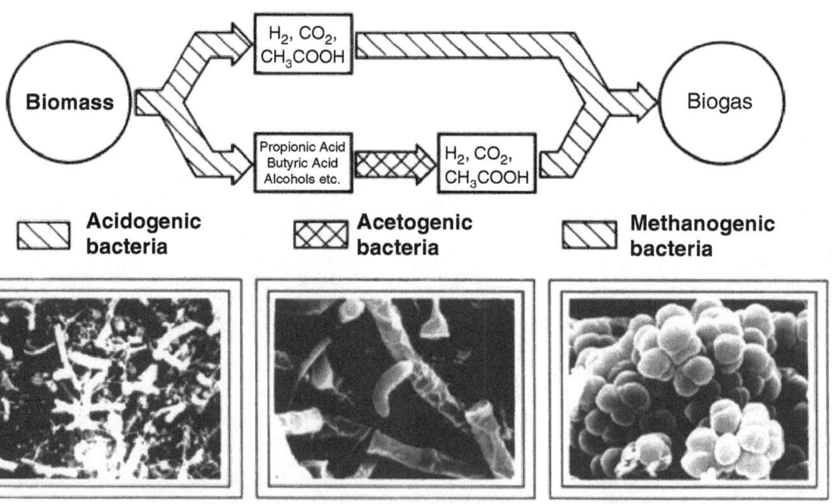


**Figure 5.1.** Schematic representation of the course of anaerobic methane generation from complex organic substances showing scanning electron micrographs of individual microorganisms involved.

acid formation, with the simultaneous generation of hydrogen and carbon dioxide, a step referred to as acetogenesis. It is at this point that methane formation can commence. Indeed, tracer experiments have shown that ca. 70% of the methane produced in the overall process comes from the degradation of acetic acid (Smith and Mah, 1978), whereas the remainder arises from the reaction of hydrogen with carbon dioxide. The complete degradation chain is in reality much more complicated and requires a balanced cooperation among the various microorganisms present, ensuring that there will be neither a shortage nor an accumulation of intermediates.

## 5.3. THERMODYNAMICS

As shown in Figure 5.1, the anaerobic degradation of organic matter proceeds via a series of intermediate steps. In terms of both thermodynamics and kinetics, further degradation may be severely limited if intermediates are allowed to accumulate. The concentration of hydrogen takes on a regulative role by influencing the distribution of metabolic products formed by the fermentative bacteria (Hungate, 1970; Thauer et al., 1977; Wolin, 1976). This point is emphasized in Figure 5.2 and reinforced

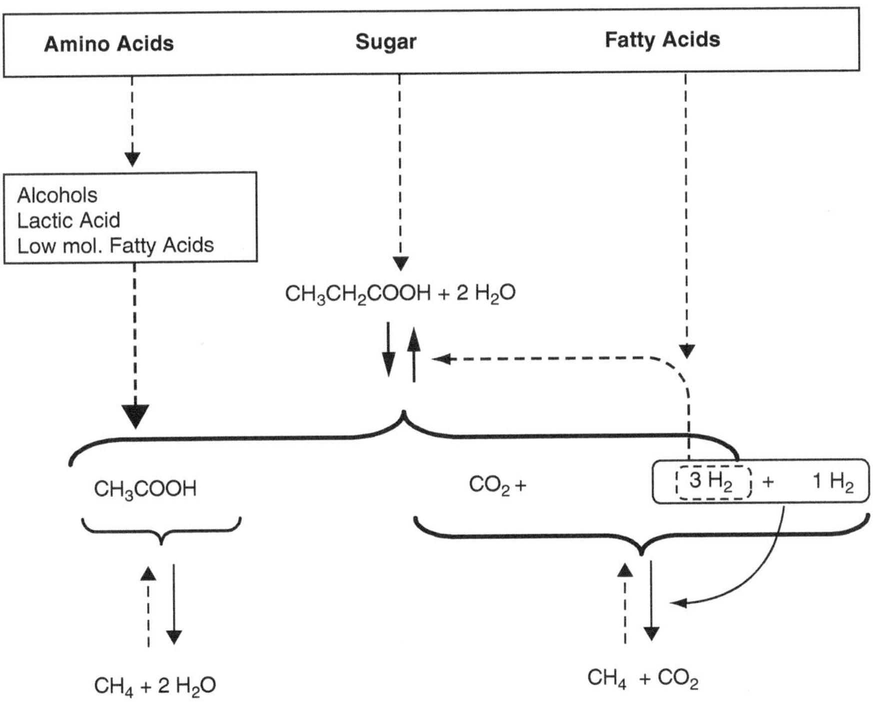


**Figure 5.2.** Methane formation from acetic acid and the combination of hydrogen and carbon dioxide: the influence of hydrogen on propionic acid degradation.

by the equations in Scheme 5.1, involving the products of acidogenesis (acetic acid, propionic acid, and butyric acid) and the hydrogen produced from them.

While the decomposition of acetic acid is independent of the partial pressure of hydrogen, the degradation of propionic acid and butyric acid would be inhibited by a high partial pressure of hydrogen. At the same time, an adequate hydrogen partial pressure is necessary for the formation of methane from hydrogen and $CO_2$. This results in a narrow ''thermodynamic window'' within which the decisive reactions are able to proceed simultaneously (Figure 5.3). Consequently, there must be a close

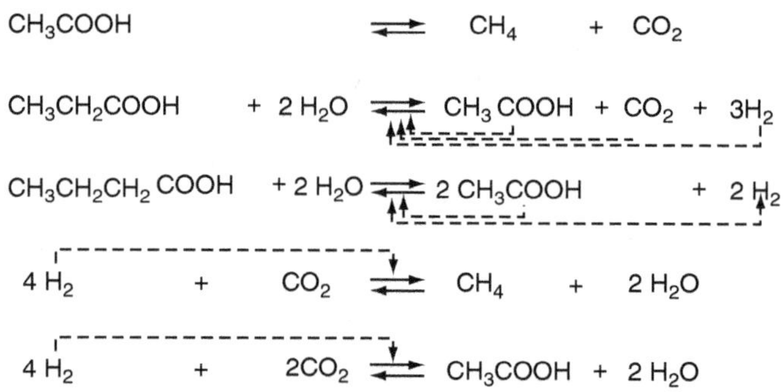


**Scheme 5.1.** Acidogenesis products and methane formation.

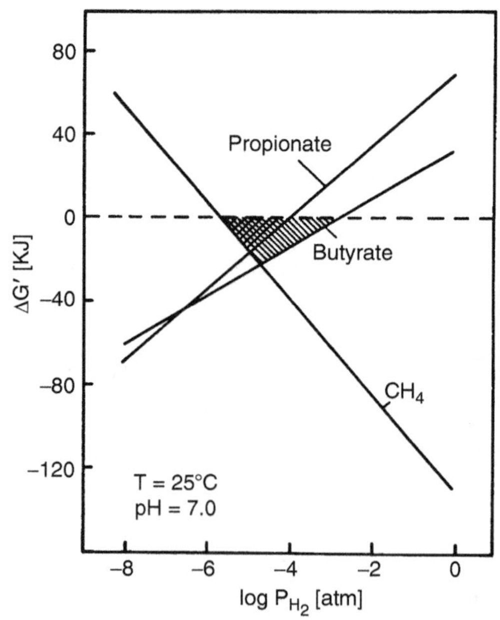


**Figure 5.3.** ''Thermodynamic window'' for simultaneous methane formation from propionic acid, butyric acid, and $H_2/CO_2$ as a function of hydrogen partial pressure.

$$C_6H_{12}O_6 + 4H_2O \rightarrow 2CH_3COO^- + 2HCO_3^- + 4H^+ + 4H_2,\ \Delta G_R^0 = -206.3\,kJ$$
$$C_6H_{12}O_6 + 2H_2 \rightarrow 2CH_3CH_2COO^- + 2H_2O + 2H^+,\ \Delta G_R^0 = -358.1\,kJ$$
$$C_6H_{12}O_6 + 2H_2O \rightarrow 2CH_3CH_2COO^- + 2HCO_3^- + 3H^+, + 2H_2,\ \Delta G_R^0 = -254.6\,kJ$$
$$C_6H_{12}O_6 + 2H_2O \rightarrow 2C_2H_5OH + 2HCO_3^- + 2H^+,\ \Delta G_R^0 = -225.9\,kJ$$
$$C_6H_{12}O_6 \rightarrow 2CH_3CHOHCOO^- + 2H^+,\ \Delta G_R^0 = -198.3\,kJ$$

**Scheme 5.2.** Fermentative conversions of carbohydrates in the presence of hydrogen.

relation between microorganisms that form hydrogen and those that instead utilize it (interspecies hydrogen transfer). Hydrogen transfer presupposes a spatial proximity on the part of the relevant microorganisms. In fact, if the population density is inadequate, vigorous stirring can actually interfere with this synergistic effect.

Figure 5.3 shows that propionic acid is poorly degradable in the presence of even a low partial pressure of hydrogen. If hydrogen utilization in a mixture culture is not sufficiently high, accumulation is observed, especially of propionic acid (and the wastewater reactor ''turns acidic''). Since the activity of methanogenic microorganisms decreases significantly at a pH below 5, a vicious circle thus is likely to result.

However, the relationships are much more complex, because it is not only propionic acid already present that may serve as an input to acetogenesis, but also additional propionic acid derived from prior steps in the context of acidogenesis. The equations at Scheme 5.2 show that fermentative conversion of carbohydrates in the presence of hydrogen is directed mainly toward propionic acid, because this step links to the greatest change in the free energy ($\Delta G_R^0$: $-358.1\,kJ/mol$).

## 5.4. REACTOR DESIGN

As a result of stepwise degradation involving various bacterial populations with the complex interactions linking fermentative acid formation with methanogenesis, it has been suggested that these two individual steps have to be separate in space by the use of a multistage reactor combination (Ghosh and Conrad, 1975; Massey and Pohland, 1978; Pohland and Kang, 1971). The most widely applied variant is a two-stage mode of operation, in which acidogenic microorganisms are most active in the first stage and mainly methanogenic microorganisms in the second (Pipyn and Verstraete, 1981; Cohen et al., 1982). The latter is usually implemented in the form of an upflow anaerobic sludge blanket or fixed-bed reactor.

Today the trend in anaerobic treatment of highly contaminated wastewater is clearly toward a two-stage process design. The generation time of microorganisms in the first stage is clearly shorter than that in the second stage, and therefore incorporation after the first stage of a sedimenter with partial biomass return is sufficient to maintain stable conditions in the first stage. Nevertheless, in a large-scale operation it is extremely important to ensure effective separation of preacidification biomass from the inlet of the second stage to prevent solids loading either in the methanogenic stage or at the outlet of the plant complex as a whole.

## 5.5. BIOMASS RETENTION/RECYCLE

Achieving high performance in a reactor is not simply a function of the biomass activity. Because of the autocatalytic factor, the biomass concentration is important as well. The biomass concentration cannot be adjusted independently of the wastewater residence time in the bioreactor. Given the long generation time of anaerobic microorganism, important components of the mixed population are always lost along with the partially treated water even at a residence time of several days if no precautions are taken with respect to biomass retention or external separation and return. In general, this requires increasing the particle size and/or the density difference relative to the liquid phase.

The most important methods for decoupling the residence times for liquid substrate and biomass in anaerobic wastewater treatment via internal retention are illustrated in Figure 5.4.

A selection in favor of either flocculating or granule-forming microorganisms that sediment easily in a bioreactor is still accomplished in most cases by empirical optimization of the reaction conditions. Factors that influence the formation and maintenance of anaerobic granules include the liquid upflow velocity and hydraulic retention time, the characteristics of the substrate, the characteristics of seed sludge, the addition of polymers and cations ($Ca^{2+}$, $Mg^{2+}$, $Fe^{2+}$, $Fe^{3+}$, $Al^{3+}$), reactor temperature, and pH (Yu et al., 2001a, b). This principle of biomass retention is the basis for the upflow anaerobic sludge bed (UASB) reactor and the expanded granular sludge bed (EGSB) reactor (Lettinga et al., 1980; Pette and Versprille, 1981; Schmidt and Ahring, 1996; Yan and Tay, 1997; Liu and Tay, 2002; Liu et al., 2003).

Free biomass or pelletized biomass can be separated with the help of filtration techniques, depending on the pore size of the filter, although such filters easily become clogged (Lee et al., 2001; Fuchs et al., 2003). Moreover, activity losses must be anticipated due to the mechanical shear stress involved in recirculation of a microbial suspension.

One way to achieve an increase in the particle size, with a corresponding improvement in the separability of the liquid face, is to immobilize the biomass on a carrier material (vandenBerg and Lentz, 1979; vandenBerg and Kennedy, 1981; Kennedy and vandenBerg, 1982; Huysman et al., 1983; Szendrey, 1984; Wulfert and Weiland, 1985). Evidence for the potential of this approach to microbial retention is available from nature, because most microorganisms in the environment are present in a fixed state. Nevertheless, adsorption presents the risk that the biomass might become detached as a consequence of shear forces.

The colonization of macroporous carriers with high surface areas lowers the risk of biomass detachment, while at the same time substantially increases the

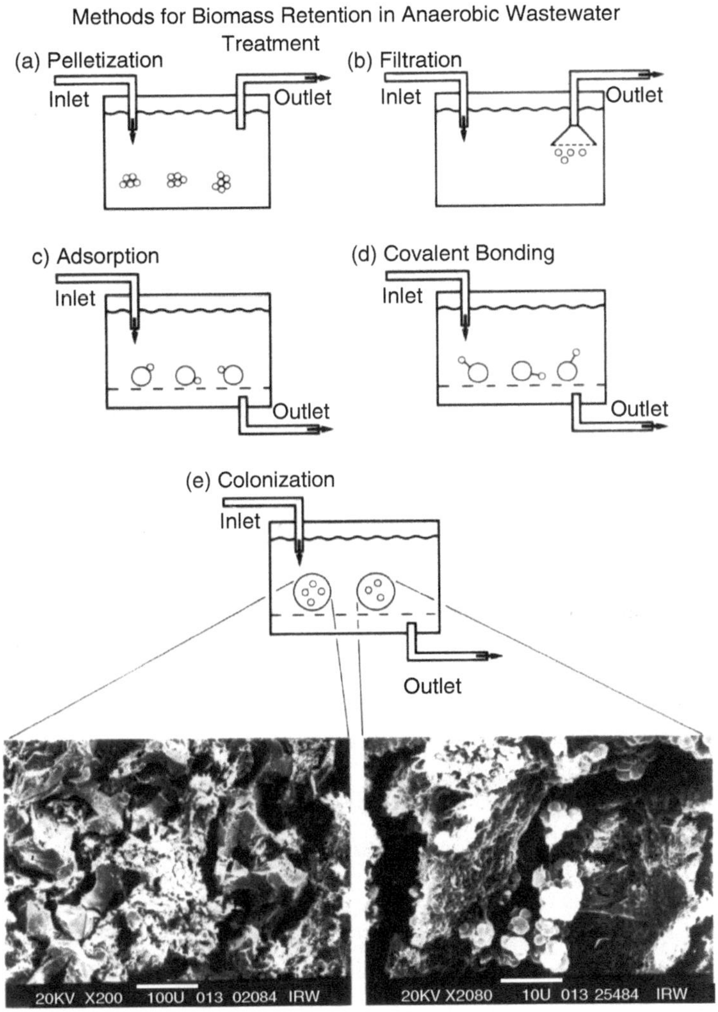


**Figure 5.4.** Schematic representation of various approaches to biomass retention: (a) Flocculation–pelletization; (b) filtration; (c) immobilization through adsorption; (d) immobilization through covalent bonding; (e) immobilization through colonization on a macroporous carrier. Scanning electron micrographs and show an open-pored sintered glass colonized by anaerobic bacteria.

amount of biomass that can be accumulated per unit volume (Aivasidis and Wandrey, 1985, 1986, 1987, 1988, 1989, 1990; Aivasidis, 1989; Aivasidis et al., 1992). Proper adjustment of the pore size of the carrier material permits the flow partially to penetrate the interior of the particles without also detaching the biomass. This has the further advantage that it facilitates the rapid removal of biogas as it is formed. A summary of common reactor configurations based on immobilized microorganisms is provided in Figure 5.5.

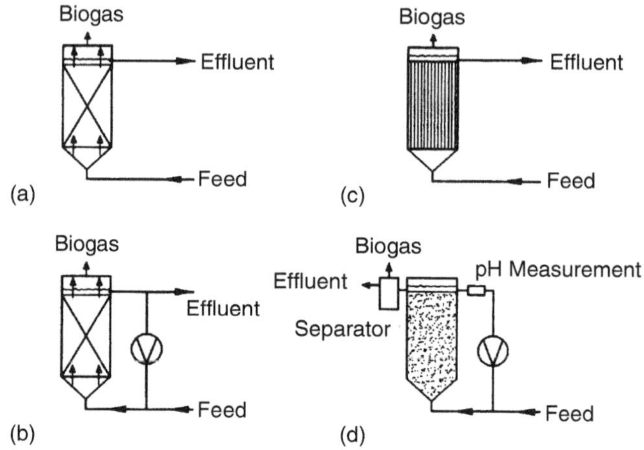


**Figure 5.5.** Reactors for the immobilization of microorganisms in anaerobic wastewater treatment: (a) fixed bed reactor; (b) fixed bed loop reactor; (c) fixed film reactor; and (d) Fluidized bed reactor.

In the case of external biomass recycling, sedimentation is achieved either by flocculation induced by chemical additives or by increasing the separation area with the help of lamella separators. This technique is also known as the anaerobic contact process (vandenBerg and Lentz, 1978; Schlegel and Kalbskopf, 1981). It is also possible to increase the density difference by flotation (Zlokarnik, 1985) or to increase the gravitational field by centrifugation. The principles of external biomass recycling are illustrated schematically in Figure 5.6.

### 5.5.1. Upflow Anaerobic Sludge Bed Reactors

The UASB process has been successfully applied on a large variety of industrial effluents, originating mainly from the agrofood industry, beverage industry, alcohol distilleries, but also from the pulp and paper industry and chemical industry. A simplified flow diagram of the UASB process is presented in Figure 5.7. Wastewater flows vertically through the system starting at the bottom. Microbial granules sediment on the reactor floor, forming a sludge bed. Enrichment of the biomass in the reactor is made possible by a three-phase separation system, designed especially to retain sludge flocs and granules and for separating and releasing the biogas from the system. Apart from granular sludge, a considerable amount of flocculated sludge is formed as well, which in turn constitutes a basis for the formation of new pellets.

A distributor on the floor of the reactor is used for uniform liquid feeding and it must necessarily meet special design requirements (Lettinga and Pol, 1991). In order to achieve an optimal contact between sludge and wastewater, channeling of the wastewater through the sludge bed, the formation of dead corners in the reactor, as well as uneven distribution due to clogging of the inlet pipes should be avoided.

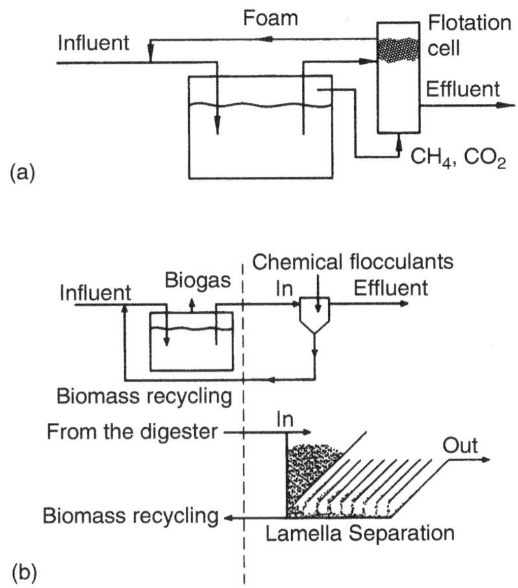


**Figure 5.6.** Schematic representation of methods for biomass recycling in the anaerobic contact process: (a) foam removal through flotation; and (b) separation by external chemical or physical sedimentation.

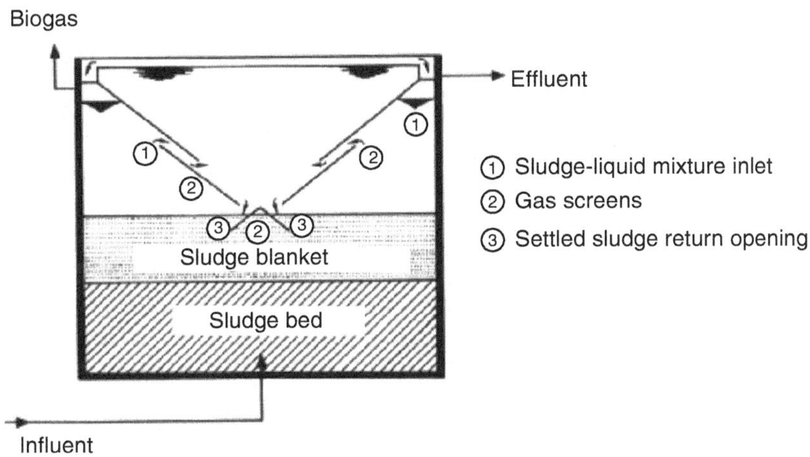


**Figure 5.7.** Schematic representation of the upflow anaerobic sludge bed (UASB) process.

For partially soluble and soluble wastewaters, the operational upflow velocity, in UASB reactors with granular sludge, is approximately 1–3 m/h, while velocities <0.5 m/h are applied for flocculent type of sludge (Lettinga and Pol, 1991). A modification of the UASB reactor in which higher superficial velocity is applied (5–10 m/h) is the EGSB reactor. Because of the higher upflow velocities, mainly granular sludge is retained in an EGSB system (Lettinga, 1996). As a result, the contact between the wastewater and the sludge is higher compared to UASB reactors, resulting in the maximum achievable loading rate.

Recirculation of partially purified wastewater to the UASB reactor is required only in the case of high COD concentrations in the wastewater in order to avoid pronounced concentration gradients as a function of height. The introduction of recycling can minimize in several cases the total operational costs (savings in alkaline addition), due mainly to the efficient use of the internal buffering capacity (Romli et al., 1994).

The process chain is generally enhanced by a preceding preacidification step that simultaneously ensures hydraulic equilibration. Special care should be given for the acidifying biomass to be separated prior to entering the UASB reactor compartment. Moreover, suspended solid, either poorly degradable or nondegradable, should be prevented from entering the UASB reactors because an irreversible sharp drop in the specific methanogenic activity occurs, while granule formation is influenced in a negative manner.

From the large-scale wastewater treatment plants, which were designed for anaerobic operation—1,215 units until 2001 (Frankin, 2001)—more than 70% were conceived as UASB and EGSB plants. This speaks well for the idea, but there also are disadvantages to be considered. For example, the process of microbial pellet formation is still very empirical and it is neither applicable to all types of wastewater nor uniformly reproducible. The formation of granular sludge may occasionally require several months or even years and sometimes it never takes place; moreover, existing pellets may redissolve and be washed out of the system due to a change in process conditions (Hickey et al., 1991). For this reason, pellet sludge is extracted from existing plants for the purpose of ''seeding'' newly commissioned UASB reactors and material may be transported overseas to this end.

An examination of the utilization efficiency of the existing reactor volume shows that only 25–30% of the space is actually devoted to bioconversion, while the rest functions simply as a phase-separating device. The biosolids concentration approach 75–85 g/L in the granular sludge region but the COD removal rate (at a maximum space yield) approximates 6–14 kg/($m^3$d) as a result of hindered mass transport. The minimum realizable hydraulic residence time (HRT) in a UASB reactor is 2–6 h. However, such short residence times are actually achieved only with lightly contaminated wastewater (1–2 kg/$m^3$). An HRT of 8–20 h is required for a moderately concentrated effluent (COD 3–20 kg/$m^3$).

### 5.5.2. Fixed-Bed Loop Reactors

The principle underlying this type of reactor combines the advantages of fixed-bed and fluidized-bed reactors, but avoids several of the disadvantages of both (Figure 5.8). Most of the liquid responsible for the upward flow in a fixed-bed loop reactor (Aivasidis and Wandrey, 1985, 1986, 1987, 1988, 1989, 1990; Aivasidis et al., 1992; Aivasidis, 1989) leaves the fixed-bed and returns once again through an

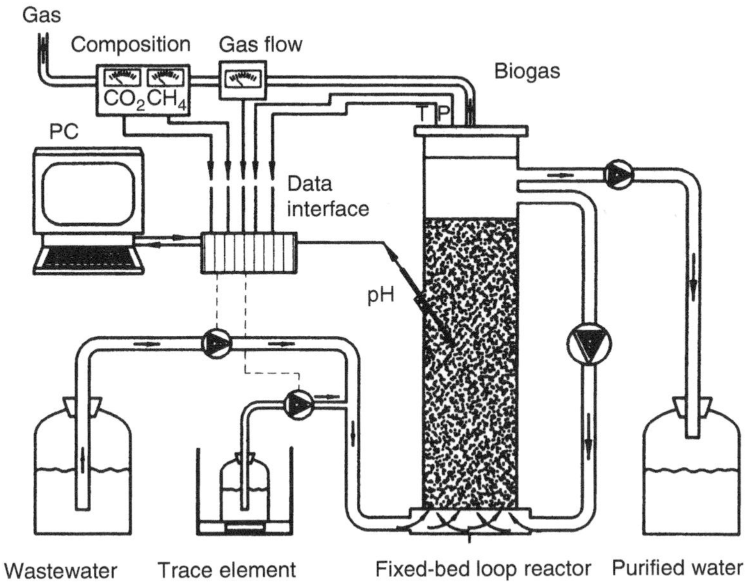


**Figure 5.8.** Schematic illustration of a fixed-bed loop reactor.

external loop, where it is mixed with fresh influent. Circulation is gradually increased until concentration gradients no longer play a role. Establishing the correct rate of circulation requires that biomass accumulation in the intergranular volume also be taken into account. This in turn means that the void volume must be neither rinsed entirely free of biomass nor allowed to become clogged. In this way it is possible to establish a very high biomass concentration. Thus, despite the low propagation rate of anaerobic microorganisms, provision must be made for periodically washing out excess sludge by briefly increasing the circulation rate. Simultaneously, small-suspended solid particles not trapped in the course of physical pretreatment of the wastewater, and which may have accumulated in the fixed-bed loop over time, are also flushed out.

This particular anaerobic-reactor concept has become especially familiar in the literature in conjunction with the use of macroporous carrier materials for microbial immobilization with three-dimensional colonization (Figure 5.9). Conventional carrier material manufactured for microbial adsorption and biofilm formation have the main disadvantage of biomass detachment due to inevitable hydrodynamic shear forces and particle–particle interactions during collisions (abrasion). Microbial colonization on high-surface-area macroporous carrier materials avoid the risk of biomass detachments, while at the same time the amount of biomass that can be accumulated per unit reactor volume increases substantially. Reticulated sintered glass with adjustable material properties, such as porosity, pore size distribution, and shape, proved especially suitable. Application of the latter material in combination

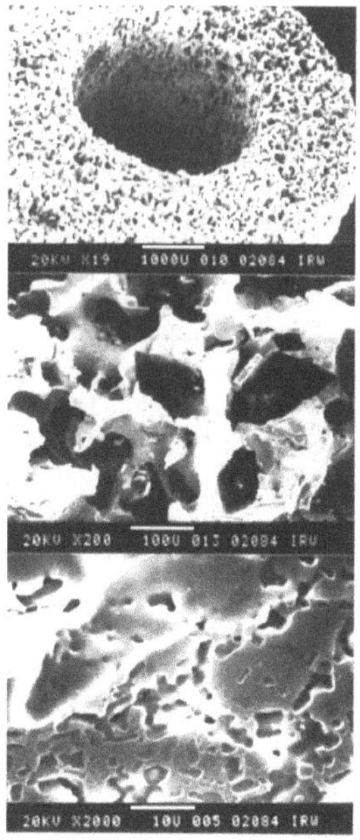


**Figure 5.9.** Scanning electron micrographs at various magnifications showing an open-pored sintered glass that has been formed into Rasching rings.

with fixed-bed loop reactors led to the development of a biogas high-performance process for the treatment of highly organically polluted wastewaters. The technique itself results in an increase in particle size as well as in the density difference between the particles and the fluid, making it possible to decouple the residence time for the biocatalyst and the fluid in the reactor without the addition of chemicals (e.g., for carrier fixation, flocculation, or flotation) or supplementary energy (e.g., for filtration or centrifugation) and without the need for constructing additional facilities (e.g., sedimenters or lamellar separators). COD removal rates of 80–160 kg/(m$^3$d) have been obtained in laboratory and pilot-scale reactors, while the removal rate—even at full scale—ranged from 30 to 80 kg/(m$^3$d).

## 5.6. EXAMPLES OF APPLICATION

### 5.6.1. UASB Reactor for the Fruit Canning Industry

#### 5.6.1.1. Introduction

The fruit canning industry is of great importance for Greece and for other Mediterranean countries with major export activities. The industrial branch of canned fruits employs increased quantities of freshwater, both for fruit washing at different processing steps and for sanitizing the production plant and the equipment. Since most of the water used is not employed for consumptive use, significant quantities of wastewater are generated. The latter contains mainly organic substances, sugars and carbohydrates, in soluble, colloidal, and suspended form. In Table 5.1 wastewater characteristics from peach and pear canning processes are presented. The data presented show that the peeling and the blanching streams are major contributors to the total organic contamination of the raw wastewater.

#### 5.6.1.2. Wastewater Treatment

In the literature several studies have been conducted with UASB reactors treating fruit canning wastewater. Some results from these studies are summarized in Table 5.2.

In order to study the anaerobic digestion of original fruit canning wastewater, a pilot plant was constructed and operated with wastewater originating from the overall streams of a Greek fruit cannery during its annual campaign. The objective of the study was to obtain reliable information about the UASB performance and sufficient data in order to design a full-scale installation and predict it performance. Wastewater batches were sampled daily, analyzed, and used for the study. Details about the experimental design can be found elsewhere (Diamantis et al., 2005). A summary of the UASB performance and operational characteristics during the study period is given in Table 5.3.

*Table 5.1.* Typical wastewater characteristics originating from a Greek fruit cannery (Diamantis et al., 2003)

| | Parameter units | COD total mg/L | COD soluble mg/L | pH – | Temperature °C | Electrical conductivity mS/cm |
|---|---|---|---|---|---|---|
| | Peach canning wastewater | 3,582 | 2,973 | 9.53 | 24.0 | 1.785 |
| Peach peeling | Caustic bath to sewer | 29,800 | 25,550 | 12.23 | 45.0 | 198.400 |
| Peach peeling | High pressure water sprays | 4,350 | 3,340 | 11.25 | 22.4 | 1.811 |
| Peach peeling | Fruit washing | 10,400 | 9,500 | 9.64 | 22.8 | 1.514 |
| | Pear canning wastewater | 3,178 | 3,044 | 7.40 | 17.6 | 1.680 |
| Pear peeling | Mechanical peeling and coring | 3,909 | 3,705 | 7.25 | 16.5 | 0.930 |
| Pear peeling | Fruit washing 1st stage | 511 | nd | 7.88 | 16.3 | 0.913 |
| Pear peeling | Fruit washing 2nd stage | 38 | nd | 7.87 | 15.9 | 0.916 |
| | Pear flume conveying | 1,188 | 1,075 | 7.92 | 16.8 | 0.910 |
| | Pear blanching | nd[a] | 54,680 | 4.42 | 83.5 | 1.667 |
| | Can cooling | 16 | nd | 7.24 | 20.2 | 0.933 |

[a] nd, Not determined.

*Table 5.2*. Continuous studies with UASB reactors concerning the treatment of fruit processing wastewater

| Feed characteristics | | | | | | | | | |
|---|---|---|---|---|---|---|---|---|---|
| Origin | COD (mg/L) | pH | Reactor (volume) | $T^a$ °C | $L_{RS}$ kg/(m$^3$ d) | $U_S$ % | HRT h | COD out mg/L | Reference |
| Diluted lye-peeling effluent | 4,500 | 8.5 | Pilot UASB (2.3 L) | 35 | 4.5 | 90–93 | 24 | 315–400 | Sigge et al. (2001) |
| Diluted apple juice | 2,500 | 7 | Pilot UASB (9 L) | 30 | 13–14 | > 85 | 4.8–5.4 | 250–290 | Dold et al. (1987) |
| | | | | 25 | 7–8 | > 80 | 7–8 | 170–350 | |
| Apple processing wastewater | 2,500–5,000 | 7 | Pilot UASB (9 L) | 30 | 13–14 | > 85 | 8 | – | |
| | | | | 25 | 7–8 | > 80 | 12 | – | |
| Fruit canning effluent | 265–2,200 | – | Full scale UASB | – | 0.53–2.2 | > 85 | 12–24 | – | Oliva et al. (1995) |
| Fruit canning effluent | 2,300–4,000 | 8.5 | Pilot UASB (2.3 L) | 35 | 2.3–4.0 | > 88 | 24 | – | Trnovec and Britz (1998) |
| | 4,000 | 8.5 | | 35 | 11 | > 88 | 8 | – | |
| Fruit juice industry effluent | 2,500–5,000 | – | Pilot UASB | 24–30 | 1.9–17.6 | > 80 | – | – | Austermann-Haun et al. (1997) |
| | 1,500–2,400 | – | Full scale UASB (591 m$^3$) | 26 | 1.5–4.4 | 82–88 | 13–24 | – | |
| | 820–1,600 | – | | 26 | 0.4–0.8 | 87–90 | 46 | – | |
| | 900–1,400 | – | | 26 | 0.7–1.4 | 87–90 | 24–44 | – | |
| Fruit cocktail canning effluent | 4,000 | 7.5 | Pilot UASB (50 L) | 35 | 9.2 | 81 | 10 | – | Britz et al. (1999) |
| Fruit and vegetable processing wastewater | 200–2,500 | 6.9–7.7 | Full scale UASB (88 m$^3$) | 27–29 | 0.3–2.5 | 61–88 | 24 | 70–300 | Campos et al. (1986) |
| | | | | 28–30 | 0.4–1.3 | 43–85 | 12 | 60–280 | |
| | | | | 29 | 0.6–3.3 | 55–82 | 7.5 | 80–260 | |

[a] T, Operational temperature; $L_{RS}$, organic loading rate; $U_S$, COD removal; HRT, hydraulic residence time.

*Table 5.3.* Operational characteristics of the pilot scale UASB reactor

| Parameter | Diluted peach pulp | Peach canning wastewater | Pear canning wastewater |
|---|---|---|---|
| Testing period (d) | 0–37 | 53–113 | 164–180 |
| HRT (h) | 9.23 (± 4.14) | 8.71 (± 2.71) | 9.41 (± 2.17) |
| Temperature | 34.8 (± 0.6) | 28.2 (± 1.7) | 31.2 (± 1.5) |
| Hydraulic loading [$m^3/(m^3d)$] | 3.0 (± 1.0) | 3.0 (± 0.9) | 2.7 (± 0.8) |
| $L_{RS}$[$kg/(m^3d)$] | 5.5 (± 2.3) | 8.0 (± 2.8) | 5.1 (± 3.0) |
| $L_{RS}$[$kg/(m^3_{VSS}d)$] | 21.4 (± 8.7) | 26.7 (± 8.7) | 24.2 (± 14.5) |
| COD influent (mg/L) | 1,795 (± 376) | 2,638 (± 520) | 1,965 (± 763) |
| COD out total (mg/L) | 284 (± 87) | 470 (± 163) | 383 (± 191) |
| $U_S$ total (%) | 83.9 (± 5.4) | 81.4 (± 8.7) | 80.8 (± 5.2) |
| COD out soluble (mg/L) | 110 (± 60) | 310 (± 144) | 206 (± 75) |
| $U_S$ soluble (%) | 93.4 (± 2.6) | 87.7 (± 6.6) | 88.5 (± 2.7) |
| Biogas production rate [$m^3/(m^3d)$] | 1.80 (± 0.65) | 2.38 (± 0.76) | 1.62 (± 0.81) |
| Biogas selectivity [$m^3/kg$] | 0.40 (± 0.11) | 0.47 (± 0.18) | 0.41 (± 0.17) |
| $CH_4$ in biogas (%) | 66.5 (± 9.0) | 70.8 (± 9.1) | 84.8 (± 6.5) |
| $CH_4$ selectivity [$m^3/kg$] | 0.28 (± 0.07) | 0.34 (± 0.14) | 0.33 (± 0.13) |
| Sludge activity [$g_{CODr}/(g_{VSS}d)$][a] | 0.30 (± 0.12) | 0.35 (± 0.14) | 0.32 (± 0.20) |

[a] The sludge bed volume was continuously monitored. The mean VSS concentration was 50 g/L.

The mean hydraulic residence time (HRT) of the UASB reactor was maintained between 8 and 10 h for the testing period. The organic loading rate ($L_{RS}$) showed great variation during the study period, between 2.0 and 16.0 $kg/(m^3d)$ mainly due to the daily fluctuation of the incoming wastewater COD values. The influent COD concentration varied from 1,000 to 4,000 mg/L, but the effluent COD values were consistently below 800 mg/L and 600 mg/L for total and soluble COD, respectively. Reactor start-up after 14 months of sludge inactivation was accomplished within 5–7 d at $L_{RS}$ 6 $kg/(m^3\ d)$ and 37°C. After short-term pauses (1–5 d) reactor start-up was completed within less than 24 h.

The UASB reactor can be operated efficiently with fruit canning wastewater at $L_{RS}$ between 10 and 15 $kg/(m^3\ d)$ at HRT between 5 and 10 h. At the following conditions COD removal efficiency ($U_s$) is higher than 80% at operational temperatures between 25 and 35°C. The efficiency and stability of the UASB reactor during treatment of fruit canning wastewater was satisfactory. The results of the present study together with the operational results from previous studies suggest feasibility of anaerobic treatment for the studied wastewater.

## 5.6.2. A Fixed-Bed Loop Reactor for the Alcohol Distillation Industry

### 5.6.2.1. Introduction

The primary raw materials for ethyl alcohol manufacturing processes in Greece are molasses, mainly from beets (around 70%) but also from figs, raisins, and wines

(30%) (Vlissidis and Zouboulis, 1993). Production of alcohol from molasses consists of diluting it with water to a sugar content of about 15%, enriching this solution with nutrients (ammonium salts and phosphates), adding a yeast suspension to the mash, fermenting the mash, and distilling off the alcohol.

The composition of the wastewater from a molasses distillery is given in Table 5.4. They consist chiefly of cooling and condenser waters, but also contains spent mash, wastes from the washing of yeast, and wastes from washing of rooms and equipment. The most concentrated constituent in the wastes is spent mash (slops) obtained as a distillation residue from the stills. They have brown color caused by the humus present in the molasses. They are slightly acidic and turbid because of the large concentration of colloids, the content of settleable matter being fairly small.

In order to study the treatability of alcohol distillery wastewater and to optimize anaerobic digestion processes for these wastewaters a series of pilot studies were conducted including the operation of:

1. A single-stage fixed-bed loop reactor (methanization reactor).
2. A two-stage fixed-bed loop reactor (acidification and methanization reactor).
3. A two-stage fixed-bed loop reactor with methanization cascade (acidification and two methanization reactors in series).

### 5.6.2.2. Wastewater Treatment

In all cases, before being fed into the reactor, the water was passed through a sedimentation unit to separate suspended solids and inert particles. Open-pored sintered glass in the form of Rasching rings (50% porosity, 60–120 μm pore-size distribution) was used as the carrier for the fixed-bed reactors. The anaerobic reactors were operated at pH–static at 6.4–6.7 and the operational temperature was 37°C. More details about the experimental setup and wastewater characteristics can be found elsewhere (Aivasidis et al., 1989).

*Table 5.4.* Composition (g/L) of the distillery slops (Vlissidis and Zouboulis, 1993)

| Parameter | Molasses | Raisins | Wines | Figs |
|---|---|---|---|---|
| pH | 4.3 (± 0.5) | 3.2 (± 0.5) | 4.2 (± 2) | 3.6 (± 0.5) |
| BOD | 27.5 (± 9.5) | 30 (± 2.2) | 16.3 (± 1.5) | 20.4 (± 5.5) |
| COD | 55.5 (± 25) | 57.5 (± 15) | 27.5 (± 2.1) | 35.4 (± 15) |
| Total Solids | 95 (± 25) | 73.2 (± 22) | 36.6 (± 0.69) | 48.5 (± 18) |
| Organic solids | 55 (± 15) | 44.5 (± 15) | 32.2 (± 6.2) | 32 (± 12) |
| Volatile acids | 1.95 (± 0.185) | 3.2 (± 0.35) | 1.1 (± 0.6) | nd |
| Total N | 4.75 (± 0.15) | 0.75 (± 0.15) | 0.65 (± 0.12) | 0.88 (± 0.18) |
| Total P | nd[a] | 0.22 (± 0.06) | nd | 0.17 (± 0.05) |
| Sulfates | 3.5 (± 1.05) | 0.48 (± 0.15) | 0.12 (± 0.055) | 0.9 (± 0.18) |

[a] nd; Not determined.

5.6.2.3. Influence of Single-Stage and Two-Stage Methanization Cascade

The investigation focused on mean hydraulic residence times (HRT) of 5–16 h. Although the COD loading rates in the single methanization reactor and in the cascade at given residence times were identical, significant differences occurred looking at the individual cascade stages. COD effluent values from the cascade arrangement were in any case lower than those from the effluent data of the single-stage methanization. The clear superiority of the process involving a two-stage methanization cascade, in relation to the single stage, is especially apparent from curves representing the residual part of COD and the COD conversion as a function of the HRT (Figure 5.10).

During the study period, significant amounts of propionic acid accumulated in the single-methanization stage compared to the effluent of the cascade. Additionally, the $H_2$ concentration in the biogas of the single methanization stage was in the range of 250–350 ppm, the corresponding concentration in the second cascade stage was always below 100 ppm.

5.6.2.4. Comparison of Single-Stage and Two-Stage Process Operation

In the case of a single methanization (fixed-bed loop) reactor the major aim was to operate the process in a COD loading rate ($L_{RS}$) comparable to the two-stage treatment. The most important degradation results are shown in Table 5.5.

In general, it can be established that both the single- and two-stage process can be operated satisfactorily up to an $L_{RS}$ of approximately 40 kg/(m$^3$d). However, if the degradation performance is considered in detail, then significant differences result indicating the superiority of the two-stage process. It becomes clear from the dependence of COD elimination on $L_{RS}$ (Figure 5.11) that improved degradability of the spent mash is achieved by phase separation.

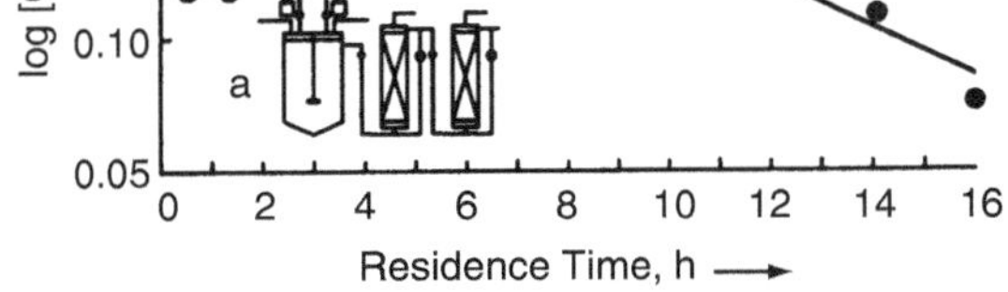


**Figure 5.10.** Change of remaining COD as a function of residence time during anaerobic treatment of distillery residues in a two-stage process with and without cascade connection.

*Table 5.5.* Degradation results from the single stage treatment of distillery mashes in the fixed bed loop reactor ($COD_o$: 33.8 kg/m$^3$)

| $L_{RS}$(kg/(m$^3$d)) | HRT (h) | $U_S$ (%) | $R_{RS}$[a] (kg/(m$^3$d)) |
|---|---|---|---|
| 45 | 17 | 60 | 27 |
| 37 | 22 | 79 | 29 |
| 34 | 23 | 79 | 27 |
| 27 | 31 | 86 | 22 |
| 23 | 35 | 89 | 20 |
| 31 ($COD_o$: 55.3 kg/m$^3$) | 42 | 84 | 26 |

[a] $R_{RS}$, Volumetric COD removal rate.

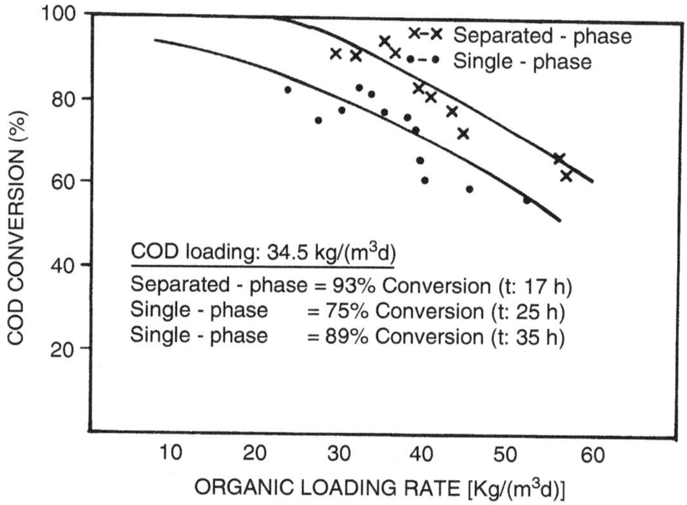


**Figure 5.11.** Influence of the COD loading rate on the COD conversion in the single and two-stage process.

In conclusion, a single-stage treatment of alcohol distillery wastewater in a fixed-bed loop reactor is in any case possible in the $L_{RS}$ range of 27–45 kg/(m$^3$ d). However, if a continuously operated stirred tank reactor is connected in front of a fixed-bed loop reactor for methanization in order to implement acidification and phase separation processing under defined conditions, then both improved degradation (higher COD conversion) as well as greater process stability can be achieved in comparison to the single-stage process under comparable conditions. A further improvement on COD conversion can be obtained when two-stage methanization cascade are used for the same total methanization reactor volume.

### 5.6.3. Anaerobic Treatment for the Potato Starch Industry

In discussing the wastes from the potato products industry the manufacture of potato starch will be considered. Water pollution problems related to the potato starch industry are serious. Starch plants operate seasonally, the season lasting for about 100 days from September to January, but sometimes also in the spring (Koziorowski and Kucharski, 1972). The wastewater is divided into three groups:

1. Effluent from flumes and potato washing.
2. Protein waters originating from the centrifuges or pulp screens, the settling tanks, the dewatering processes, etc.
3. Cooling waters from vapor condensers and vacuum driers.

The effluent in the first group is mainly polluted with suspended matter such as soil particles, sand, and clay; it may also contain very small potatoes or pieces of potato, sprouts, straw, potato leaves, bits of root, etc. The water from washing and transport of the potatoes also contains some dissolved organic compounds leached from damaged potatoes. When the potatoes begin to rot in the store, or if they are frozen, the concentration of organic material in these waters is increased. Protein waters contain mainly organic compounds such as carbohydrates. They are very turbid and when fresh usually have a weakly alkaline reaction. After some time, they become acidic due to lactic and butyric acid fermentation, especially if they are very diluted. Besides dissolved organic compounds, protein waters also contain a relatively large amount of insoluble compounds, such as small starch grains, parts of cells, tiny pieces of potato seed buds, root fibers, and leaves. Effluents in the third group (cooling water) are practically unpolluted.

From the data quoted by a number of authors, the COD of starch industry wastewater appears to cover a fairly wide range, from 1,000 to 12,000 mg/L. Some physicochemical characteristics of starch industry wastewater, potato processing facilities effluent, and several production processes are given in Table 5.6.

Pilot studies with fixed-bed loop reactors for the treatment of potato starch wastewater were conducted prior the design of a full-scale industrial plant (Table 5.7). During these studies COD conversion was greater than 78% for both configurations with influent COD in the range of 16–18 g/L and effluent values between 3.5 and 4.6 g/L. These results were achieved at HRT equal to 12–14 h for the acidification and the methanization reactor each. The $L_{RS}$ of the anaerobic reactor was 26–30 kg/(m$^3$ d).

According to the results presented above, a full-scale anaerobic reactor was constructed for a potato starch industry based on the fixed-bed loop reactor process. A two-stage process designed as a significant part of the COD load consists of proteins and carbohydrates, which require a hydrolysis step prior to methanization. The wastewater streams were mixed into the entrance tank of the treatment plant and introduced in the acidification stage, which comprised a continuous stirred tank reactor. The total flow was about 2,000 m$^3$/d with a 17,000 mg/L COD

*Table 5.6.* Physicochemical characteristics of effluents from a potato processing and starch plants

| Parameter units | COD mg/L | $BOD_5$ mg/L | pH – | TSS mg/L | TKN mg N/L | Total P mg P/L | Reference |
|---|---|---|---|---|---|---|---|
| Raw wastewater tapioca starch plant | 13,940 (± 350) | 12,780 (± 500) | 4.2 (± 0.4) | 9,130 (± 3,010) | 85–250 | 25–48 | Rajbhandari and Annachatre (2004) |
| Potato chips wastewater | 5,940–7,058 | 2,729–3,415 | 6–7 | 3,346–3,569 | 118–223 | 28–34 | El-Gohary et al. (1999) |
| Potato steam peeling | 27,400 | – | – | 13,200 ($mg_{COD}$/L) | – | – | Rintala and Lepisto (1999) |
| Potato blanching | 39,600 | – | – | 8,300 ($mg_{COD}$/L) | – | – | Rintala and Lepisto (1999) |
| Potato chips wastewater | 7,293 | 5,450 | 4–10 | 1,300 | – | – | Hadjivassilis et al. (1997) |
| Starch plant | – | 6,130–6,670 | 6.2–7.0 | 700–2,050 | 136–435 | – | Kozioroswski and Kuchaski (1972) |
| Potato starch plant | 900–1,400 | 1,100–1,500 | | 2,500–13,000 | 90–250 | 20–30 | Koziorowski and Kucharski (1972) |
| Potato and vegetable products | 7,500 | – | 4.5 | 1,500 ($mg_{COD}$/L) | 50–200 | 10–50 | Zoutberg and Eker (1999) |
| Potato and corn crisps | 4,500 | – | 5–9 | 3,075 ($mg_{COD}$/L) | 20–70 | 2–10 | Zoutberg and Eker (1999) |

*Table 5.7.* Experimental data on anaerobic treatment of potato starch wastewater in laboratory-scale and pilot-scale designs compared to the full-scale design (estimated data)

| | Test unit | Pilot plant | Full scale |
|---|---|---|---|
| Acidification | | | |
| Reactor volume ($m^3$) | 0.045 | 3.4 | 1,000 |
| HRT (h) | 24 | 14 | 12 |
| COD influent ($kg/m^3$) | 17.5 | 17.5 | 16.6 |
| Methanization | | | |
| Reactor volume ($m^3$) | 0.018 | 3.4 | 1,200 |
| HRT (h) | 30 | 13 | 14 |
| $L_{RS}$ ($kg/(m^3d)$) | 30 | 32 | 26 |
| COD out ($kg/m^3$) | 3.5 | 3.8 | 4.6 |
| $U_S$ (%) | 80 | 78 | 73 |

*Table 5.8.* Comparison of plant design evaluation and full-scale operational data

| Parameter | Design value | Actual value |
|---|---|---|
| Hydraulic load ($m^3/d$) | 2,064 | 2,460 (+19%) |
| COD influent (mg/L) | 16,600 | 24,400 (+47%) |
| COD load (t/d) | 34.6 | 60 (+73%) |
| COD out (mg/L) | 4,600 | 5,600–6,200 |
| $U_S$ (%) | 73 | 75 |
| COD elimination (t/d) | 25.3 | 45.3 (+79%) |

concentration at temperature of 37°C. The acidification reactor volume was 1,000 $m^3$ corresponding to an HRT of about 11–12 h.

The effluent of the first stage after biomass separation through sedimentation entered the methanization stage. The unit has a modular design and consists of four parallel operating fixed-bed loop reactors that contain volcanic stone and have a common recirculation of the liquid. The volume of each methanization reactor is 300 $m^3$ (1,200 $m^3$ total). The $L_{RS}$ of the methanization reactor is approx. 26–29 $kg/(m^3\ d)$ at HRT of about 13–14 h. The cell dry mass in the methane stage during steady-state operation is about 25 $kg/m^3$.

The start-up of the acidification stage in the full-scale plant took place within a few days due to the very short doubling time of the biomass present. However, the effluent COD of the methanization stage was far below the operational values during the first operational weeks. The start-up phase was completed within 6 weeks. During full-scale operation, a hydraulic as well as a COD overloading of the anaerobic treatment plant occurred (Table 5.8). The hydraulic overload was up to 20% higher and the COD overload was up to 73% higher in comparison to the design. The COD of the centrifuged samples was consistently below 4,600 mg/L but the homogeneous samples were in the range of 5,600 to 7,500 mg/L. Nevertheless the plant could handle this overloading at least with respect to conversion. COD conversion was 79% higher with respect to design values. This was mainly due to steadily increasing biomass concentration in the fixed-bed loop reactor.

The case study described above demonstrates that it is possible to scale up the anaerobic bioreactors for the treatment of complex organic pollutants. Careful experiments in laboratory and pilot-scale studies can provide sufficient data to design a large-scale system and to predict its performance. Crucial to the successful operation of the large scale is the ability to generate adequate amounts of pre-adapted biomass that can then be used to inoculate and start-up the large-scale reactor system.

## 5.7. CONCLUSIONS

The anaerobic treatment of wastewater represents another example of the fact that innovations may result from the joint application of biological and technological principles. Anaerobic wastewater treatment is a key technology for the treatment of wastewater from the food industry, where organically high-polluted wastewater is generated. It has the advantage that water reclamation is preformed without oxygen supply. The excess sludge produced is about one tenth compared to aerobic processes and biogas is produced as a result of the degradation of the organic compounds by anaerobic bacteria. The biomass developed consists of high-quality soil amendment compared to the less stabilized sludge produced during aerobic treatment. Moreover, it is possible to maintain the anaerobic biomass active for significant periods without feed, which an issue that is especially important for seasonal operating industries. In the case of highly contaminated wastewater, anaerobic treatment not only is energy self-sufficient, but also capable of providing excess energy for use in other applications.

Wastewater treatment plants should be regarded as an integral part of the overall production plant and should be designed and handled in an integrated fashion. Wastewater treatment is an industrial process associated with one of the largest levels of mass throughput known, and for this reason it has provided a major impetus to further developments in bioprocess technology in general.

## 5.8. REFERENCES

Aivasidis A., 1989, Process development for anaerobic wastewater treatment using porous carrier for microbial colonization, *Bioprocess Eng.*, **4**:19.

Aivasidis, A., and Wandrey, C., 1985, Process and apparatus for the continuous anaerobic decomposition of organic compounds, Forschungzentrum Julich, *US Patent 4532042*.

Aivasidis, A., and Wandrey, C., 1986, Forschungzentrum Julich, *DE Patent 3345691*.

Aivasidis, A., and Wandrey, C., 1987, Forschungszentrum Julich, *US Patent 4670140*.

Aivasidis, A., and Wandrey, C., 1988, Recent developments in process and reactor design for anaerobic wastewater treatment, *Water Sci. Technol.*, **20**:211.

Aivasidis, A., and Wandrey, C., 1989, Development and practical implementation of a high-performance biogas process for purification of wastewater with extensive organic contamination, *Chem. Ing. Tech.*, **61**:484.

Aivasidis, A., and Wandrey, C., 1990, Development and scale up of a high-rate biogas process for treatment of organically polluted effluents, *Ann. N.Y. Acad. Sci.*, **589**:599.

Aivasidis, A., Wandrey C., and Hilla, E., 1989, Studies on reaction techniques concerning reactor design for the anaerobic degradation of complex substrates with the example of the methanation of effluents in the fermentation industry, *Bioprocess Eng.*, **4**:63.

Aivasidis, A., Wandrey, C., and Kiefer, W., 1992, Forschungszentrum Julich, Schott Patent Glaswerke, *US Patent 5096814.*

Annachhatre, A.P., and Amatya, P.L., 2000, UASB treatment of tapioca starch wastewater, *J. Environ. Eng.-ASCE*, **126**:1149.

Austermann-Haun, U., Seyfried, C.F., and Rosenwinkel, K.H., 1997, UASB-reactor in the fruit juice industry, *Water Sci. Technol.*, **36**:407.

Britz, T.J., Trnovec, W., van Schalwyk, C., and Roos, P., 1999, *Enhanced granulation in Upflow Anaerobic Sludge-Bed Digesters (UASB) by process induction and microbial stimulation*, WRC Report No 667/1/99, Water Research Commission, Pretoria, South Africa.

Campos, J.R., Foresti, E., and Camacho, R.D.P., 1987, Anaerobic wastewater treatment in the food processing industry: two case studies, *Water Sci. Technol.*, **18**:87.

Cohen, A., Vandeursen, A., Vanandel, J.G., and Breure, A.M., 1982, Degradation patterns and intermediates in the anaerobic-digestion of glucose – experiments with C-14-Labeled substrates, *A. Van Leeuw. J. Microb.*, **48**:337.

Diamantis, V.I., Parisis, A., and Aivasidis, A., 2003, Anaerobic/ aerobic treatment of peach canning wastewater, *Proc. 8th Int. Conf. Environmental Science and Technology*, 8–10 September, Lemnos Island, Greece.

Diamantis, V.I., Pavlidou, E., and Aivasidis, A., 2005, UASB treatment of fruit canning wastewater: pilot-scale investigations, *Environ. Eng. Management J.*, **4**:339.

Dold, P.L., SamSoon, A., Palmer, I.H., and Marais, G.v.R., 1987, Anaerobic treatment of an apple processing wastewater, *Water Sci. Technol.*, **19**:237.

El-Gohary, F.A., Nasr, F.A., and Aly, H.I., 1999, Cost-effective pre-treatment of food-processing industrial wastewater, *Water Sci. Technol.*, **40**:17.

Frankin, R.J., 2001, Full-scale experiences with anaerobic treatment of industrial wastewater, *Water Sci. Technol.*, **44**:1.

Fuchs, W., Binder, H., Mavrias, G., and Braun, R., 2003, Anaerobic treatment of wastewater with high organic content using a stirred tank reactor coupled with a membrane filtration unit, *Water Res.*, **37**:902.

Ghosh, S., and Conrad, J.R., 1975, Anaerobic processes, *J. Water Pollut. Con. F.*, **47**:1278.

Hadjivassilis, I., Gajdos, S., Vanco, D., and Nicolaou, M., 1997, Treatment of wastewater from the potato chips and snacks manufacturing industry, *Water Sci. Technol.*, **36**:329.

Hickey, R.F., Wu, W.M., Veiga, M.C., and Jones, R., 1991, Start-up, operation, monitoring and control of high-rate anaerobic treatment systems, *Water Sci. Technol.*, **24**:207.

Hungate, R.E., 1970, Formate as an intermediate in bovine rumen fermentation, *J. Bacteriol.*, **102**:398

Huysman, P., vanMeenen, P., vanAssche, P., and Verstraete, W., 1983, Factors affecting the colonization of non porous and porous packing materials in model upflow methane reactors, *Biotechnol. Lett.*, **5**:643.

Kennedy, K.J., and vandenBerg, L., 1982, Stability and performance of anaerobic fixed film reactors during overloading at 10–35°C, *Water Res.*, **16**:1391.

Koziorowski, B., and Kucharski, J., 1972, *Industrial waste disposal*, 1st Ed., Pergamon Press, Wydawnictwa Naukowo-Techniczne.

Lee, S.M., Jung, J.Y., and Chung, Y.C., 2001, Novel method for enhancing permeate flux of submerged membrane system in two-phase anaerobic reactor, *Water Res.*, **35**:471.

Lettinga, G., 1996, Sustainable integrated biological wastewater treatment, *Water Sci. Technol.*, **33**:85.

Lettinga, G., and Pol, L.W.H., 1991, UASB-process design for various types of wastewaters, *Water Sci. Technol.*, **24**:87.
Lettinga, G., Van Velsen, A.F., Hobma, S.W., de Zeeuw, W., and Klapwijk, A., 1980, Use of the upflow sludge blanket (UASB) reactor concept for biological wastewater-treatment, especially for anaerobic treatment, *Biotechnol. Bioeng.*, **22**:699.
Liu, Y., and Tay, J.H., 2002, The essential role of hydrodynamic shear force in the formation of biofilm and granular sludge, *Water Res.*, **36**:1653.
Liu, Y., Xu, H.L., Yang, S.F., and Tay, J.H., 2003, Mechanisms and models for anaerobic granulation in upflow anaerobic sludge blanket reactor, *Water Res.*, **37**:661.
Massey, M.L., and Pohland, F.G., 1978, Phase separation of anaerobic stabilization by kinetic controls, *J.Water Pollut. Con. F.*, **50**:2204.
Oliva, L.C.H.V., Zaiat, M., and Foresti, E., 1995, Anaerobic reactors for food processing wastewater treatment: established technology and new developments, *Water Sci. Technol.*, **32**:157.
Pette, K.C., and Versprille, A.I., 1981, Application of the UASB-concept for waste water treatment, In: *Anaerobic Digestion*, D.E. Hughes et al., eds, Elsevier Biomedical Press, Amsterdam.
Pipyn, P., and Verstraete, W., 1981, Lactate and ethanol as intermediates in 2-phase anaerobic-digestion, *Biotechnol. Bioeng.*, **23**:1145.
Pohland, F.G., and Kang, S.J., 1971, Anaerobic processes, *J. Water Pollut. Con. F.*, **46**:1129.
Rajbhandari, B.K., and Annachatre, A.P., 2004, Anaerobic ponds treatment of starch wastewater: case study in Thailand, *Bioresource Technol.*, **95**:135.
Romli, M., Greenfield, P.F., and Lee, P.L., 1994, Effect of recycle on a two-phase high-rate anaerobic wastewater treatment system, *Water Res.*, **28**:475.
Rintala, J.A., and Lepisto, S.S., 1997, Pilot scale thermophylic anaerobic treatment of wastewaters from seasonal vegetable processing industry, *Water Sci. Technol.*, **36**:279.
Schlegel, S., and Kalbskopf, K.H., 1981, Treatment of liquors from heat-treated sludge using the anaerobic contact process, In: *Anaerobic Digestion*, D.E. Hughes et. al., eds, Elsevier Biomedical Press, Amsterdam.
Schmidt, J.E., and Ahring, B.K., 1996, Granular sludge formation in upflow anaerobic sludge blanket (UASB) reactors, *Biotechnol. Bioeng.*, **49**:229.
Sigge, G.O., Britz, T.J., Fourie, P.C., Barnardt, C.A., and Strydom, R., 2001, Use of ozone and hydrogen peroxide in the post treatment of UASB treated alkaline fruit cannery effluent, *Water Sci. Technol.*, **44**:69.
Smith, M.R., and Mah, RA., 1978, Growth and methanogenesis by Methanosarcina strain 227 on acetate and methanol, *Appll. Environ. Microb.*, **36**:870.
Szendrey, L.M., 1984, Start up and operation of the Bacardi Corporation anaerobic filter, *Environ. Prog.*, **3**:222.
Thauer, R.K., Jungermann, K., and Decker, K., 1977, Energy – conservation in chemotrophic anaerobic bacteria, *Bacteriol. Rev.*, **41**:100.
Trnovec, W., and Britz, T.J., 1998, Influence of organic loading rate and hydraulic retention time on the efficiency of a UASB bioreactor treating a canning factory effluent, *Water SA*, **24**:1147.
vandenBerg, L., and Kennedy, K.J., 1981, Support materials for stationary fixed film reactors for high-rate methanogenic fermentations, Biotechnol. Lett., **3**:165.
vandenBerg, L., and Lentz, C.P., 1978, Factors affecting sedimentation in the anaerobic contact fermentation using food-processing wastes, *Proc. Ind. Waste Conf. 33rd*, pp. 185–193.
vandenBerg, L., and Lentz, C.P., 1979, Comparison between up and down flow anaerobic fixed film reactors of varying surface-to-volume ratios for the treatment of bean blanching waste, *Proc. Ind. Waste Conf. 34th*, pp. 319–325.
Vlissidis, A., and Zouboulis, A.I., 1993, Thermophylic anaerobic digestion of alcohol distillery wastewaters, *Bioresource Technol.*, **43**:131.
Wolin, M.J., 1976, Microbial formation and utilization of gases, Verlag Goltze, Gottingen, pp.141–150.
Wulfert, K., and Weiland, P., 1985, Operating performance of fixed-bed reactors for biomethanization of effluent in the fermentation industry, *Chem. Ing. Tech.*, **57**:472.

Yan, Y.G., and Tay, J.H., 1997, Characterization of the granulation process during UASB reactor start-up, *Water Res.*, **31**:1573.

Yu, H.Q., Fang, H.H.P., and Tay, J.H., 2001a, Enhanced sludge granulation in upflow anaerobic sludge blanket (UASB) reactors by aluminium chloride, *Chemosphere*, **44**:31.

Yu, H.Q., Tay, J.H., and Fang, H.H.P., 2001b, The roles of calcium in sludge granulation during UASB reactor start-up, *Water Res.*, **35**:1052.

Zlokarnik, M., 1985, New flotation techniques for the separation and thickening of sludge in biological wastewater purification, *Chem. Ing. Tech.*, **57**:710.

Zoutberg, G.R., and Eker, Z., 1999, Anaerobic treatment of potato processing wastewater, *Water Sci. Technol.*, **40**:297.

# 6

# Fermentation of Distiller's Wash in a Biogas Plant

Anselm Gleixner

## 6.1. INTRODUCTION

Biogas plants for energy recovery of distillery residues have been in use for more than 20 years. Primarily distiller's wash but also rectification wastewater, surplus yeast, and superposed raw materials can be used in a biogas plant. The anaerobic fermentation is applicable to the gasification of distiller's wash made out of potatoes, grain, maize, and also molasses, which was the object of very successful experiments (Gleixner and Spradau, 1996).

Biogas plants for distiller's wash are sophisticated and also operate outstandingly with distiller's wash as monosubstrate. The anaerobic fermentation is a natural process that can positively be supported by means of technical and technological measures. The involved microorganisms are among the oldest creatures on earth. A highly efficient and trouble-free operation of the biogas plant is guaranteed by consideration of the natural rules. The energy generation from distiller's wash is an interesting alternative to the usual feed conversion, particularly if the feed market does not provide sufficient revenues.

## 6.2. SOME BASICS FOR A SUCCESSFUL OPERATION OF THE MONOFERMENTATION

A good knowledge of nature-given processes in the biogas plant is a necessary requirement for an unobstructed operation. The degradation of organic substances in a biogas plant occurs in four principal steps:

1. Hydrolysis
2. Acidogenesis

ANSELM GLEIXNER • INNOVAS Innovative Energie und Umwelttechnik, Margot-Kalinke-Straße 9, D-80939 München, Germany.

3. Acetogenesis
4. Methanogenesis

It is an essential requirement that those successive or parallel running biochemical individual fermentation processes of the fermentation of distiller's wash and other crop material are combined in two different phases, means the hydrolysis phase (step 1 + 2) and the methanization phase (step 3 + 4).

Experience shows that the common concentrations of distiller's wash lying between 6 and 9% of dry substance possess a hydrolysis time of 3 to 6 days. This depends on the raw material, the disintegration, and the consistency of the distiller's wash. Therefore distiller's wash from potatoes needs less time for hydrolysis than distiller's wash from grain. The holding time of standardized distiller's wash in the fermenter should be between 20 and 25 days.

Table 6.1 is taken from a typical analysis sheet of a very well operating biogas plant of a distillery. It shows the development of fatty acids and free fatty acids in hydrolysis (first phase) and their degradation in methanogenesis (second phase) (Pesta, 2005).

Hydrolysis is the limiting factor of biogas production. The more fatty acids, primarily acetic acid, are generated in the acidogenesis, the higher is the degradation rate and biogas production at the next steps. That means that only substances provided in the phase of hydrolysis and acidogenesis can be decomposed in acetogenesis and methanogenesis into biogas through bacteria.

The results and experiences of monofermentation of distiller's wash can be transferred to monofermentation of any plant material, e.g., biogenous waste or renewable primary products. The following principles concerning a good and reliable operating of a monofermentation plant should be taken into consideration:

- A preferably high cell disintegration should be established. The greater the surface of the particles, the better the degradation and therefore the yield. The cell disintegration can be thermally, mechanically, enzymatically, or chemically.
- It is not possible to ferment lignin, wood, and woody crude fiber. If cellulose and hemicellulose are wrapped with lignin, the anaerobic degradation can

*Table 6.1.* Example of fatty acid composition and degradation of a biogas plant

| | Fatty acids | | | | | | | | |
|---|---|---|---|---|---|---|---|---|---|
| Phase | C-2 acetic | C-3 propionic | i-C-4 isobutyric | C-4 butyric | i-C-5 isovaleric | C-5 valeric | C-6 caproic | C-7 oenanthic | Sum |
| Hydrolysis (mg/L) | 6,857 | 2,606 | 104 | 2,612 | 145 | 1,413 | 1,722 | 666 | 16,125 |
| Methanolysis (mg/L) | 109 | 16 | 5 | 22 | 5 | 30 | 28 | 14 | 229 |
| Degradation (%) | 98.4 | 99.4 | 95.2 | 99.2 | 96.6 | 97.3 | 98.4 | 97.9 | 98.6 |

only be carried out with great effort. Therefore these substrate constituents should be removed via separation technologies and not added to the fermenter feeding.

- The biogas plant should be a two-phase model with a separation of the hydrolysis and the methanogenesis phase (not to be mistaken with commonly used agricultural systems of main fermenter and postfermenter).
- The fermenting substrate preferably should be homogeneous and properly mixed for pumping.
- These are the requirements to receive a good specific biogas yield by volume and time as well as a safe line operation.
- No supplement of liquid manure is necessary in the fermentation of plants and plant parts. The output is even higher without liquid manure.

In this context the application of special enzymes, which are able to split up the lignin husk of crude fiber making the sugar of cellulose available, is very interesting for an up to 20% rise of energy output (Bäuml, 2005). In contrast to that the higher costliness, the additional mechanistic buildup and the slightly higher internal energy demand should be taken into account.

As a matter of course the biology works also in a single phase, but very long retention times and a low organic load in the system must be accepted to reach an optimal balance in the successive degradation steps. Otherwise, it results in rapid acidification and a total breakdown of the digestion process.

## 6.3. ENERGY POTENTIAL

Fats, proteins, and carbohydrates are utilizable in biogas production, whereas lignin is not anaerobically degradable. Gas production and amount of methane (gas quality) also depend on the material composition. With the knowledge of the composition, the theoretic achievable amount of biogas as well as the $CH_4$ and $CO_2$ content can be calculated according to the Buswell formula (Wellinger et al., 1991). On the basis of these values and of the known composition of the fermenting substrates, the biogas amount and thus the achievable energy output can be estimated.

Table 6.2 shows a comparison of the composition of different substrates, taken from a "DLG-feed composition table." It can vary depending on the distillery procedure between 4.5 and 16%. For better comparing the content of dry substance is standardized at 7.0% DS.

These gas yields can be achieved only with a two-phase high-capacity biogas plant with accurately harmonized processing. Ordinary, single-phase agricultural biogas plants with the traditional inserting and stirring technique are not able to realize such degradation rates and gas yields from distiller's wash.

*Table 6.2.* Typical biogas yields from various distiller's wash, sustainable biomass crops (NaWaRo), and liquid manure for comparison

| Substrate | DS/ODS (%) | Raw fat (g/kg TS) | Proteins (g/kg TS) | Carbo-hydrates (g/kg DS) | Crude fiber (g/kg DS) | Specific gas production rate ($m^3$/kg ODS) | Biogas per 1 $m^3$ mash ($m^3/m^3$) | $CH_4$-content |
|---|---|---|---|---|---|---|---|---|
| Distiller's wash from potatoes | 7.0/85 | 17 | 285 | 451 | 80 | 0.60 | 36 | 75% |
| Distiller's wash from wheat | 7.0/88 | 67 | 362 | 416 | 97 | 0.65 | 38 | 68% |
| Distiller's wash from rye | 7.0/88 | 54 | 431 | 406 | 56 | 0.68 | 40 | 60% |
| Distiller's wash from maize | 7.0/94 | 82 | 297 | 466 | 104 | 0.70 | 45 | 70% |
| Maize silage mash[a] | 7.0/95.5 | 41 | 98 | 743 | 73 | 0.62 | 41 | 63% |
| Grass silage mash[a] | 7.0/91.4 | 53 | 219 | 558 | 85 | 0.58 | 38 | 66% |
| Liquid manure swines[b] | 7.0/86 | 70 | 180 | 450 | 160 | 0.50 | 30 | 68% |
| Liquid manure cattle[b] | 7.0/80 | 30 | 150 | 410 | 210 | 0.48 | 27 | 64% |

[a] After separation of not degradable crude fiber.
[b] Not separated liquid manure.

## 6.4. SYSTEM CONCEPT

Due to its operating procedure, a distillery is ideally suited to the application of biogas technology for regenerative production of energy (heat and electricity). Thus the distiller's wash is already optimally solubilized for yeast fermentation, the substrate has a constant composition, the biogas plant does not need internal energy for substrate preheating, and the distillery itself is a great energy consumer (Figure 6.1).

Biogas generated from distiller's wash can be used like natural gas. From the beginning of biogas utilization in distilleries, biogas was also used directly together with natural gas or instead of fuel oil in steam boiler. With the first German energy supply law of 1990, the application of block heat and power plants became state of the art.

With the changes in the new version of the Act on Granting Priority to Renewable Energy Sources (EEG), energy production became more lucrative. Distiller's wash from an agricultural distillery in compliance with the spirits monopoly law is classified as renewable resource, and therefore a bonus of 6 ct €/$kWh_{el}$ is paid additionally to the basic payment. A second bonus of 2 ct €/$kWh_{el}$ can be accessed for combined heat and power generation, if the surplus heat is implemented in the distilleries production process (N.N., 2004).

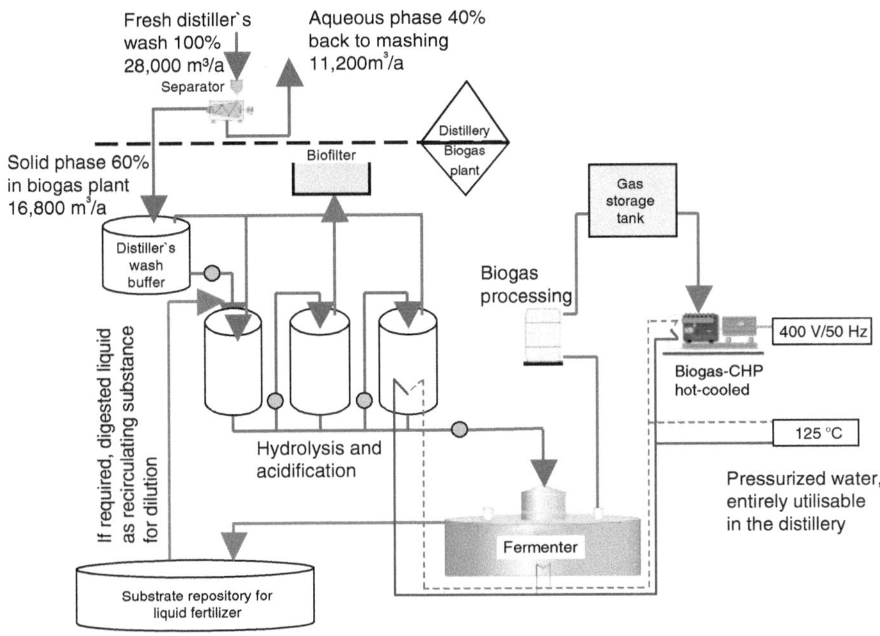


**Figure 6.1.** Procedures of a distiller's wash biogas plant considering as example for 28,000 hl distillery (Gleixner, 2004).

In principle the following biogas utilization is possible:

- Direct combustion in the steam boiler for production of process steam, together with natural gas or instead of fuel oil (multifuel burner).
- Combustion in a combined heat and power unit (CHP) with high-temperature cooling system gas motor for the generation of electricity and process heat as compression water or as low pressure steam or CHP with uncoupling of exhaust gas heat for steam generation.
- Combustion in a combined heat and power unit with gas turbine for generation of electricity and steam/process heat (only at larger plants).
- Application of biogas in fuel cell, generation of electricity, and process heat.

Figure 6.2 gives an impression of the ideal feed and energy flow in modern distilleries.

With the generated biogas, it is possible to substitute up to 90% of the process energy (heat and electricity) of the distillery by using in-house production residues. The distillery should possess some preconditions for the self-supply with energy from distiller's wash. These conditions are also the operational requirements for a profitable alcohol production:

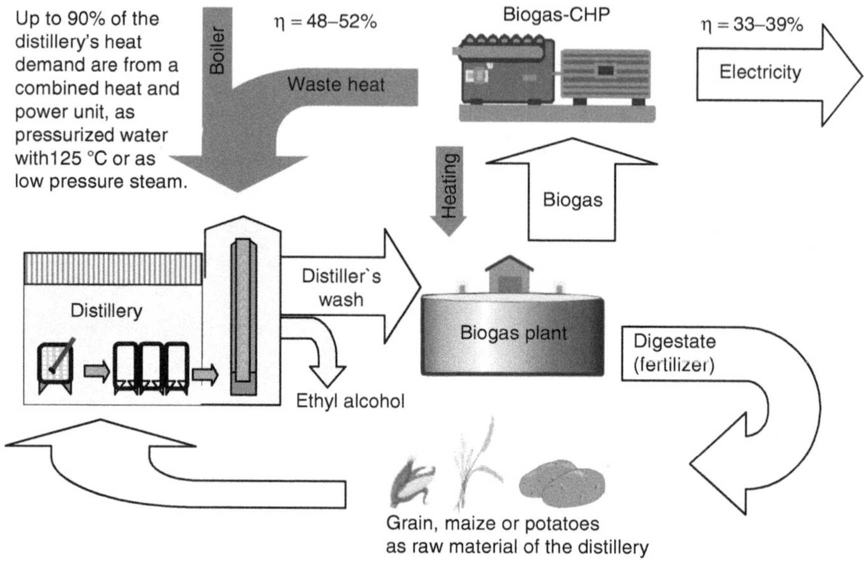


**Figure 6.2.** Self-supply of a distillery with biogas from distiller's wash (Gleixner, 2004).

- a constant production over 24 h (three shifts), 6–7 days per weeks
- year-round operation, 350 days p.a.
- constant energy consumption
- relatively constant substrate spectrum, no high load variations

## 6.5. ENERGY UTILIZATION

Considering a 28,000 hl distillery as an example, the possible energy utilization is described and explained:

| | |
|---|---|
| • Supposed amount of distiller's wash[1,2]: | 16,800 t/a (= 48 t/d) |
| • Distiller's wash composition: | 11% DS |
| | there from 89.5% ODS |
| | about 4,885 kg ODS/d |
| • Digester volume of the biogas plant: | 1,700m$^3$ |
| • Load factor of the fermenter: | 2.92 kg ODS/m$^3$d |
| • Specific biogas yield[3]: | 600 L biogas/kg ODS |

[1] In comparison to the distillery, the feeding of biogas plant happens constantly. The amount of distiller's wash is distributed on 7 days per week, 24 hours per day. Thus there are 350 days of operation per year.

[2] Separated amount after distiller's wash recirculation.

[3] With safety reduction.

| | |
|---|---|
| • Producible biogas: | ca. 2,950 $m^3/d$ |
| | approx. 1,032,500 $m^3/a$ |
| • Biogas quality: | $>70\%$ $CH_4$ |
| • Primary energy amount of biogas: | approx. 20,500 $kWh_{prim.}/d$ |
| | approx. 7,190,000 $kWh_{prim.}/a$ |

Because of the realignment of the law for the priority of renewable energy, there are some significant improvements of the grants for feeding the grid.

| | |
|---|---|
| • Grants for feeding the grid according to the EEG (status quo: 2004): | 11.5 ct.€/kWh up to 150 kW |
| | 9.9 ct.€/kWh above 150 kW |
| | 10.6 ct.€/kWh mixed price |
| • Additional renewable primary products bonus: | 6.0 ct.€/kWh |
| • Additional combined heat and power generation bonus[4]: | 2.0 ct.€/kWh |
| • Possible grant for feeding the grid: | 18.6 ct.€/kWh |

The preceding rate calculation refers to the first year of commencement of the EEG. In biogas plants set into operation to a later date, the primary grant for feeding the grid is reduced about 1.5% a year in each case.

A high-temperature cooling-system combined heat and power unit is used to recover the lost heat also in the distillery. In these aggregates the heat is available as pressured water of 125°C. An alternative are combined heat and power units (CHP) that use the exhaust temperature (400°C) with special exhaust gas heat exchangers. Approximately half of the surplus heat from CHP can be used for steam generation.

With an annual operational performance of 8000 hours of operation, a CHP with 330 kW can be operated:

| | |
|---|---|
| • Combined heat and power unit: | 330 kW (electric power) |
| • Efficiency $\eta_{electr.}$: | 36% |
| • Producible electricity: | approx. 7,380 $kWh_{electr.}/d$ |
| | approx. 2,583,000 $kWh_{electr.}/a$ |

## 6.6. INTERNAL ENERGY CONSUMPTION, OPERATION EXPENSES

The internal energy consumption of a biogas plant for distiller's wash is relatively low. The internal consumption of electricity of the biogas plant is about 200,000

[4] Especially distilleries have a very good requirement for the combined heat and power generation bonus, because the biogas plant in principle has no internal heat demand. The distiller's wash is already available warm.

$kWh_{el}/a$. Heat energy demand is negligible, if distiller's wash with temperatures of 40 to 50°C is provided.

The staff employment for the operation of a biogas plant for distiller's wash is very low due to the possible automation level. Two hours for supervision are calculated.

- Proportionate labor costs: ca. 14,000 €/a
- Maintenance and repairs costs: ca. 50,000 €/a
- Diverse equipment, analytics, etc.: ca. 5,000 €/a

Capital charges for outside resources and depreciation of the plant are added. Costs for supply of distiller's wash and for removal of digesting residues have not been calculated. Both costs are part of raw material and production costs of the distillery.

## 6.7. THE DIGESTION RESIDUES

The digestion residues of agricultural distilleries are collected, stored, and used as fertilizer during plant growth. The degradation of organic ingredients in the biogas process resulted in the low viscosity (approx. 2–3% DS) of the distiller's wash. The mineral ingredients like N, P, K are conserved to a large extent in the fermenting residues. Studies of well-known experts showed that the effect of fertilizer from digested distiller's wash is improved compared to undigested fresh distiller's wash.

Distiller's wash has to be well digested to lose the typical off odors and to obtain a shelf life. Degradation rates over 85% can be achieved with well operating, two-stage plants, enabling an open storage without unpleasant odors. This method depicts an ecologically optimal solution: Digested distiller's wash is brought back to the fields as fertilizer; in turn the fertilizing digestate is metabolized by the plants, closing the recycling circle.

In bigger companies with 500 $m^2$ distiller's wash daily, the direct utilization of liquid fertilizer will soon approach logistic limits. According to the law of fertilizing and the relevant regulations deploying digestate or fertilizers on fields in winter is not allowed. In the meantime very huge storage capacities would be required. Additionally the impact of nitrogen to the fields is extremely limited. Due to the fact that in bigger distilleries raw materials are usually purchased nationwide, transportation costs for deploying digestate onto these fields, where the raw material comes from, would have a negative impact on the calculation.

After the digestion process it is practicable to separate digested distiller's wash into a slurry and a water phase. The slurry phase is stored in a concentrated state and needs less storage volume. The water phase is either treated aerobically, purified with microfilters, or vaporized and drained off as runoff ditch water. Thereby the surplus water can be finally cleaned to the quality of process water, provided it is economically advantageous. The slurry also can be further dried and conditioned, so that it can be stored and deployed with low costs.

## 6.8. EXAMPLES FOR BIOGAS PLANTS

As already mentioned, long-time experiences with biogas plants in distilleries are available. There are big plants in South Africa and overseas, which already were operating in the 1960s, processing wastewater from distilleries (Temper et al., 1986).

One of the first operating biogas plants in Germany was a plant of Mr. Spradau in Twistringen-Heiligenloh. This plant was designed by Schraufstetter and is the archetype of INNOVAS biogas plants.

1. Biogas plant Twistringen-Heiligenloh
   - 20 years in operation, from 1982 to the closure of the distillery 2002;
   - Digester volume $1 \times 300\,m^3$, $1 \times 700\,m^3$;
   - Distiller's wash from potatoes was used as substrate;
   - More than $10 \times 10^6\,m^3$ biogas was produced.
2. Biogas plant Mellinghausen
   - The plant had operated for 10 years, from 1984 to 1994 (closure of distillery);
   - Digester volume 2,200 $m^3$, used substrate was distiller's wash from potatoes;
   - For 3 months per year the plant was partly shut down and then started up trouble-free within days.
3. Biogas plant Altheim
   - Operating since 1999;
   - Digester 1,200 $m^3$;
   - Biogas plant operates with distiller's wash in winter and pulp in summer;
   - Long-time heat utilization in distillery and drying plant.

## 6.9. CONCLUSION

The utilization of distiller's wash and/or vegetable parts for monofermentation in a biogas plant is a fully developed and reliable technology with long-time operating experiences. With a well-planned operating concept, the profitability of the plant is assured.

Distilleries are the best example. The generated biogas substitutes up to 90% of the energy demand for fossil fuels and therefore determines the autarky of the distillery. The independence of imponderability of the international energy market is the most important criterion for the biogas application and alcohol production. Small and medium organic ethyl alcohol producers have a chance to compete in this market, if they possess a biogas plant for energetic utilization of distiller's wash.

The only alternative is the utilization of distiller's wash: the drying of distiller's wash for fodder production, which should not be evaluated at this point. Due to the fact that the energy input for the drying process is very high and that there is always an energy surplus in the biogas process, an energetic use of distiller's wash by anaerobic digestion should be beyond all questions.

## 6.10. REFERENCES

Bäuml, M., and COWATEC GmbH, 2005, Der Einsatz des hydrolytischen Enzyms MethaPlus zur Optimierung von Biogasanlagen, technical discussion (biomasse 2005), CARMEN, Straubing.

Gleixner, A., 2004, Brennerei und Biogas, Bioethanol- und Energieerzeugung in mittelständischen landwirtschaftlichen Brennereien, technical report.

Gleixner, A., and Spradau, L., 1996, 12 Jahre Biogasanlagen in Brennereien – Ein Erfahrungsbericht, in: *Handbuch für die Brennerei- und Alkoholwirtschaft, Brennereikalender 1996*, Beck, M., ed., Zimmermann Druck, Balve.

N.N., 2004, Gesetz für den Vorrang Erneuerbarer Energien (Erneuerbare-Energien-Gesetz – EEG), Federal Ministry for the Environment, Nature Conservation and Nuclear Safety, Berlin.

Pesta, G., 2005, Analysis Report Biogas Plant Altheim, technical report, Technische Universität München, Chair for Energy and Environmental Technologies of the Food Industry.

Temper, U., Pfeiffer, W., Steiner, A., Mücke, J. V., and Carozzi, A., 1986, Anaerobic waste water treatment - results of a literature review, in: *Proceedings of Aqua '86 - Symposium on anaerobic waste water treatment*, Amsterdam.

Wellinger, A., Baserga, U., Edelmann, W., Egger, K., and Seiler, B., 1991, *Biogas-Handbuch, Grundlagen – Planung – Betrieb landwirtschaftlicher Anlagen*, Wirz, Aarau.

# 7

# Brewery and Winery Wastewater Treatment: Some Focal Points of Design and Operation

António G. Brito, João Peixoto, José M. Oliveira, José A. Oliveira, Cristina Costa, Regina Nogueira, and Ana Rodrigues

## 7.1. INTRODUCTION

Environmental issues are a critical factor for today's industry competitiveness. Indeed, the society and the individual consumers could set a common framework for companies' commitment and engagement regarding environment protection. Redesigning the process, recovering by-products, or reusing effluents are some of the possible actions towards an ecoefficient strategy. Nevertheless, a point remains crucial in such a mission: the ability to defend natural ecosystems from polluted wastewaters. For such a purpose, a wastewater treatment plant that maximizes removal efficiency and minimizes investment and operation costs is a key factor.

Brewery and winery are traditional industries with an important economic value in the agrofood sector. In 2003, the total beer production in the European Union (18 countries) was $344 \times 10^5$ m$^3$, being recorded at 1,800 breweries with 110,000 employees. If Norway, Switzerland, and Turkey are also included, those numbers rise to $358 \times 10^5$ m$^3$, 1,839 units and 117,000, respectively. The excise revenue from the beer industry in all these countries reaches over $8800 \times 10^6$€ (The Brewers of Europe, 2004).

ANTÓNIO G. BRITO, JOÃO PEIXOTO, JOSÉ M. OLIVEIRA, REGINA NOGUEIRA, AND ANA RODRIGUES • University of Minho, School of Engineering – Center of Biological Engineering, Campus de Gualtar, 4710-057 Braga, Portugal. JOSÉ A. OLIVEIRA • Adega Cooperativa de Ponte da Barca, Lugar de Agrelos, 4980-601 Ponte da Barca, Portugal. CRISTINA COSTA • Unicer SA, Leça do Balio, Matosinhos, 4466-955 S. Mamede de Infesta, Portugal.

The worldwide wine production is $261 \times 10^5$ m$^3$ (data from 2002), of which 69% is from Europe, 18% from America, 5% from Asia, 4% from Africa, and 4% from Oceania. The worldwide wine consumption (2002) is $228 \times 10^5$ m$^3$, distributed by Europe (68%), America (20%), Asia (7%), Africa (3%), and Oceania (2%) (OIV, 2002).

This chapter intends to present some key points on design and operation in wastewater treatment of the brewery and winery industries. Therefore, an introduction of the industrial processes is first presented and then wastewater characteristics and treatment processes are discussed. Finally, the experience of a collaborative effort between the University of Minho and two industrial companies, Unicer SA and ACPB (Adega Cooperativa de Ponte da Barca) is presented in order to address some practical problems of wastewater systems design and operation. Unicer SA and ACPB are very important players in their field of activity: Unicer has the major share of the beer market in Portugal and ACPB is a very well-known producer of wine with appellation of origin *Vinho Verde*.

## 7.2. BREWERY AND WINERY INDUSTRIES: AN OVERVIEW

### 7.2.1. Brewing Processes

Beer is a soft drink obtained through alcoholic fermentation, using selected yeasts of the genera *Saccharomyces*, of wort prepared from malt cereals, mainly barley, and other amylaceous or sugar-based raw materials, to which were added hop flowers, or their derivatives, and adequate water. Figure 7.1 shows a typical technological process.

A mass balance is depicted in Figure 7.2, which represents water and energy inputs, and also the outputs respecting residues and subproducts, liquid effluents, and air emissions. Residues similar to urban residues, simple industrial residues, glass, paper, cardboard, plastic, oils, wood, biological sludge, green residues, etc., are classified as solid wastes; surplus yeast and spent grains are considered subproducts. Brewer's spent grains are generally used for the production of low-value composts, livestock feed, or disposed of in landfill as waste (Jay et al., 2004). Alternatively, the spent grains can be hydrolyzed for the production of xylo-oligosaccharides (probiotic effect), xylitol (sweetener), or pentose-rich culture media (Carvalheiro et al., 2004, 2005; Duarte et al, 2004).

### 7.2.2. Winemaking Processes

Wine is the product obtained from the total or partial alcoholic fermentation of fresh grapes, whether or not crushed, or of grape must. Producing wine requires the implementation of a biotechnological sequence involving several unit operations. Although some few products are added to the must and/or wine, several residues are rejected, either as liquid or solid wastes. White wine is normally produced by

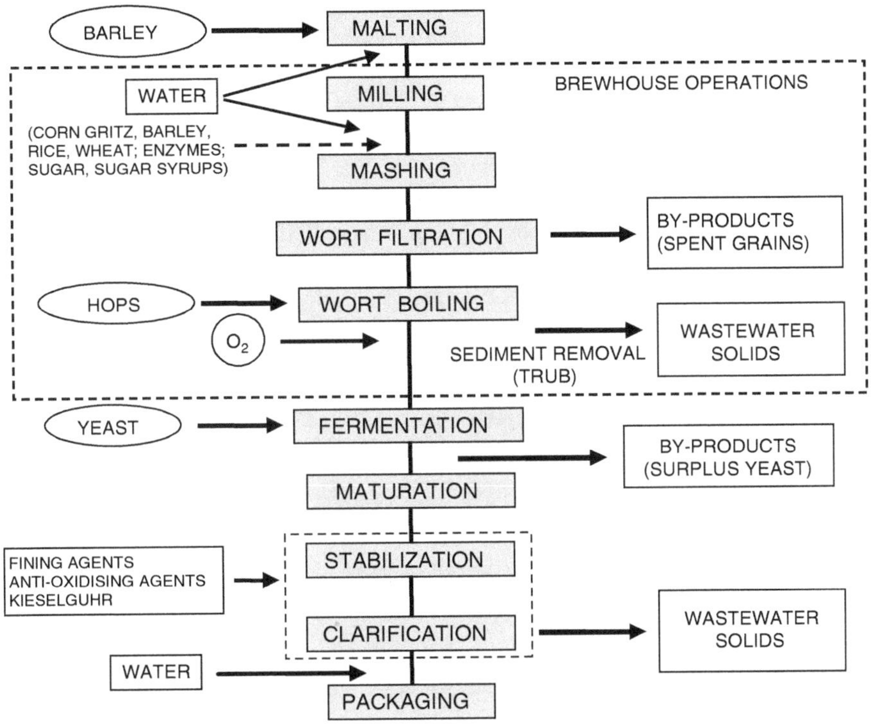


**Figure 7.1.** Technological process in breweries (adapted from Unicer SA and Varnam and Sutherland, 1994).

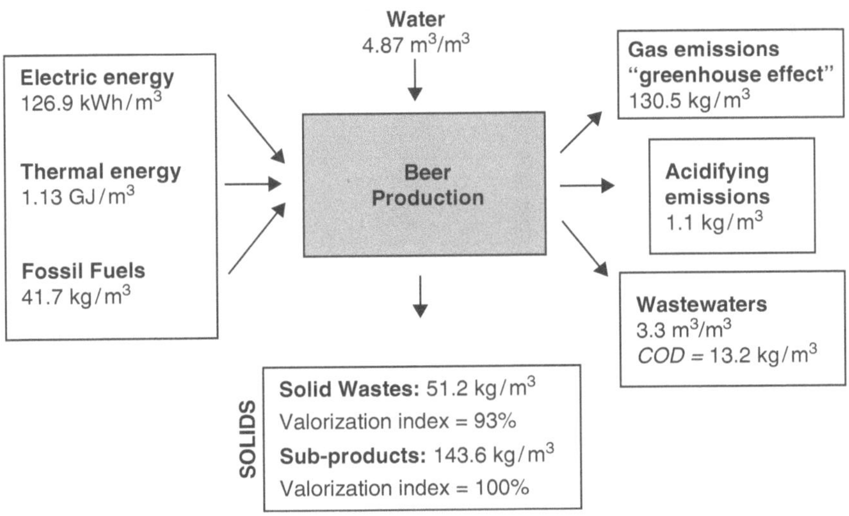


**Figure 7.2.** Mass balance applied to Unicer SA breweries representing specific values, i.e., values per cubic meter of produced beer (Unicer SA, 2005).

the fermentation of a clarified must, which is obtained after grape stem removal, pressing of the resulted grape berries, and subsequent clarification. The production of red wines is usually conducted in non-clarified musts, prepared after grape stem removal, and crushing of grape clusters. Musts also can be fermented in the presence of grape stems. After fermentation, wines must be clarified and stabilized, chemically and microbiologically, before bottling. Figure 7.3 shows a schematic process, applied at ACPB, to produce wines (*Vinho Verde*). These wines follow the ordinary winemaking process, but ageing is avoided, in order to preserve the original freshness and fruity characteristics.

Wineries, distilleries, and other grape-processing industries annually generate large volumes of wastewater. This mainly originates from various washing

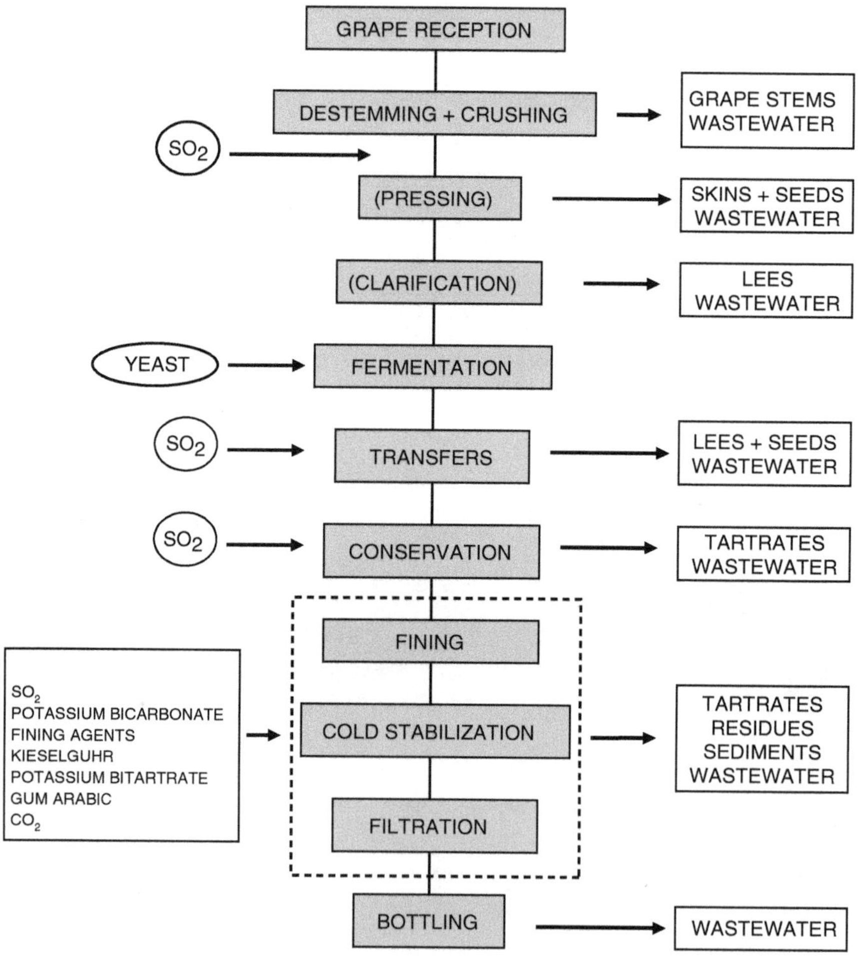


**Figure 7.3.** Technological process adopted at ACPB wine cellar.

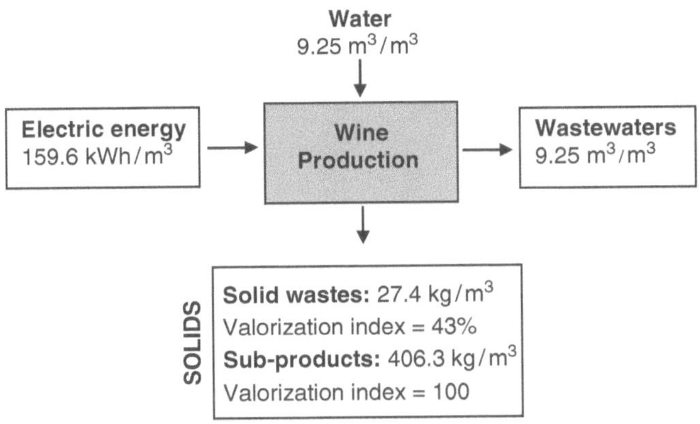


**Figure 7.4.** Mass balance applied to ACPB winery representing specific values, i.e., values per cubic meter of produced wine (2004). Losses of water by evaporation were neglected.

operations during the crushing and pressing of grapes, as well as rinsing of fermentation tanks, barrels, and other equipment or surfaces (Petruccioli et al., 2000). Over the year, volumes and pollution loads greatly vary in relation to the working period (vintage, racking, bottling) and to the winemaking technologies used, e.g., in the production of red, white, and special wines (Rochard, 1995; Anon, 1996).

A mass balance of wine production is depicted in Figure 7.4, which represents water and energy inputs and also the outputs respecting residues and subproducts, as well as liquid effluents. Simple municipal and some industrial residues (glass, paper, cardboard, plastic, wood, and filtration earths) but also yeasts, grape stems, pomace, and lees should be recycled and valorized whenever possible.

Yeasts cannot be used in animal dietary because they have high contents of polyphenols and may contain some residues coming from treatments; they only can be composted with pomace. However, pomace, seeds, lees, effluents resulting from tartar removal, and wine rests can be valorized to produce compounds with adding value like alimentary colorant E163, alimentary oil, tartaric acid, 1,3-propanediol, and dihydroxy-acetone (Bourzeix et al., 1998). On the other hand, the grape stems can be composted, the final compost being used as organic soil amendment and the grape pomace can be sold to distilleries.

## 7.3. WASTEWATER TREATMENT

### 7.3.1. Brewery Industry

#### 7.3.1.1. Wastewater Characterization

The composition of brewing effluents can fluctuate significantly as it depends on various processes that take place within the brewery, but the amount of wastewater produced depends on the water consumption during the process. In general, water consumption per volume of produced beer attains 4.7 $m^3/m^3$ (Carlsberg, 2005), but it should be pointed that the wastewater to beer ratio is often 1.2 $m^3/m^3$ to 2 $m^3/m^3$ less because part of the water is disposed of with by-products and lost by evaporation (Drissen and Vereijken, 2003).

Organic components in brewery effluent are generally easily biodegradable and mainly consist of sugars, soluble starch, ethanol, volatile fatty acids, etc., leading to a BOD/COD[1] ratio of 0.6 to 0.7. The effluent solids consist of spent grains, kieselguhr, waste yeast, and ''hot'' trub. The pH levels are determined by the amount and the type of chemicals used at the CIP (clean in place) units (e.g., caustic soda, phosphoric acid, nitric acid). Nitrogen[2] and phosphorous levels mainly depend on the handling of raw material and the amount of spent yeast present in the effluent. High phosphorous levels also can result from the chemicals used in the CIP unit. Table 7.1 summarizes some of the most important environmental parameters.

#### 7.3.1.2. Treatment Processes

Different environmental and socioeconomics criteria can be considered when deciding on a wastewater treatment plant for a brewery industry. The aim is to select a process that is flexible enough to cope with large fluctuations in organic load and

*Table 7.1.* Characteristics of some industrial brewery effluents including Unicer's

| Parameter / benchmark per unit | Brewery effluent composition | | |
|---|---|---|---|
| | Unicer | Typical[a] | Opaque beer[b] |
| COD (mg/L) | 800–3,500 | 2,000–6,000 | 8,240–20,000 |
| BOD (mg/L) | 520–2,300 | 1,200–3,600 | |
| TSS[c] (mg/L) | | 200–1,000 | 2,901–3,000 |
| TS[c] (mg/L) | | | 5,100–8,750 |
| T (°C) | 30–35 | 18–20 | 25–35 |
| pH | 6.5–7.9 | 4.5–12 | 3.3–6.3 |
| Nitrogen (mg/L) | 12–31 | 25–80 | 0.0196–0.0336 |
| Phosphorous (mg/L) | 9–15 | 10–50 | 16–24 |
| (Water/beer) ($m^3/m^3$) | 4.87 | | |
| (Liquid effluent/beer) ($m^3/m^3$) | 3.3 | 2–8 | |
| (COD/beer) ($kg/m^3$) | 13.2 | 5–30 | |
| (BOD/beer) ($kg/m^3$) | | 2–20 | |
| (TSS/beer) ($kg/m^3$) | | 1–5 | |

[a] Driessen and Vereijken (2003).
[b] Parawira et al. (2005).
[c] TS, TSS, Total solids, total suspended solids.

[1] BOD, Biochemical oxygen demand; COD, chemical oxygen demand (mass of $O_2$ per volume).
[2] N, Nitrogen mass concentration (mass of N per volume). $NO_3^-$-N, $NO_2^-$-N, $NH_4^+$-N Nitrate, nitrite, and ammonia mass concentration as mass of N per volume.

characteristics of such wastewaters, while keeping capital and operating costs as low as possible. Because organic matter concentration in brewery effluent is significant, a high input of energy for aeration is required. Another factor is the amount of waste sludge generated from aerobic metabolism, which also needs to be handled and disposed of. Both increase the cost of operation of the treatment system. Therefore, anaerobic processes are preferred for the purpose of brewery wastewaters pretreatment because energy is saved and sludge disposal costs are minimized. When discharging into surface water bodies, anaerobic pretreatment combined with subsequent aerobic posttreatment for organic or nutrient removal is considered to be the best solution (Rodrigues et al., 2001; Nogueira et al., 2002).

Several types of anaerobic reactors can be applied to brewery wastewater treatment. However, the upflow anaerobic sludge blanket (UASB) reactor clearly accounts for the most usual full-scale systems (Batston et al., 2004; Parawira et al., 2005). The upflow mode of operation induces the development of a characteristic biological self-aggregation process without addition of support material. The resulting biofilm structure is usually denominated ''granules'' and is the main factor for their high biomass concentration and biological activity (Brito et al., 1997a). The expanded granular sludge bed (EGSB) reactor is a tower reactor using granular anaerobic sludge, identical to UASB reactors, built with tank heights of 12 m to 16 m. The internal circulation (IC) reactor also uses granular anaerobic sludge and is built with higher tank heights (up to 24 m). Whereas the EGSB and UASB reactors separate the biomass, biogas, and wastewater in a one-step three-phase-separator located in top of the reactor, the IC reactor is a two-staged UASB reactor design. The lower UASB receives extra mixing by an internal circulation, driven by its own gas production. While the first separator removes most of the biogas, turbulence is significantly reduced, allowing the second separator to effectively separate the anaerobic sludge from the wastewater. The loading rate of the IC reactor, as COD, is typically twice as high as the UASB reactor (15 kg $m^{-3}d^{-1}$ to 30 kg $m^{-3}d^{-1}$). Another positive factor resulting from the applied high hydraulic upflow velocities is the selective washout of brewery solids, like kieselguhr, trub, and yeast.

In order to meet stringent requirements of surface water quality, an aerobic polishing step is necessary after the anaerobic pretreatment. Sequencing batch reactors (SBR) are well suited for such purpose (Brito et al., 1997b; Rodrigues et al., 2004). The SBR is a periodic process that performs multiple biological reactions in non–steady-state conditions. Biomass retention throughout the introduction of a decanting step and the ease of automation are additional advantages for using SBR technology (Rodrigues et al., 1998). Nevertheless, some other interesting experiences regarding aerobic processes can be named. Selected examples are jet loop reactors (Bloor et al., 1995), fluidized bed bioreactor (Ochieng et al., 2002) and membrane bioreactors (Cornelissen et al., 2002). It should be noted that membrane bioreactors deserve a special attention within the brewing industry. Their market share can increase in the next few years, including in the anaerobic concept (Ince et al., 2000).

### 7.3.2. Winery Industry

#### 7.3.2.1. Wastewater Characterization

Winemaking is seasonal with high activity in autumn (at north hemisphere), which corresponds to vintages and fermentations, a notoriously less important activity in spring on the occasion of transfers (racking period) and filtrations, and a weak activity during winter and summer. Winery effluents contain four types of principal pollutants:

- Subproduct residues: stems, seeds, skins, lees, sludge, tartar, etc.;
- Loss of brut products: musts and wines occurred by accidental losses and during washings;
- Products used to wine treatments: fining agents, filtration earths, etc.;
- Cleaning and disinfection products used to wash materials and soils.

Musts and wine constituents are present in wastewaters, in variable proportions: sugars, ethanol, esters, glycerol, organic acids (e.g., citric, tartaric, malic, lactic, acetic), phenolic compounds (coloring matter and tannins), and a numerous population of bacteria and yeasts. They are easily biodegradable elements, except for polyphenols (60 mg/L to 225 mg/L) which make this biodegradation more difficult and requiring an adapted flora. Effluents have a pronounced demand in nitrogen and phosphorous, with a $BOD_5/N/P$ relation often near 100/1/0.3 (Torrijos and Moletta, 1998). Additionally, effluents have a daily great variability, in both quantity and quality, making evaluation of daily pollution complex. Generally, the production of 1 $m^3$ of wine generates a pollution load equivalent to 100 persons. The pH is usually acidic but, punctually, it may display basic values, on the occasion of the cleaning operations (with alkaline products and organochlorides) and on the occasion of chemical detartaration.

Rejected volumes per volume of produced wine vary from one wine cellar to another, with extreme values comprised between 0.1 $m^3/m^3$ and 2.4 $m^3/m^3$. For the ratio of water consumption to produced wine, 1.0 $m^3/m^3$ is the rule of thumb, while Pévost and Gouzenes (2003) refer to values between 0.3 $m^3/m^3$ and 2.5 $m^3/m^3$. Table 7.2 shows some examples of winery effluents, main characteristics. Washing operations carried out during different winemaking steps, which are at the origin of the rejection of fully charged wastewaters, can be distributed as follow:

- During vintage preparation: washing and disinfection of materials;
- During grape reception: washing of reception materials (hoppers, destemmers, crushers, presses, dejuicers, conveyors and transport pumps); cleaning the floors, with or without addition of cleaning products;
- During vinifications: rinsing of fermentation and clarification vats; cleaning the floors, with or without addition of cleaning products;
- During transfers: rinsing vats after transfers; cleaning the floors, with or without addition of cleaning products;
- During filtrations: rinsing kieselguhr and earth filters.

*Table 7.2.* Examples of effluent composition (mean or range values) of four different wineries, including that of ACPB

| | | Wine cellar[a] | | |
|---|---|---|---|---|
| | ACPB | A[b] | B[b] | C[c] |
| Production ($m^3$/year) | 250 | 730 | 3,000 | 6,000 |
| pH | 5.7 | 4.9 | 4.7 | 4.0–4.3 |
| COD (mg/L) | 1,200–10,266 | 5,200 | 14,150 | 9,240–17,900 |
| BOD (mg/L) | 130–5,320 | 2,500 | 8,100 | 5,540–11,340 |
| TSS (mg/L) | 385–5,200 | 522[e] | 1,060 | 1,960–5,800 |
| TVS[d] (mg/L) | — | — | 742 | 81–86% of the TSS |
| Total N (kjeldahl) (mg/L) | 12–93 | 61 | 48.2 | 74–260 |
| Total P (mg/L) | 23 | 25 | 5.5 | 16 to 68 |

[a] Torrijos and Moletta (1998).
[b] Vintage period, mean value after 24 h.
[c] Extreme values.
[d] TVS, Total volatile solids.
[e] After primary sedimentation.

### 7.3.2.2. Treatment Processes

The criteria for selecting an anaerobic or an aerobic biological treatment are identical in brewery and winery industries. Like in the brewery industry, the winery wastewaters are characterized by their high content on organic biodegradable compounds. In this case, the anaerobic technology is the most economical bioprocess due to lower running costs for aeration and sludge processing. However, as previously mentioned for the brewery case, the anaerobic conversion is generally insufficient to attain the effluent quality required for discharge in surface waters. Therefore, the anaerobic treatment should be followed by an aerobic system, if the option of cotreatment of the winery wastewaters in a (aerobic) municipal wastewater treatment plant is not available. Despite such rule, in the case of small wine industries where the minimization of investment costs is the key factor and only one biological process may be considered, the option must be an aerobic process if the objectives for effluent quality are high. Obviously, the financial burden of an aerobic operation is not so heavy in the case of a low wastewater flow.

Organic matter is essentially in soluble form. Therefore, a static sedimentation unit is not an option for significant concentration reduction. Besides, an important fraction of the suspended matters is easily removed by settling (seeds, tartaric salts, filtration earths). Another focal point is the removal of inorganic suspended solids from such type of wastewaters because the abrasive solids used in precoated filters can damage mechanical equipment. Furthermore, many biological processes face difficulties for treating nonsoluble wastewaters: a pretreatment step using screening and/or sedimentation is then mandatory.

The anaerobic process shows a very good reliability for winery wastewaters. The COD/N/P ratio is appropriate for anaerobic bacteria and the seasonal activity is not a problem for process start-up. The anaerobic digesters are generally heated to reach the mesophilic range (but psychrophilic conditions are possible) and it is advisable to measure alkalinity routinely in order to avoid a sudden pH drop in one-stage processes. All anaerobic technologies can be applied for treating winery wastewaters. Among them, two of the most promising ones are granular UASB reactors and the anaerobic sequencing batch reactor (aSBR). An interesting approach is reported by Keyser et al. (2003) who evaluated three UASB reactors with the aim of tailoring granules for the treatment of winery wastewater, a novel ecotechnological approach. One reactor was seeded with granular sludge enriched with *Enterobacter sakazakii* and a 90% COD removal at hydraulic retention time of 24 h could be reached. This performance compares favorable with a second reactor seeded with brewery granules that achieved 85% COD removal and with a third one seeded with municipal sludge, which showed problems and had continuously to be reseeded. Ruíz et al. (2002) operated an anaerobic sequencing batch reactor at an organic loading rate, as COD, around 8.6 kg/($m^3$d) with soluble COD (sCOD) removal efficiency greater than 98%, hydraulic retention time of 2.2 d and a specific organic loading rate, as COD/VSS (volatile suspended solids), of 0.96 g/(g d). Anaerobic filters and completely mixed reactors are also used in the winery industry, but fewer systems are under construction now.

As stated before, aerobic technologies are well suited for the depollution of wastewaters from wineries, if their running costs are not decisive. Sequencing batch reactors are becoming the most popular, since Torrijos and Moletta (1997) used them to treat a winery wastewater and reported a 95% sCOD elimination, and a nitrogen and phosphorous removal of 50% and 88%, respectively. These results could be generalized and the simplified automation and the possibility of coping with load fluctuations are decisive SBR advantages. Nevertheless, other different designs are currently available. Eusébio et al. (2004) have operated jet-loop reactors, Andreottola et al. (2005) performed the treatment of a winery wastewater applying a two-stage fixed bed biofilm reactor, and Coetzee et al. (2004) have implemented a pilot-scale rotating biological contactor. The seasonal operation of wineries may be a problem for aerobic biological systems leading to decreased sludge settleability, floc disintegration, and increased solids in the treated effluent (Chudoba and Pujol, 1996). Therefore, in order to work efficiently, even during those temporary overloading periods, the plant has to be oversized. This strategy is rather costly, because such a plant has to run below its nominal capacity during a major part of the year.

In small wineries, simplified systems of low energy consumption—lagoons, constructed wetlands, land spreading/irrigation—are also scenarios for effluent treatment or polishing, but a landscape integration is sought and large areas of land should be available (Bustamante et al., 2005). The feasibility of such an approach depends on external factors that restrain a generalized use, namely, meteorological, hydrogeological, and soil and biomass characteristics. Therefore, the engineering of a specific biological treatment process for wineries wastewater, including the selection of ancillary equipment, should be decided on a case by case basis, as stated by Rochard and Kerner (2004).

## 7.4. CASE STUDY 1: BREWING WASTEWATER TREATMENT

The brewery industry Unicer SA has in operation a UASB reactor (1600 m$^3$) for the industrial wastewater treatment. The start-up of UASB reactors often relies on a massive inoculation with biomass already in pellets/granules (Nollet et al., 2005), representing an additional cost for the brewery industry. Indeed, the Unicer SA reactor was inoculated with granular sludge imported from a paper factory in Spain. A 70% to 80% COD removal is generally recorded in the UASB process. In spite of such efficiency, the final COD and ammonium nitrogen levels are above the threshold values prescribed by legislation for wastewater discharge in surface waters. On the other hand, due to the anaerobic digestion process, the carbon concentration in the UASB effluent is very low, imposing difficulties on conventional postdenitrification processes. Therefore, as depicted in Figure 7.5, several steps were performed. First, there was the formation of anaerobic granules in a lab-scale UASB reactor using dispersed biomass as inoculum and the industrial wastewater from Unicer SA as substrate. Second, the feasibility of SBR technology for the posttreatment of the effluent from the UASB reactor was assessed. For the posttreatment of the brewery wastewater, two different SBR strategies for nitrogen removal were considered. One was based on an aerobic–anoxic sequence and the other one comprised a predenitrification step, that is, an anoxic–aerobic–anoxic sequence. In both tests, SBR performance and biological kinetics were evaluated.

### 7.4.1. UASB Operation for the Formation of Biomass Granules

Nonaggregated biomass from an anaerobic digester used in the stabilization of activated sludge was tested for granulation. The operational protocol was based on the selection of aggregate-forming bacteria, mainly focused on the acetotrophic *Methanothrix spp*, by favoring the washout of nonaggregated biomass (Hulshoff Pol, 1989). In order to attain such an objective, the loading rate was increased when acetate concentration was lower than 50 mg/L, a value near the half saturation

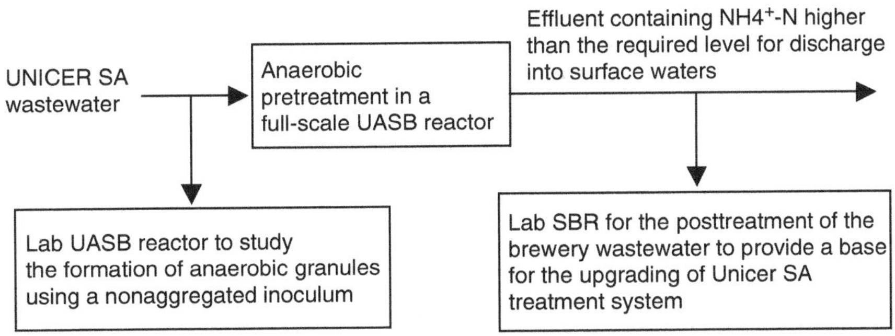


**Figure 7.5.** Schematic diagram of the goals of the present chapter.

constant of *Methanothrix spp.* During the first 3 months, the treated effluent was partly recycled to increase the hydraulic load. The operating temperature in the UASB reactor was 35°C. The pH ranged from 6.5 to 7.9.

Figures 7.6 and 7.7 show the operational conditions and results of the UASB reactor, namely, $B_V$ (volumetric organic load, organic matter mass concentration, as COD, per time unit), COD (influent and effluent), and COD removal efficiency. The granular activity sustained the application of high $B_V$, up to 20 kg/($m^3$ d), with average COD removal efficiencies of 80%. The objective of granulation process was successfully achieved, but a 6-month period of operation was necessary. The sedimentation velocity of aggregated biomass attained 40 m/h to 50 m/h and the SVI (sludge volume index) was 10 mL/g. The TS and TVS in granules amounted to 114 kg/$m^3$ and 87 kg/$m^3$. Figure 7.8 shows a scanning electron microscopy (SEM) picture of the granules, obtained at the end of operation.

The feasibility of UASB reactor start-up based on an inoculation with non-aggregated biomass was demonstrated for the treatment of brewery industry

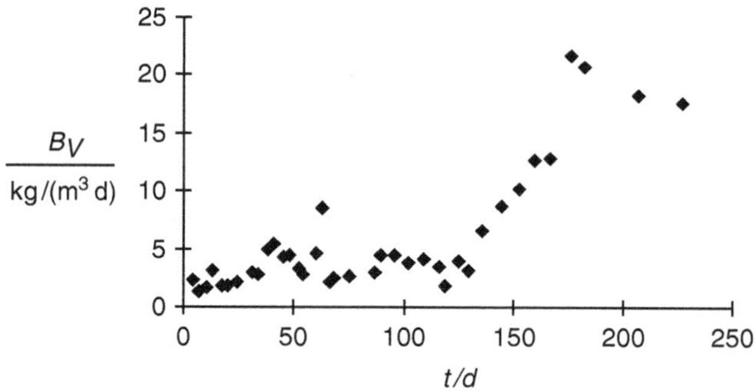


**Figure 7.6.** Organic load applied to the UASB reactor.

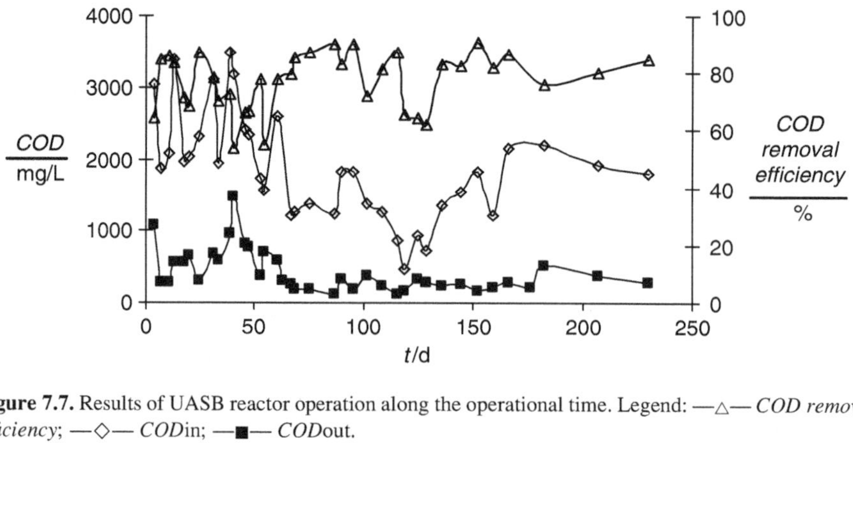


**Figure 7.7.** Results of UASB reactor operation along the operational time. Legend: —△— *COD removal efficiency*; —◇— *COD*in; —■— *COD*out.

**Figure 7.8.** SEM photograph of the biomass after granulation.

*Table 7.3.* Brewery wastewater composition after UASB pretreatment

| Parameter | Range |
|---|---|
| pH | 7.5–8.0 |
| Total COD, tCOD (mg/L) | 400–2,000 |
| Soluble COD, sCOD (mg/L) | < 470 |
| Soluble organic carbon (C) (mg/L) | 60–83 |
| $NH_4^+$-N (mg/L) | 23–87 |
| $NO_2^-$-N (mg/L) | 0–1.2 |
| $NO_3^-$-N (mg/L) | 0–3 |
| Soluble P (mg/L) | 8–20 |
| TSS (mg/L) | 320–1,440 |

wastewaters, concerning organic matter elimination. However, an ammonification process occurred, $NH_4^+$-N in the effluent ranging between 23 mg/L and 87 mg/L, while the influent $NH_4^+$-N was just 12 mg/L to 29 mg/L. Therefore, a further nitrogen removal process was necessary in order to attain effluent thresholds for discharge into surface waters.

### 7.4.2. SBR Operation for the Posttreatment of the Brewery Wastewater

The average composition of the UASB effluent collected at Unicer SA brewery is shown in Table 7.3. The bench-scale SBR was operated in the typical sequence of fill, react, settle, and draw. Two SBR operating strategies were tested during the present study. Their main features are summarized in Figure 7.9. The SBR operational conditions are described in Table 7.4. The biomass inoculum was a grab sample collected in a municipal activated sludge plant of the extended aeration type (around 90% of the inoculum), supplemented with an inoculum of *Alcaligenes denitrificans* and nitrifying microorganisms. The average MLVSS (mixed liquor volatile suspended solids concentration, mass per volume) during the experimental assays was 1690 mg/L. Solids sampling represented the only biomass wastage

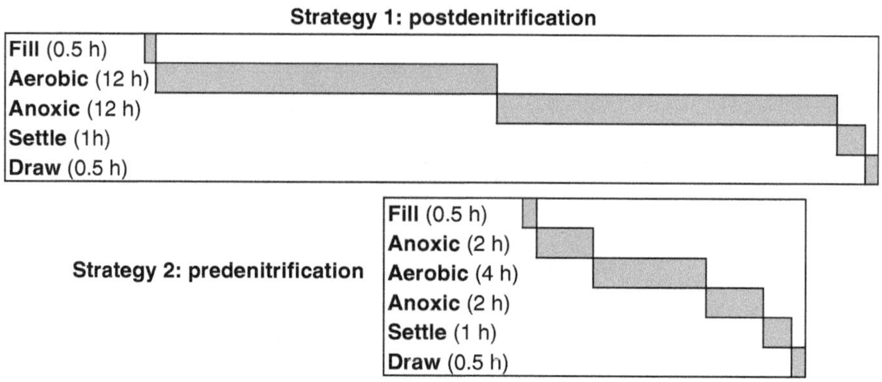


**Figure 7.9.** SBR operational strategies.

*Table 7.4.* SBR operating conditions

| | Strategy 1 | Strategy 2 |
|---|---|---|
| Reaction sequence | Aerobic (12 h) /anoxic (12 h) | Anoxic (2 h)/aerobic (4 h) /anoxic (2 h) |
| Total cycle time (h) | 26 | 10 |
| Working volume (L) | 2.9 | 1.7 |
| Volumetric replacement (%) | 60 | 30 |
| HRT[a] (d) | 1.9 | 1.2 |
| DO (in the aerated phase) (mg/L) | 3.7 | 2.8 |
| N after Fill (mg/L) | 30–45 | 20–28 |
| Nitrogen (N) load per volume [kg/(m$^3$d)] | 0.040 | 0.086 |
| Nitrogen (N) load per VSS[b] [kg/(kg d)] | 0.024 | 0.051 |

[a] HRT, Hydraulic retention time.
[b] VSS, Volatile suspended solids.

carried out along the experimental work. Therefore, the sludge age was rather long, being estimated as 37 d. The concentration of nitrogen compounds in the treated effluent is depicted in Figure 7.10.

Strategy 1 was characterized by the use of an aerated phase (dissolved oxygen concentration, mass of $O_2$ per volume, DO = 3.7 mg/L) followed by an anoxic phase (see Figure 7.11). Complete nitrification took place, during the aerated phase, ammonium and nitrites being removed from the anaerobically pretreated effluent. The maximum observed SDR (specific denitrification rate, mass of nitrogen, N, per *VSS* per time) was 0.165 kg/(kg d). However, the nitrogen removal efficiency was 50%, resulting in an effluent $NO_3^-$-N above 15 mg/L (the value prescribed by the legislation for discharge in surface waters is 11 mg/L). Moreover, a nitrogen balance in the liquid phase showed that $NO_3^-$-N at the end of the aerobic phase (15 mg/L to 20 mg/L) was roughly 50% lower than the theoretically one expected according to reaction stoichiometry. Biomass yield was not sufficient to fill this gap.

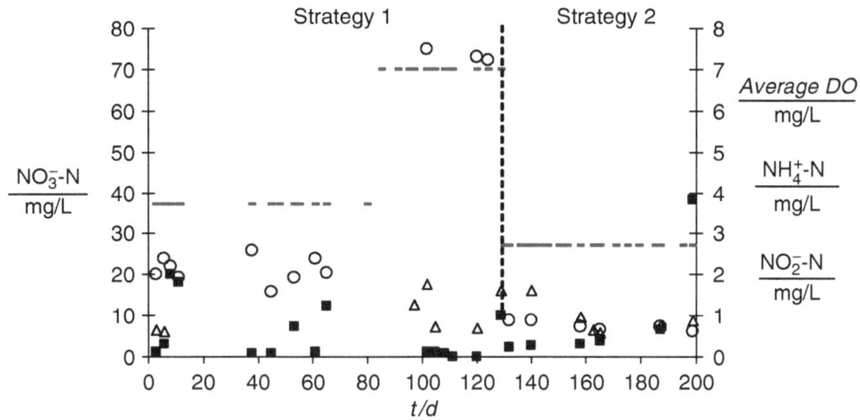


**Figure 7.10.** Long-term ammonium, nitrite, and nitrate effluent concentrations and average DO levels in the aerated phase. Legend: —△— $NH_4^+$-N; —■— $NO_2^-$-N; —○— $NO_3^-$-N; —-— *Average DO*.

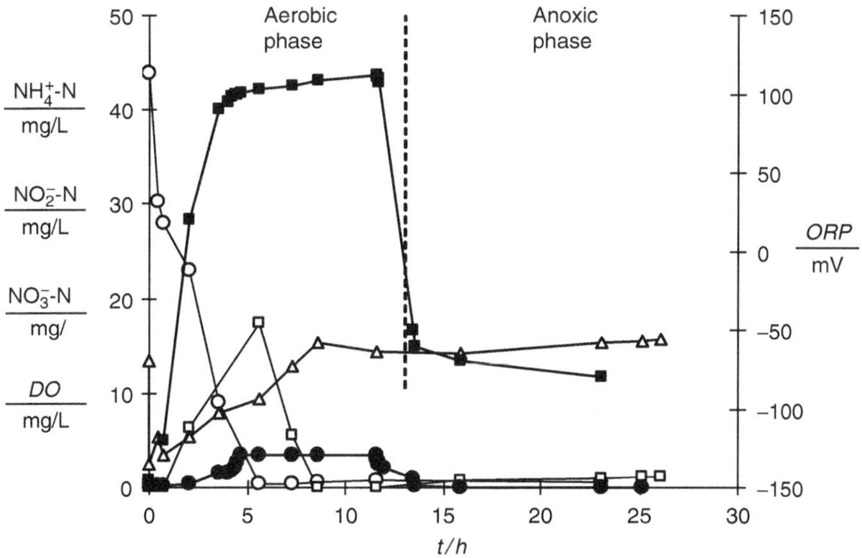


**Figure 7.11.** ORP (oxidation–reduction–redox–potential) values and DO, ammonium, nitrite, and nitrate nitrogen concentrations in the bulk liquid, along a typical SBR cycle, in strategy 1 ($DO = 3.7$ mg/L during the aerated phase). Legend: —■— *ORP*; —●— *DO*; —○— $NH_4^+$-N; —□— $NO_2^-$-N; —△— $NO_3^-$-N.

Consequently, the data indicate the occurrence of a significant denitrification process during the aerated phase.

As declared above, the maximum observed SDR was 0.165 kg/(kg d). An explanation for such phenomenon relies on oxygen limitations within microbial flocs providing the oxygen-free conditions for heterotrophic denitrifying bacteria activity (van Loosdrecht and Heijnen, 1993). Such hypothesis was tested setting a DO of 7 mg/L during the aerated phase (Figure 7.12). In fact, at such high DO, denitrification did not occur during the aerobic period, confirming that there was an oxygen limitation when the bulk liquid DO was 3.7 mg/L.

The strategy 2 involved a predenitrification step and thereafter the aerated phase (DO = 2.8 mg/L) and the anoxic phase. In Figure 7.13 the behavior of nitrogen compounds and the ORP and DO profiles during a typical SBR cycle are shown.

The overall experimental results (Figure 7.10) demonstrated that the most appropriate strategy for nitrogen removal in order to achieve the legal compliance for wastewater discharge in surface waters was the anoxic–aerobic–anoxic sequence, with DO = 2.8 mg/L in the aerated period and a volumetric replacement of 30% (strategy 2). Under such conditions, the maximum observed SNRR (specific nitrogen removal rate) had the value 0.038 kg/(kg d) and $NO_3^-$-N in the effluent was lower than 8 mg/L.

Thus, this strategy optimizes the energy requirements for aeration with an appropriate effluent quality for discharge in surface waters. The denitrification during the final anoxic phase (after the aerobic period) was practically meaningless in all runs.

The C/N ratio in the UASB effluent had an average value of 0.8 and carbon requirements for complete nitrogen removal were not satisfied: the stoichiometric

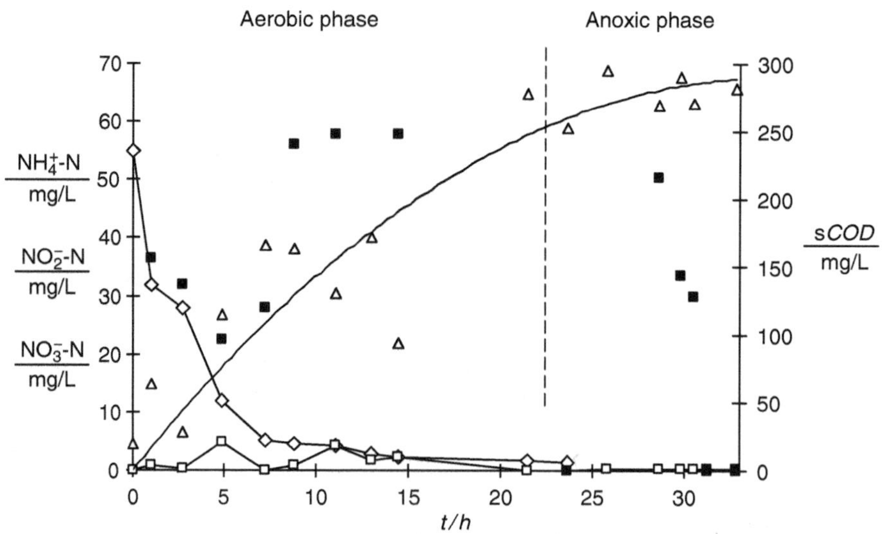


**Figure 7.12.** Ammonium, nitrite, nitrate, and soluble COD in the bulk liquid along a typical SBR cycle in strategy 1 (during the aerated phase: DO = 7 mg/L). Legend: —◇— $NH_4^+$-N; —□— $NO_2^-$-N; —△— $NO_3^-$-N; —■—*sCOD*.

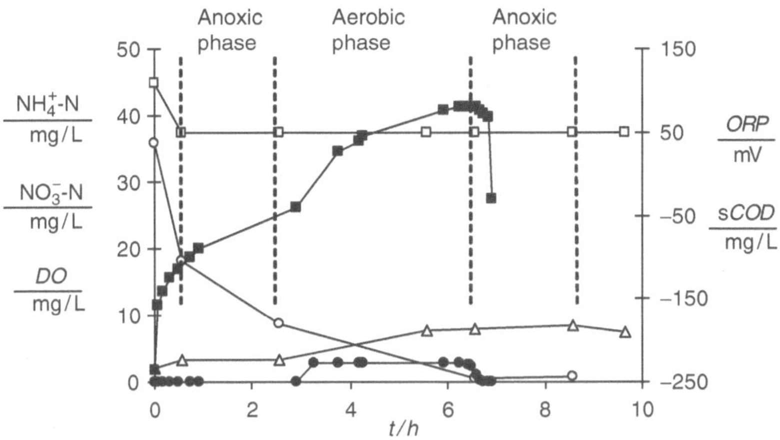


**Figure 7.13.** Variation of ORP values, DO, soluble COD, and nitrogen compounds concentration in the bulk liquid along a typical SBR cycle, during strategy 2. Legend: —■— *ORP*; —●— *DO*; —□— *sCOD*; —○— $NH_4^+$-N; —△— $NO_3^-$-N.

C/N ratio, using an easily degradable carbon source like acetate, must be 1.25. The results obtained when acetate was used to increase the mass C/N ratio to 1.3, during the anoxic phase, leading to a complete nitrate removal, confirmed that the soluble carbon source was limiting denitrification. Simultaneous nitrification and denitrification was detected during the aerated phase at DO = 2.8 mg/L and 3.7 mg/L (Figures 7.11 and 7.13). On the other hand, denitrification was inhibited during the aerated period when the bulk liquid DO was raised to 7 mg/L (Figure 7.12). The redox potential was kept within the range + 100 mV to −240 mV, in response to oxygen concentration along each cycle. The ORP provides information about the process regime and can be used to control the duration of the denitrification phase (Demoulin et al., 1997). However, the typical breakpoint, the "nitrate knee," that appears in the ORP curve at $NO_3^-$-N close to zero could not be observed (Figures 7.11 and 7.13). Due to the simultaneous nitrification–denitrification and carbon limitations, nitrate was always present, even if at low concentrations, along the entire operating cycle. An improvement of the biological floc settleability was noticed along SBR operation. Soon after the start-up, flocs became larger and the SVI decreased from 200 mL/g down to 115 mL/g. A concomitant decrease of TSS in the treated effluent was observed, attaining only 30 mg/L at the end of the experimental period.

## 7.5. CASE STUDY 2: WINERY WASTEWATER TREATMENT

The start-up and optimization of the wastewater treatment process in the wine industry are presented using the ACPB case study. The aimed optimization was based on two operational strategies during the periods of high daily flows and organic loads.

### 7.5.1. Full-Scale Sequencing Batch Reactor Operation

A schematic diagram of the WWTP (wastewater treatment plant) is presented in Figure 7.14.

A sand remover is located at the beginning of the WWTP in order to remove solid materials, including diatomaceous earth, from the wastewater. If necessary, the correction of the pH value is made in a 300 $m^3$ equalization tank, throughout $CaCO_3$ addition. Inside the SBR, two superficial jet aerators with swing-arms are installed, with a capacity of oxygenation of 5.5 kg/h each.

The SBR was inoculated with domestic sewage, with VSS = 540 mg/L. Nutrients were supplied throughout the SBR feeding with domestic wastewater produced at ACPB. The ratio $BOD_5/N/P$ was 100/10/0.4. The excess sludge from the SBR was conducted to a gravity thickener. According to the working conditions, nine operational phases were identified during the first year of operation of the SBR (Table 7.5). Each phase was either related to a working period of the wine industry (with a typical effluent composition), or to changes in the SBR operating conditions in order to increase the treatment efficiency. In order to deal with the high effluent volumes generated during the vinification and racking periods, the SBR was fed twice a day, representing a reduction of 50% in the reaction time. The time-based SBR schedule is depicted in Table 7.6, describing the operation with one and two cycles per day. Table 7.7 summarizes the general operating conditions of the SBR, considering the operation with one cycle per day.

The pH values of the effluent were in the range between 7 and 8. Nevertheless, during the vinification period, pH values of 3 were detected in the equalization tank. The temperature of the effluent ranged between 15°C and 25°C. In general, high COD removal efficiencies were detected, despite the $B_V$ changes (Figure 7.15), but during the vinification and racking periods, $B_V$ increase led to a significant decrease in the COD removal efficiency, due to oxygen limitations.

In fact, $B_V$, which was usually in the range between 0.5 kg/($m^3$ d) and 1.5 kg/($m^3$ d), reached, in this period, the averaged value of 2.5 kg/($m^3$ d). As a

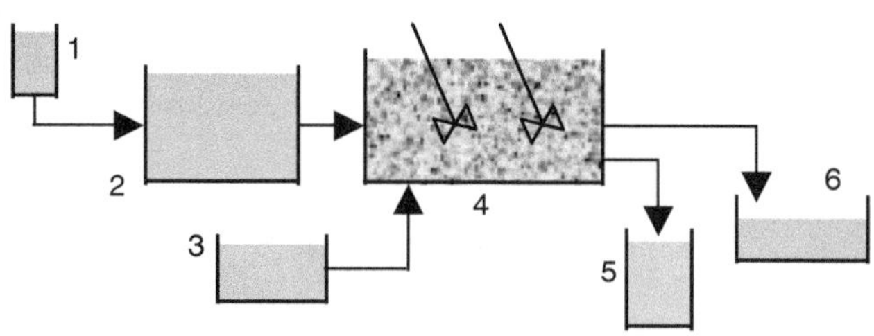


**Figure 7.14.** Schematic diagram of the WWTP from ACPB: 1, Sand remover; 2, equalization/neutralization tank; 3, septic tank; 4, biological unit (SBR); 5, sludge thickener; 6, river.

*Table 7.5.* Characterization of the different phases of WWTP operation

| Operational phase | Operational reactor phases | Working period at the winery | Cycles per phase | Cycles per day |
|---|---|---|---|---|
| 1 | Start-up | — | 65 | 1 |
| 2 | Operation | Washing operations and bottling | 37 | 1 |
| 3 | Operation | Vinification | 10 | 1 |
| 4 | Operation | Vinification and racking | 29 | 1 |
| 5 | Operation | Bottling | 24 | 1 |
| 6 | Operation | Second racking | 42 | 1 → 2 |
| 7 | 1st sludge purge | — | 26 | 2 |
| 8 | 2nd sludge purge | — | 33 | 2 |
| 9 | Biomass recirculation | — | 27 | 1 |

*Table 7.6.* Time-based SBR schedule for operation with one and two cycles per day

| | Aerated fill | Aerated react | Settle | Draw |
|---|---|---|---|---|
| 1 cycle per day | 0.5 h | 21 h | 2 h | 0.5 h |
| 2 cycles per day | 0.5 h | 10 h | 1 h | 0.5 h |

*Table 7.7.* SBR general operating conditions (one cycle per day)

| Parameter | Value |
|---|---|
| Working volume ($m^3$) | 150 |
| Volumetric replacement (%) | 17 |
| HRT (d) | 5.7 |
| VSS (g/L) | 2.5–4.5 |
| $B_V$ [kg/($m^3$d)] | 0.5–2.5 |
| Applied specific load (as VSS) [kg/(kg d)] | 0.26–0.57 |

consequence, the biomass concentration increased significantly and the oxygen supply was not enough to fit the needs, resulting in tCOD values of 5000 mg/L in the discharge, despite the higher COD removal rates. A washout of the biomass was observed for VSS higher than about 4.5 g/L, leading to an increase in the final effluent total COD and TSS. In fact, for VSS higher than 4.5 g/L, the biomass exhibited a low sedimentation capability, due to the high sludge age (45 d), leading to SVI >120 mL/g. The results obtained showed that SVI values should not exceed 80 mL/g, in order to maintain a good performance of the biological reactor. Toward the goal of increasing the SBR performance during the vinification and racking periods and in order to account for the high daily flow and organic load of the industrial effluent, two operational strategies were tested (Figure 7.16).

The first strategy, based on the operation of the SBR with two cycles per day (resulting in a 50% decrease in HRT, from 7.4 d to 3.7 d, and therefore in the

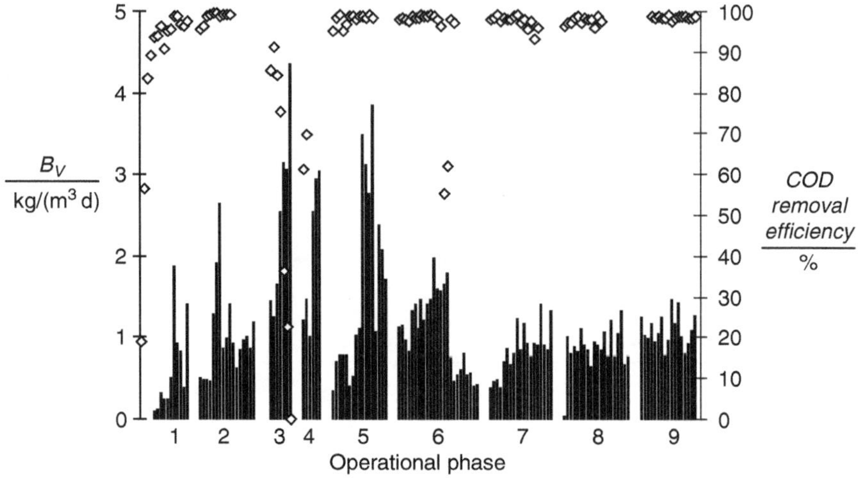


**Figure 7.15.** COD removal efficiency as a function of the applied volumetric load. Legend: —◇— COD removal efficiency; —■— $B_V$.

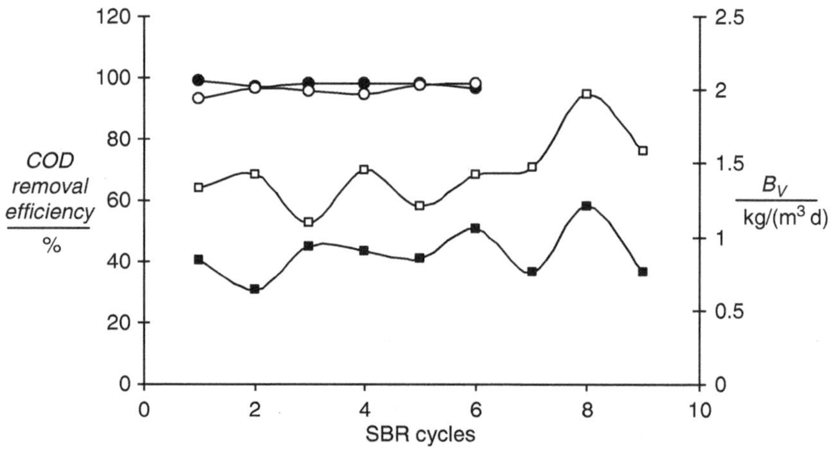


**Figure 7.16.** *COD* removal efficiency according to the applied organic load for different operational strategies (1 and 2 cycles per day). Legend: —○— Efficiency (1 cycle); —●— Efficiency (2 cycles); —■— $B_V$ (1 cycle); —□— $B_V$ (2 cycles).

duplication of $B_V$), was tested during the bottling period (average tCOD in the equalization tank of 4000 mg/L). At this time, the winery wastewater comes mainly from the washing operations and from the cooling processes, leading to high daily wastewater flows. The second strategy was used when $B_V$ was high [above 1.5 kg/(m$^3$d)] and consisted of the recirculation of biomass from the SBR to the equalization tank, and the use of an additional aeration system in both units in order to provide the oxygen needed for the organic matter biodegradation. The biomass recirculation to the equalization tank and the aeration of the medium allowed the beginning of the biodegradation processes at this stage, thus reducing the organic load applied to the SBR.

The results of the present study showed the suitability of a SBR designed on the basis of averaged values of organic matter concentration and effluent flow, by changing the operational strategy during the vinification and racking periods. In fact, during the periods of high organic load (vinification and racking periods), the additional oxygen supply led to a significant improvement in the WWTP performance, in terms of COD elimination. During the rest of the year, the COD removal efficiency was always higher than 90% (Figure 7.16), despite the operation of the SBR with one or two cycles per day, according to the industrial wastewater daily flow.

## 7.6. CONCLUSION

Brewery and winery industries are small and medium enterprises but with a significant social and economic value. Therefore, their sustainability policy requires wastewater treatment systems with the best performance; the fact is that well-known processes and technologies are available for such a purpose. The experience obtained at Unicer SA and ACPB demonstrated that the technological solutions are quite site specific—in their case, UASB and sequencing batch reactors were very appropriate—and highlighted that a good operation requires bioengineering knowledge. However, it also requires continuous effort to minimize costs and maintain the best quality and service.

## 7.7. ACKNOWLEDGMENTS

The authors are deeply thankful to Gerd Teunissen, Patrícia Moreira, and Agostinha Castro for their contribution during design, operation, and analytical control of ACPB WWTP and UASB lab-scale reactor. We also want to offer our recognition to Luís Melo for his pertinent research suggestions.

## 7.8. REFERENCES

Andreottola, G., Foladori, P., Nardelli, P., and Denicolo, A., 2005, Treatment of winery wastewater in a full-scale fixed bed biofilm reactor, *Water Sci. Technol.* **51**(1):71.

Anon, 1996, La vendemmia 1996 nei paesi UE, *Ind. Bev.* **25**:640.

Batston, D. J., Keller, J., and Blackall, L. L., 2004, The influence of substrate kinetics on the microbial community structure in granular anaerobic biomass, *Water Res.* **38**:1390.

Bloor, J. C., Anderson, G. K., Willey, A. R., 1995, High rate aerobic treatment of brewery wastewater using the jet loop reactor, *Water Res.* **29**(5):1217.

Bourzeix, M., Escudier, J. L., and Mourgues, J., 1998, Produits de diversification, in: *Œnologie – Fondements Scientifiques et Technologiques*, Flanzy C. (Coordonnateur), Éditions Tec & Doc, Paris, pp. 1143–1179.

Brito, A. G., Rodrigues, A. C., and Melo, L. F., 1997a, Granulation during the start-up of a UASB reactor used in the treatment of low strength wastewaters, *Biotechnol. Lett.* **19**(4):363.

Brito, A. G., Rodrigues, A. C., and Melo, L. F., 1997b, Feasibility of a pulsed sequencing batch reactor with anaerobic aggregated biomass for the treatment of low strength wastewaters, *Water Sci. Technol.* **35**(1):193.

Bustamante, M. A., Paredes, C., Moral, R., Moreno-Caselles, J., Pérez-Espinosa, A., Pérez-Murcia, M. D., 2005, Uses of winery and distillery effluents in agriculture: characterisation of nutrient and hazardous components, *Water Sci. Technol.* **51**(1):145.

Carlsberg/Carlsberg Breweries A/S Environmental Report 2003 and 2004, 2005, Copenhagen (February 1, 2005); http://www.carlsberg.com/

Carvalheiro, F., Duarte, L. C., Lopes, S., Parajó, J. C., Pereira, H., and Gírio F. M., 2005, Evaluation of the detoxification of breweries spent grain hydrolysate for xylitol production by *Debaryomyces hansenii* CCMI 941, *Process Biochem.* **40**:1215.

Carvalheiro, F., Esteves, M. P., Parajó, J. C., Pereira, H., and Gírio F. M., 2004, Production of oligosaccharides by autohydrolysis of brewer's spent grain, *Bioresour. Technol.* **91**:93.

Chudoba, P., and Pujol, R., 1996, Activated sludge plant facing grape harvest period – a case study, *Water Sci. Technol.* **34**(11):25.

Coetzee, G., Malandra, L., Wolfaardt, G.M., Viljoen-Bloom, M., 2004, Dynamics of a microbial biofilm in a rotating biological contactor for the treatment of winery effluent, *Water SA* 30(3):407.

Cornelissen, E.R., Janse, W., Koning, J., 2002, Wastewater treatment with the internal MEMBIOR, *Desalination* **146**:463.

Demoulin, G., Goronsky, M. C., Wutscher, K., and Forsthuber, E., 1997, Co-current nitrification/denitrification and biological P-removal in cyclic activated sludge plants by redox controlled cycle operation, *Water Sci. Technol.* **35**(1):215.

Driessen, W., and Vereijken, T., 2003, Recent developments in biological treatment of brewery effluent, *The Institute and Guild of Brewing Convention*, Livingstone, Zambia, March 2–7.

Duarte, L. C., Carvalheiro, F., Lopes, S., Marques, S., Parajó, S., and Gírio, F. M., 2004, Comparison of two post-hydrolysis processes of brewer's spent grain autohydrolysis liquor to produce a pentose-containing culture media, *Appl. Biochem. Biotechnol.* ABAB Symposium Series, pp.1041–1058.

Eusébio, A., Petruccioli, M., Lageiro, M., Federici, F., Duarte, J. C., 2004, Microbial characterisation of activated sludge in jet-loop bioreactors treating winery wastewaters, *J. Ind. Microbiol. Biotechnol.* **31**:29.

Hulshoff Pol, L. W., 1989, The phenomenon of granulation of anaerobic sludge, *Ph.D. Thesis*, Agricultural University Wageningen, The Netherlands.

Ince, B. K., Ince, O., Sallis, P. J., Anderson, G. K., 2000, Inert COD production in a membrane anaerobic reactor treating brewery wastewater, *Water Res.* **34**(16):3943.

Jay, A. J., Parker, M. L., Faulks, R., Smith, A. C., Wilde, P. J., Faulds, C. B., and Waldron, K. W, 2004, A systematic micro-dissection of brewer's spent grain, in: *Total Food – Exploiting Co-products – Minimizing Waste*, Waldron, K., Faulds, C., and Smith, A., eds., Norwich, UK, pp. 150–156.

Keyser, M., Witthuhn, R. C., Ronquest, L.-C., Britz, T. J., 2003, Treatment of winery effluent with upflow anaerobic sludge blanket (UASB) – granular sludges enriched with *Enterobacter sakazakii*, *Biotechnol. Lett.* **25**:1893.

Nogueira, R., Melo, L. F., Purkhold, U., Wuertz, S., and Wagner, M., 2002, Nitrifying and heterotrophic population dynamics in biofilm reactors: effects of hydraulic retention time and the presence of organic carbon, in: *Modern Scientific Tools in Bioprocessing*, P. Wilderer and S. Wuertz, eds., Ch. 10, Elsevier, Amsterdam, pp. 469–481.

Nollet, H., Van de Putte, I., Raskin, L., and Verstraete, W., 2005, Carbon/electron source dependence of polychlorinated biphenyl dechlorination pathways for anaerobic granules, *Chemosphere* **58**:299.

Ochieng, A., Ogada, T., Sisenda, W., Wambua, P., 2002, Brewery wastewater treatment in a fluidised bed bioreactor, *Journal Hazard. Mater. B* **90**:311.

OIV – International Organisation of Vine and Wine/Situation and Statistics of the World Vitiviniculture Sector 2002, Paris (February 1, 2005); http://www.oiv.org/.

Parawira, W., Kudita, I., Nyandoroh, and M. G., Zvauya, R., 2005, A study of industrial anaerobic treatment of opaque beer brewery wastewater in a tropical climate using a full-scale UASB reactor seeded with activated sludge, *Process Biochem.* **40**:593.

Petruccioli, M., Duarte, J. C., and Federici F., 2000, High rate aerobic treatment of winery wastewater using bioreactors with free and immobilized activated sludge, *J. Biosci. Bioeng.* **90**(4):381.

Pévost, M., and Gouzenes, E., 2003, Le traitment des effluents vinicoles du bassin adour garonne, *Adour Garonne. Revue de l'Agence de l'Eau*, **86**:1.

Rochard, J., 1995, Nature et conséquences de la pollution vinicole: réduction de la charge polluante, *Ind. Bev.* **24**:16.

Rochard, J., Kerner, S., 2004, Traitement d'épuration des effluents vinicoles, *Revue des Œnologues* **113**:45.

Rodrigues, A. C., Brito, A. G., and Melo, L. F., 1998, Fate of phosphorus concentration in a SBR designed for nitrogen removal, in: *2nd European Symposium on Biochemical Engineering Science*, Porto, Portugal, September 16–19, pp. 343.

Rodrigues, A. C., Brito, A. G., and Melo, L. F., 2001, Post-treatment of a brewery wastewater using a Sequencing Batch Reactor, *Water Environ. Res.* **73**:45.

Rodrigues, A. C., Brito, A. G., and Melo, L. F., 2004, Biological treatment of hydrocarbon slurries by a Sequencing Batch Biofilm Reactor (SBBR), in: *Proceedings of the Leading-Edge Conference on Water and Wastewater Treatment Technologies*, Prague, Czech Republic, June 1–4.

Ruíz, C., Torrijos, M., Sousbie, P., Lebrato-Martinez, J., Moletta, R., Delgenes, J.P., 2002, Treatment of winery wastewater by anaerobic sequencing batch reactor, *Water Sci.Technol.* **45**(10):219.

The Brewers of Europe/Beer Facts 2003, 2004, Brussels (February 1, 2005); http://www.brewersofeurope.org/.

Torrijos, M., Moletta, R., 1997, Winery wastewater depollution by sequencing batch reactor, *Water Sci. Technol.* **35**(1):249.

Torrijos, M., and Moletta, R., 1998, Effluents vinicoles et filières de traitement, in: *Œnologie – Fondements Scientifiques et Technologiques*, Flanzy C. (Coordonnateur), Éditions Tec & Doc, Paris, pp. 1263–1284.

Unicer SA/Sustainability Report 2004, 2005, Matosinhos (February 1, 2005); http://www.unicer.pt/

van Loosdrecht, M. C. M., and Heijnen, J. J., 1993, Biofilm bioreactors for wastewater treatment, *TIBTECH*, **11**:117.

Varnam, A. H., and Sutherland, J. P., 1994, Alcoholic beverages: I. Beer, in: *Beverages – Technology, Chemistry and Microbiology*, Vol. **2**, Food Products Series, 1st ed., Chapman & Hall, London, pp. 296–361.

# 8

# Olive Mill Wastewater Treatment

Evagelia Tsagaraki, Harris N. Lazarides, and Konstantinos B. Petrotos

## 8.1. INTRODUCTION

The cultivation of olive trees and the production and use of olive oil has been a well-known and established practice in the Mediterranean region for more than 7000 years. The consumption of olive oil is rapidly increasing worldwide, due to its high dietetic and nutritional value. According to the IOOC (2004), the production of olive oil increased from 1.85 million tons in 1984 to 3.17 million tons in 2003 (70% increase) (Table 8.1).

There are approximately 750 million productive olive trees worldwide, 98% of them located in the Mediterranean region, where more than 97% of olive oil is produced. The three major olive oil producers worldwide are Spain, Italy, and Greece, followed by Turkey, Tunisia, and to a lesser extent Portugal, Morocco, and Algeria. The data presented in Figure 8.1 reflect the importance of the olive oil sector in the Mediterranean area and consequently the magnitude of the problems related with the disposal of large amounts of wastes produced during olive oil production.

The traditional press extraction method as well as the continuous three-phase decanter process, which is most widely used for the production of olive oil, generate three products: olive oil (20%) and two streams of waste: a wet solid waste (30%) called "crude olive cake" or "olive husk" and an aqueous waste called "olive mill wastewater" or "olive mill effluent" or "alpechin" (50%). The solid waste (crude olive cake) is the residue that remains after the first pressing of the olives and is a mixture of olive pulp and olive stones. At present, olive husk is processed in seed oil factories in order to extract the small amount of oil remaining in the waste. Both crude and exhausted olive cake can be used as solid fuels (due to their high heating

EVAGELIA TSAGARAKI, HARRIS N. LAZARIDES, AND KONSTANTINOS B. PETROTOS • Aristotle University of Thessaloniki, Department of Food Science and Technology, 54006, Thessaloniki, Greece.

*Table 8.1.* Olive oil production in the last decade in thousand metric tones (International Olive Oil Council, 2004)

| 1994 | 1995 | 1996 | 1997 | 1998 | 1999 | 2000 | 2001 | 2002 | 2003 | 2004 |
|---|---|---|---|---|---|---|---|---|---|---|
| 1,846 | 1,736 | 2,595 | 2,466 | 2,403 | 2,375 | 2,566 | 2,826 | 2,494 | 3,165 | 2,766 |

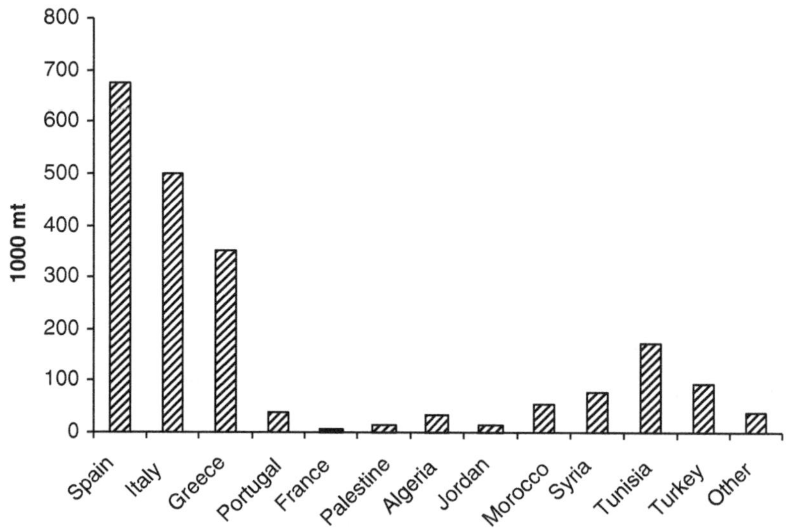


**Figure 8.1.** Olive oil-producing countries (International Olive Oil Council, 2004).

value), for animal feed supplement, or return to the olive grove as mulch. While economic concerns regarding the profitability of seed oil production now are being questioned, the problem of disposal of olive husk is adequately solved by one of these three alternatives.

On the other hand, as far as olive mill wastewater (OMWW) treatment and disposal is concerned, the situation is much more complicated in practice. Attempts to alleviate the problem, especially in the major olive oil-producing countries, are more than 50 years old; yet, there has been little success in finding an environmentally friendly and economically viable solution to be generally adopted.

## 8.2. THE DIMENSIONS OF THE PROBLEM

OMWW is a mixture of vegetation water and soft tissues of the olive fruit and the water used in the various stages of the oil extraction process, i.e., water added during centrifugation, water from filtering disks, and from washing rooms and equipment. It contains olive pulp, mucilage, pectin, oil, etc., suspended in a relatively stable emulsion.

The annual world OMWW production is estimated from 7 to over 30 million $m^3$. Although the quantity of the waste produced is still much smaller than other types of waste (i.e., domestic sewage) and its production is seasonal, the contribution of OMWW to environmental pollution is important, because of some ''peculiarities'' of the case that have to do with both the chemical synthesis of the waste and some aspects of the current situation in the olive oil sector.

### 8.2.1. Problems Arising from OMWW Synthesis

As far as its chemical synthesis is concerned, OMWW basic characteristics that prove its ''strong'' nature as industrial waste are:

- Strong offensive smell.
- Extremely high degree of organic pollution (COD values up to 220 g/L) and a $COD/BOD_5$ ratio between 2.5 and 5 (hardly degradable).
- pH between 3 and 5.9.
- High content of polyphenols (up to 80 g/L) which are not easily biodegradable and toxic to most microorganisms.
- High content of solid matter (total solids up to 20 g/L).

In terms of pollution effect, 1 $m^3$ of OMWW is equivalent to 100–200 $m^3$ of domestic sewage. Its uncontrolled disposal in water reservoirs leads to severe problems for the whole ecosystem and especially for the natural water bodies (ground water reservoirs, surface aquatic reservoirs, seashores, and sea). The most visible effect is discoloration, a result of oxidation and subsequent polymerization of tannins. OMWW also has a considerable content of reduced sugars, high phosphorus content, and phenolic load that has a toxic action to some organisms. Some microorganisms that metabolize sugars develop more rapidly at the expense of other living organisms. The high phosphorus content accelerates the growth of algae resulting in eutrophication. Some aquatic organisms (i.e., the river fish *Gambusia affinis* and some crustaceans) become severely intoxicated even at exposures corresponding to 1 liter of unprocessed OMWW into 100,000 liter of circulating water (Fiorentino et al., 2004).

OMWW dispersion on the ground and its subsequent metabolization (by microorganisms, insects, earthworms, etc.) to humic extracts or acids also could lead to soil enrichment with nutrients (i.e., organic matter, nitrogen, phosphorus, and potassium) and a low-cost source of water. However, OMWW high concentration of potassium affects the cation exchange capacity of the soil, leading to change of environmental conditions for soil microorganisms and consequently to changes in the fertility of the soil. Soil porosity also could be affected. Other possible negative effects include the immobilization of available nitrogen and decreased available magnesium, perhaps because of the antagonistic effect on potassium.

Finally, no land disposal of OMWW should be done without taking under consideration its severe phytotoxic and antimicrobial properties that may damage the existing crops (Cox et al., 1997; Paredes et al., 1999; Sierra et al., 2001).

The phytotoxic and antimicrobial properties of OMWW have been mainly attributed to its phenolic content and some organic acids, such as acetic and formic acid, that are accumulated as microbial metabolites during storage. Its direct application on plants inhibits the germination of different seeds and early plant growth of different vegetable species and may cause leaf and fruit abscission as well. Different types of crops show different reactions to OMWW spreading and some of them may tolerate a certain amount of OMWW during early growing stages (Rinaldi et al., 2003).

As far as its antimicrobial activity is concerned, catechol, 4-methyl-catechol, and hydroxytyrosol are its most active compounds against a number of bacteria and fungi. Several authors have reported OMWW activity against soil gram(+) spore bacteria like *Bacillus megaterium ATCC 33085, Geotrichum, Rhizopus, Rhizoctonia, Bactrocera oleae,* and *Pseudomonas syringe* (Oikonomou et al., 1994) .

These biotoxic properties of phenols in OMWW constitute a significant inhibitor of the biological processes that take place in common wastewater treatment plants. Such plants do not present the desired performance with treatment of OMWW. Thus, the treatment of straight OMWW together with domestic sewage is not economically feasible, because of serious overload of the sewage treatment plant. So, research is oriented toward more complex treatment methods that usually demand higher capital or operational costs.

### 8.2.2. Other Problematic Characteristics of OMWW

The problems mentioned above make the technological design of an OMWW treatment plant difficult. Factors that make the economic design of such a plant difficult is the intense and seasonal production of the waste (maximum 4 months each winter), the great variability both of synthesis and quantity, the high regional scattering of olive mills, and the small size of the majority of them in the olive oil-producing regions.

Because of its highly variable input and seasonal production, storage facilities for the excess quantities of waste produced during winter months should be considered during design of a treatment plant. Similar design problems would arise in holiday resorts, where the population also can increase by an order of magnitude.

Olive mills are usually small-scale enterprises that cannot afford the costs of a proper wastewater treatment unless the treatment is a very simple and cheap procedure. Most treatment technologies, however, require high investment costs and a high level of technological know-how. Thus the design of centralized treatment plants is considered more suitable to treat OMWW produced by several mills. This creates a burden to operational costs, as high transportation costs due to high geographic scattering must be taken into account. In some cases, local conditions may call for separate treatment plants. Finally, serious nuisance due to the unpleasant odors and insects from OMWW may cause a serious difficulty at finding a suitable location of a treatment plant. All these factors introduce economic, technical, and organizational constraints that vary greatly from place to place, making the adoption of an environmentally compatible approach on a wide scale very difficult.

### 8.2.3. The Current Situation

As mentioned above, in all countries that produce olive oil the direct discharge of OMWW into the receiving natural water bodies is strictly forbidden, because of its detrimental effects on ecological balance. Direct discharge of OMWW in public sewers is also formally forbidden, because it causes corrosion to the pipes and builds up sediments. Nevertheless, because of the problems mentioned in the previous chapter, the most common practice followed today from the olive mills is the illegal, direct disposal of OMWW into nearby aquatic bodies, i.e. rivers, lakes, or even the sea.

In Spain, the government subsidized and promoted the construction of storage ponds to promote evaporation during the summer period. Around 1,000 evaporation ponds were constructed, which definitely improved the water quality but caused odor problems for the inhabitants of the region.

According to Italian environmental legislation, OMWW has to undergo treatment before their release into the environment. OMWW spreading on land is permitted under controlled conditions. Taking into account the potential considerable fertilizing value of OMWW, the possibility of using them as soil conditioners is also permitted. A new regulation allowing the reuse of OMWW in agriculture is now the only state regulation in force regarding OMWW and permits use of wet solid waste from the two-phase process as fertilizers on croplands. Such spreading on land must be known to the authorities and supported by a technical report.

In Greece, small olive mills serving local communities are extensively widespread. Extensive research is underway in research organizations and universities in order to safely dispose of the produced wastewater within a maximum period of 3 months after the campaign (Azbar et al., 2004).

## 8.3. QUANTITY AND SYNTHESIS OF THE WASTE

### 8.3.1. The Effect of the Production Process

As mentioned before, one of the major problems regarding OMWW disposal is that both its quantity as well as its synthesis show great variability depending on a number of unpredictable and diverse factors such as:

- type of olives,
- area under cultivation or arable soil,

- use of pesticides and fertilizers,
- harvest time, stage of maturity,
- climate, weather conditions, and
- type of olive oil extraction process.

In modern olive mills, the most common methods for extracting olive oil from the olive paste are the following:

- Pressing (traditional or classical batch system).
- Centrifugation (continuous).
  - Three-phase decanter.
  - Two-phasc decanter.

A brief presentation of the two major continuous oil extraction systems is shown in Figure 8.2.

At the *traditional batch press process*, the quantity of added water during oil extraction is small (3–5 L/100 kg of olives processed). Thus the smallest quantity of liquid waste is produced, but the most concentrated as well. Besides, this technology gives higher levels of COD, polyphenols, and total solids compared to the waste produced from other processes. Due to the low temperature of extraction, the olive oil produced with this method is of very high quality.

During the *continuous three-phase decanter process*, the addition of warm water is required at the centrifugation stage (1.25–1.75 times more than at the press extraction) resulting in the production of increased volumes of OMWW and loss of

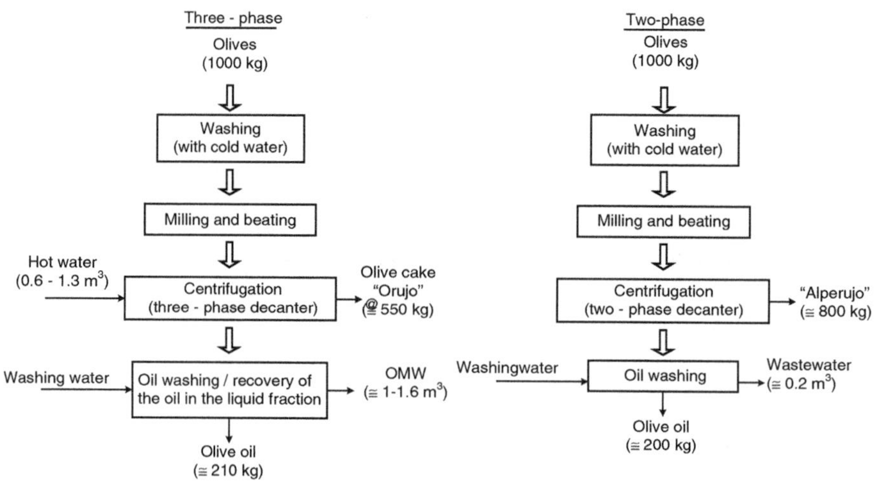


**Figure 8.2.** Comparison of the three- and two-phase centrifugation systems for olive oil extraction (Alburquerque et al., 2004).

valuable components (i.e., polyphenols) with the wastewater. The total solids content of the waste is approximately twice more than in classic press method and it is in suspended form.

The *two-phase extraction system* was developed during the 1990s in an attempt to minimize the volume of the waste produced and is widely adopted in Spain, where approximately 90% of the country's olive mills use this technology. The philosophy is the same as at the three-phase centrifugation system. The difference is that it uses no process water and delivers only two streams: olive oil and a single waste, a combination of olive husk and OMWW, i.e., a very wet olive cake, which is called ''alpeorujo.''

This modified extraction system is less complicated, consumes less energy, and yields higher-quality olive oil than the three-phase decanter. Nevertheless, although it is called ''*the ecological system*,'' because it reduces wastewater generation up to 80%, the problem remains unsolved, since a new, not yet fully characterized waste is produced with new difficulties in its treatment and disposal. Alpeorujo has much higher moisture content than olive husk (55–70% versus 20–25% for traditional press systems and 40–45% for the three-phase decanter) and a lot of polyphenols and polysaccharides that are absent from the olive husk produced with the other two processes. The high moisture content makes the drying of alpeorujo at seed oil refineries an economic burden. Polyphenols and polysaccharides cause problems to dryers because of caramelization and agglomeration effects. These problems during the seed oil extraction have forced manufacturers to perform a further purification step before drying alpeorujo, which greatly increases seed oil production costs. The new waste generated at this chemical extraction step is called ''orujillo'' and could be used as a fuel but more precise data are needed. So, the problem of alpeorujo disposal has not been fully resolved and research into new technological procedures that permit its profitable use is needed. The influence of the production process on OMWW characteristics is reflected in Table 8.2.

### 8.3.2. OMWW Synthesis

In Table 8.3 a compilation of literature data on synthesis of OMWW is presented. The most important organic constituents of OMWW are phenolic compounds, sugars, and some organic acids. Sugars vary from 1.6–4% w/v depending on the variety of olives, the climatic conditions and the production method used. The most commonly found sugars, in decreasing quantities are fructose, mannose, glucose, saccharose, sucrose, and some pentoses (Niaounakis and Halvadakis, 2004).

OMWW also contains exploitable quantities of oleanolic and maslinic acid, both acids with various biological effects (carcinogenic promoter-inhibitor effect, antihistamic effect, etc.). As far as inorganic compounds are concerned, it has high potassium content ($\approx$ 4 g/L) and notable levels of nitrogen, phosphorous, calcium, magnesium, and iron compared to other organic wastes. Main anions are $Cl^-$, $PO_4^=$, $F^-$, and $SO_4^=$ depending on the extraction process.

*Table 8.2*. Influence of the production process on OMWW characteristics

| Parameter | Press process | 3-phase | 2-phase | Reference |
|---|---|---|---|---|
| L OMWW/tn of olives | | 900–1,500 | 50–70 | Aktas et al. (2001) |
| | | 500–1,500 | | Rozzi and Malpei (1996) |
| | | 500–1,400 | | Sierra et al. (2001) |
| | 400 | 1,000 | | Improlive (2000) |
| | 400–600 | 1,000–1,200 | 85–110 | Caputo et al. (2003) |
| (% of weight of olives) | 50 | 80–110 | | Mulinacci et al. (2001) |
| pH | 4.5–5 | 4.7–5.2 | | Azbar et al. (2004) |
| | $4.5 \pm 0.3$ | $4.8 \pm 0.3$ | | Aktas et al. (2001) |
| | 4.5–5 | 4.5–5 | | Caputo et al. (2003) |
| COD (g/L) | 120–130 | 40 | | Azbar et al. (2004) |
| | $65.7 \pm 27.1$ | $103.4 \pm 19.5$ | 5–25 | Aktas et al. (2001) |
| | 125 | 50 | | Caputo et al. (2003) |
| BOD (g/L) | 90–100 | 33 | | Azbar et al. (2004) |
| | 90 | 40 | | Caputo et al. (2003) |
| TSS (%) | 0.1 | 0.9 | | Azbar et al. (2004) |
| (g/L) | 0.1 | 0.9 | | Caputo et al. (2003) |
| | $2.7 \pm 1.1$ | $27.6 \pm 5.1$ | | Aktas et al. (2001) |
| VSS (%) | 10.5 | 2.6 | | Azbar et al. (2004) |
| (g/L) | $2.5 \pm 1.1$ | $24.5 \pm 5$ | | Aktas et al. (2001) |
| TS (%) | 12 | 3 | | Azbar et al. (2004) |
| (g/L) | $44.4 \pm 13.8$ | $78.2 \pm 13.6$ | | Aktas et al. (2001) |
| Sugars (%) | 2–8 | 1 | | Azbar et al. (2004) |
| (g/L) | $2.2 \pm 1.7$ | $4.7 \pm 1.8$ | | Aktas et al. (2001) |
| | 4.5 | 1.5 | | |
| Total N (%) | 5–2 | 0.28 | | Azbar et al. (2004) |
| (mg/L) | 1.8 | 0.3 | | Caputo et al. (2003) |
| | $43.7 \pm 33.9$ | $78.8 \pm 39.6$ | | Aktas et al. (2001) |
| Polyalcohols (%) | 1–1.5 | 1 | | Azbar et al. (2004) |
| Pectin/tannin (%) | 1 | 0.37 | | Azbar et al. (2004) |
| Polyphenols (%) | 1–2.4 | 0.5 | | Azbar et al. (2004) |
| | 1.7 | 0.63 | | Caputo et al. (2003) |
| (% (w/w) dry residue) | | 24 | 20.4 | Lesage-Meesen et al. (2001) |
| (mg/L) | | 767.1 | 540.6 | Stefanoudaki-Katzouraki and Koutsaftakis (1994) |
| Oil/grease (%) | 0.03–10 | 0.5–2.3 | | Azbar et al. (2004) |

#### 8.3.2.1. Phenolic Content of OMWW

Olives are rich in simple and complex phenolic compounds, typical secondary plant metabolites, to which several antioxidant and free radical scavenging properties are attributed. During the olive oil extraction process, due to the chemical characteristics of polyphenols that are water-soluble, the partition between water and oil is different and the major proportion of these compounds goes from the olive pulp to the aqueous phase, that is the OMWW (i.e., <1% in the olive oil versus 18% in olives). OMWW phenolic content shows great variability both from the qualitative and quantitative point of view, depending on several factors, such as type of olive,

*Table 8.3*. Literature data on synthesis of OMWW

| Property | Reference | | | | | | | |
|---|---|---|---|---|---|---|---|---|
| | Azbar et al. (2004) | Niaounakis and Halvadakis (2004) | Borsani and Ferrando (1996) | Paredes et al. (1999) | Sierra et al. (2001) | Galiatsatou et al. (2002) | Eroglu et al. (2004) | Al-Malah et al. (2000) |
| pH | 3–5.9 | 4–6 | | 4.8–5.5 | 4.5–6 | 4.9–6.5 | 4.86 | 4.52 |
| Water (%) | | | 83 | | | | | |
| BOD (g/L) | 23–100 | 35–110 | | | 35–100 | 15–120 | 17.88 | 13.2 |
| COD (g/L) | 40–220 | 40–220 | | | 40–195 | 30–150 | 72.20 | 320 |
| Carbohydrates (%) | | | 2–8 | 3.37–32.91 | | 2–8 | | |
| Polyphenols (g/L) | 0.002–80 | 0.5–24 | | 1.32–3.99% | 3–24 | 1.5–2.4 | 0.13 | 3.12 |
| Fats, oils (g/L) | 1–23 | | 0.03–1% | 0.55–11.37% | 0.3–23 | 1.3 | | |
| Pectins (%) | | | 1–1.5 | | | 1–1.5 | | |
| VOC (g/L) | | 25–45 | | | | | | |
| TS (g/L) | 1–102.5 | | | | | | 42.24 | |
| SS (g/L) | | | | | | | 3.48 | 2.17 |
| N (g/L) | 0.3–1.2 | | | 0.58–1.13% | 5–15 | 0.5–2% | | |
| K (g/L) | | 4 | 0.87% $K_2O$ | 3.30–6.94% | 2.7–7.2 | | 7.81 | |
| P (g/L) | | | 0.22% $P_2O_5$ | 0.06–0.32% | 0.3–1.1 | | | |
| Ca (g/L) | | | | 0.32–0.53% | 0.12–0.75 | | 0.55 | |
| Na (g/L) | | | | 0.04–0.48% | 0.04–0.90 | | 0.41 | |
| Mg (g/L) | | | | 0.06–0.22% | 0.10–0.40 | | 0.28 | |

stage of maturity, and most important type of production process. The extraction system does not seem to qualitatively alter the phenolic composition of olive oil, but it affects its concentration, with the two-phase decanter OMWW being the richest in quantity (Lesage-Meesen et al., 2001).

More than 30 different phenolic compounds have been detected in OMWW and reported by several authors. As far as their MW is concerned, two major categories are observed.

1. Phenolic monomers, flavonoids, not autoxidated tannins, and other compounds with MW≤10kDa.
2. Medium and high MW (MW>10kDa) dark colored polymers resulting from the polymerization and autoxidation of phenolic compounds of the first group.

The color of OMWW depends on the ratio between the two groups. Their chemical synthesis divides them into three major categories:

1. Derivatives of cinnamic acid (cinnamic acid, o-, p- coumaric acid, caffeic acid, ferrulic acid).
2. Derivatives of benzoic acid (benzoic acid, protocatechuic acid).
3. β-3,4-dihydroxyphenylethanol derivatives such as tyrosol and hydroxytyrosol.

Other phenols reported to be found in OMWW are catechol, 4-methylcatechol, p-cresol, resorcinol, oleuropein, dimethyloleuropein, verbascoside, and some phenolic acids such as vanillic acid, protocatechuic acid, veratric acid, syringic acid, cinnamic acid. Among the flavonoids contained in OMWW are apigenin, cyanidin flavone, anthocyanin, luteolin, quercetin etc (Ramos-Cormenzana et al., 1996; Lesage-Meesen et al., 2001).

Hydroxytyrosol (3,4-dihydroxyphenylethanol) stands out both because it is the most abundant phenol in OMWW and because of its great bioantioxidant activity. It is produced from the enzymatic hydrolysis (by a glucosidase) of oleuropein, the major polyphenol of the olive fruit. Oleuropein degrades gradually into elenolic acid and hydroxytyrosol, as olives mature, but the majority of hydroxytyrosol quantity in OMWW is produced during the extraction of olive oil. That is why some researchers do not detect oleuropein in OMWW.

8.3.2.2. Properties of Phenols Found in OMWW

Hydroxytyrosol, 2-hydroxytyrosol, tyrosol, oleanolic acid, and maslinic acid, flavonoids, anthocyanins, and tannins that are found in OMWW are considered as natural antioxidants with considerable commercial and economic interest. The most interesting one appears to be hydroxytyrosol, a compound of high added value, due to its antioxidant and potentially beneficial (to human health) properties. Results of *in vitro* research demonstrate that hydroxytyrosol inhibits human LDL oxidation, scavenges free radicals, inhibits platelet aggregation and the production of leucotriene for human neutrofils, and confers cell protection. It also acts against both gram (+) and gram (−) bacteria. It could be used as a food preservative, in agriculture for the protection of olive trees, and in cosmetics industry in antiaging preparations (Allouche et al., 2004; Visioli et al., 1999). Also, phenolic substances are the major contributors to OMWW's antimicrobial properties.

## 8.4. SUGGESTED OMWW TREATMENT PROCESSES

As can be understood from its synthesis, OMWW possesses a double nature. It is a strong pollutant and at the same time a possible source of valuable components, such as polyphenols, flavonoids, anthocyanins, inorganic trace elements, etc., that could be isolated (removed) and economically exploited.

Research is oriented toward flexible and efficient treatment solutions that could ensure the detoxification of the waste compensating high capital and operating costs with the possibility of recovering and recycling some valuable components. According to these, the numerous treatment processes that have been proposed so far could be classified in the following categories:

- Detoxification processes.
- Processes that aim at the production of various products.
- Integrated processes aiming at energy recovery.
- Combined processes.

### 8.4.1. Detoxification Processes

These are processes that aim at ''cleaning'' the waste so as to allow its safe, subsequent disposal at water or soil reservoirs. The most important are biological and physicochemical processes.

#### 8.4.1.1. Biological Processes

Biological processes use microorganisms to break down the chemicals present in OMWW. They are divided into *aerobic* and *anaerobic* processes according to the type of the microflora used.

*Aerobic processes* can operate efficiently only if the concentration of the feed is relatively low; i.e., of the order of 1 g COD/liter. Higher concentrations can be tolerated only if the plant operates at a long hydraulic retention time or/and with high recycle ratio; both possibilities are uneconomical for a treatment plant. Also, the aerobic treatment of concentrated wastewaters yields huge volumes of excess secondary sludge that has to be removed from the system. At last, it is very difficult using aerobic processes to reach the required removal efficiency of pollutants such as polyphenols and lipids. For all the above reasons, aerobic processes are unsuitable for direct and efficient treatment of OMWW. They can be used as pretreatment or posttreatment steps to increase the efficiency of the main treatment process used.

*Anaerobic digestion* consists of a series of microbiological processes that convert organic compounds into methane and carbon dioxide. Although a pretreatment or posttreatment step is also needed, anaerobic treatment is considered most suitable for OMWW detoxification. The most important reasons for this choice are the feasibility to treat wastewaters with high organic load, such as OMWW, the low energy requirements, the production of methane that may be exploited after suitable treatment, the production of significantly less waste sludge (than aerobic processes), and the ability to restart easily after several months of shut down (Niaounakis and Halvadakis, 2004; Rozzi and Malpei, 1996).

*8.4.1.1.a. Anaerobic Processes.* Anaerobic processes are driven mostly by bacteria and have three major steps: In the first stage, anaerobic bacteria hydrolyze complex organic compounds, such as polysaccharides and polyphenols to their monomers (simple sugars and phenols, respectively). These molecules are converted into organic acids such as acetic, lactic, and formic acids and alcohol by acetogenic bacteria during the second stage of the process. In the last stage, methanogenic bacteria, which are characterized by their sensitivity to pH and temperature changes, convert the organic acids into biogas (a mixture of 60–80% methane and other gases, mainly carbon dioxide) (Sabbah et al., 2004).

Anaerobic processes are affected by temperature, retention time, pH, $H_2$ partial pressure, the chemical composition of the wastewater, and the quantity of toxic substances present. The process usually takes place under thermophilic or mesophilic conditions. Retention time varies between 10 and 35 days and pH must be controlled, because acetogenic bacteria tend to lower it and methanogenic bacteria are sensitive to pH variations.

Several technologies have been tested, including upstream anaerobic sludge blanket reactor (UASB), contact reactors, anaerobic filters (upstream and downstream), anaerobic baffled reactors (ABR), and two-stage systems that separate acidogenesis and methanogenesis processes (Azbar et al., 2004; Borja and Gonzalez, 1994; Dalis et al., 1996; Rozzi and Malpei, 1996; Zouari, 1998; Zouari and Ellouz, 1996).

UASB-type reactors and anaerobic filters are suitable for high volumetric pollution loads (5–15 kg COD/$m^3$ day). COD removals of 80% and 60–65%, respectively, have been reported but in both cases a high dilution ratio is required (1/8 and 1/5) that raises operational costs. Anaerobic filters require very little process control and 75% reduction of phenols has been reported (Dalis et al., 1996). Compared to contact reactors, greater production of methane and elimination of mechanical mixing, settling, and return of the sludge has been reported (Borja and Gonzalez, 1994). Contact reactors can operate at higher feed concentrations (up to 60 g COD/liter) with COD removal efficiencies greater than 80% but only at very low loading rates ($<5$ kgCOD/$m^3$ day) (Rozzi and Malpei, 1996).

A general problem encountered with anaerobic digestion of OMWW is that both the addition of alkali substances to neutralize pH and of substances that are sources of nitrogen such as urea or ammonia are necessary. The anaerobic microflora also shows limited efficiency in the removal of aromatics, particularly condensed tannins. Finally, scaling up these processes proves to be extremely difficult. Growth rates of anaerobic microorganisms are appreciably lower than those of aerobic ones and their metabolic degradation pathways require several different microbial populations in series which make process control and stability very delicate (Mechichi and Sayadi, 2004).

All these constraints make the use of pretreatment or posttreatment of anaerobic digestion necessary. Pretreatment methods proposed so far include dilution of the waste, gravity settling, sand filtration, centrifugation, adsorption, membrane processes, physicochemical treatments ($Ca(OH)_2$, NaOH, $Na_2CO_3$, Fenton's reagent, etc.), and aerobic degradation (Sabbah et al., 2004; Zouari, 1998).

As mentioned before, strong dilution of the waste is necessary in most of the types of anaerobic digesters. Some researchers have proposed that if OMWW is mixed with another organic effluent, definitely a more economic dilution media than water, it will also become enriched in its limiting nutrients such as nitrogen and neutralized without the addition of chemicals. At locations where the polluting load due to olive industry is comparable or lower than domestic sewage load, OMWW can be treated in conventional domestic sewage digesters if mixed with this effluent. Rozzi and Malpei (1996) and Marques (2001) studied the combined treatment of OMWW with piggery effluent, where no chemical correction was needed and 70–80% COD removal was achieved, but decolorization of the waste was not sufficient. The produced effluent could be used as irrigation water.

*8.4.1.1.b. Aerobic Processes.* The combination of aerobic and anaerobic treatment is extensively studied, as there are aerobic consortia that grow on undiluted or diluted OMWW and are capable of metabolizing and removing its aromatic compounds. In Table 8.4, some data collected from the literature are presented.

Of all the above microorganisms particular interest has been shown in the $N_2$-fixing bacteria *Azotobacter* and several white rot basidiomycetes, such as *Pleurotus*. Fungi of this kind have ligninolytic enzymes and can degrade phenolic substances of OMWW that have structural relationships with lignin. However, the majority of phenolic compounds removed are simple monomers, whereas polymerized molecules such as tannins degrade more difficultly. This happens because these compounds adsorb strongly to mycelia and extracellular enzymes so that their biodegradation is not possible.

In all cases, a certain pretreatment is necessary (dilution, thermal treatment, etc.) and the resulting effluent always needs additional treatment before it could be safely disposed off. So, aerobic processes alone are not effective enough for the detoxification of OMWW.

*Table 8.4.* Aerobic microorganisms studied for OMWW degradation

| Microorganism | Results | Comments | Reference |
|---|---|---|---|
| *Aspergillus niger* | 35–65% COD reduction | Biofertilization | Cereti et al. (2004) |
| | | | Garcia et al. (2000) |
| *Azotobacter vivelandii* | 90–96% COD reduction | Biofertilization | Piperidou et al. (2000) |
| *Yarrowia lipolytica* | 2–42% COD reduction | Production of enzymes and microbial metabolites | Lanciotti et al. (2004) |
| *Pleurotus ostreatus* | Up to 78% phenol removal | Diluted or thermally processed OMWW | Fountoulakis et al. (2002) |
| | | | Aggelis et al. (2003) |
| *Phanerochaete chrysosporium* | | Comparison concerning phenol removal capacity | Garcia et al. (2000) |
| *Aspergillus terreus* | | | |
| *Geotrichum candidum* | 65% COD removal, 75% color removal | Fresh or stored OMWW | Assas et al. (2002) |
| *Panus tigrinus* | | Removal of 4-hydroxy substituted simple phenols | D' Annibale et al. (2004) |
| *Pycnosporus coccineus* | | | Jaouani et al. (2003) |
| *Pleurotus sajor caju* | 75% color removal | | Jaouani et al. (2003) |
| *Coriolopsis polyzona* | 75% color removal | | Jaouani et al. (2003) |
| *Lentinus tigrinus* | | | Jaouani et al. (2003) |
| *Candida tropicallis* | 62.8% COD removal; 51.7% phenols removal | | Fadil et al. (2003) |
| *Pleurotus pulmonarium* | | Production of *Pleurotus basidiomata* only diluted OMWW | Zervakis et al. (1996) |
| *Pleurotus eryngii* | | | |
| *Ankistrodermus braunii* | 12% phenols reduction | Fresh OMWW or reverse osmosis fraction | Pinto et al. (2003) |
| *Scenedesmus quadric cauda (microalgae)* | | | |

8.4.1.2 Physicochemical Processes

*8.4.1.2.a. Neutralization, Precipitation/Flocculation.* These processes involve the use of additional chemicals in order to destabilize the suspended and colloidal matter of OMWW and form an insoluble solid that can be removed easily from the waste. Oil, suspended solids, COD, and BOD are decreased in this way. Destabilization of these colloids can be achieved either by reducing or increasing pH (neutralization) or by the addition of a precipitate-inducing agent (precipitation/flocculation).

Reduction of pH to the point of zero charge (pH = 2–4) has attracted little attention so far, although apart from colloids destabilization, it also is expected to contribute to the acid hydrolysis of oils to fatty acids which can be easily separated from effluents. On the contrary, the use of lime (CaO) to increase pH at about 11 has been the subject of several studies (Mitrakas et al., 1996).

By treating OMWW with lime, oil, and COD reduction, decolorization, and important reduction in odor emissions are achieved. The liquid obtained after treatment contains no phytotoxic substances and it can be treated further more easily. The major disadvantage of this process is that large quantities of sludge with high pollution load are produced leading to serious disposal problems.

The most important inorganic flocculents that have been used for OMWW treatment are ferric and ferrous chloride, ferric sulfate, and aluminium sulfate. All these reagents should not be used if the precipitated material is to be used as animal feed (Niaounakis and Halvadakis, 2004).

The processes described above, although simple and cheap, are more suitable as pre-treatment methods because the treated liquid still has a high polluting load. Considerations also arise for the disposal of the precipitated material produced.

*8.4.1.2.b. Oxidation processes.* Several oxidizing agents have been tested for OMWW treatment like hydrogen peroxide, ozone, chlorine, chlorinated derivatives (i.e., chlorine dioxide, sodium hypochloride, etc.), or a combination of them. Ozone and hydrogen peroxide systems are preferred because of their high oxidizing potential and the possibility of operating under atmospheric pressure and ambient temperatures without problematic decomposition products of the oxidizing agent (Niaounakis and Halvadakis, 2004).

In an attempt to increase oxidation rates, advanced oxidation processes have evolved (AOPs) where the combinations of oxidants as well as the combination of oxidants with ultraviolet radiation are used. They are characterized by the production of the highly oxidative $HO^{\cdot}$ radical at ambient temperature via a number of photochemical or non-photochemical pathways. This powerful radical is able to completely transform organic compounds to $CO_2$.

The principal AOPs used for OMWW treatment are Fenton's reagent reaction ($H_2O_2$ plus a ferrous salt) (Gernjak et al., 2004; Rivas et al., 2001), $O_3$ plus UV radiation (Javier-Benitez et al., 1997), $H_2O_2$ plus UV radiation, $H_2O_2/O_3$ + UV radiation and photocatalysis, where solar energy also may be used (Gernjak et al., 2004).

Most of the classic oxidation processes lack effectiveness due to either the high cost of antioxidants or the low interval of COD for which the system is suitable. AOPs manage a great COD reduction but their operating costs are considerably high. It could be said that chemical oxidation emerges as a suitable alternative, when biological degradation is not applicable.

#### 8.4.1.3 Thermal Processes

Numerous methods and variations are included in this category and what they have in common is that the concentration of OMWW is achieved either by a manmade heat source or by a natural source of thermal energy (air, sun). The most important thermal processes are evaporation, distillation, lagooning (natural evaporation), combustion, and pyrolysis.

Several distillation and evaporation processes such as vacuum, multiple effect, and flash evaporation already used in desalination and food industry have been tested on OMWW. Although these processes are claimed to reduce significantly the volume of the waste (reduction by 70–75%) great differences exist in bibliography concerning their effectiveness, because it depends on many factors such as extraction process, olive ripening, and especially storage time of the waste (Niaounakis and Halvadakis, 2004).

The main drawback of these processes is related to the posttreatment and disposal of the produced emissions: The distillate/condensate contains, apart from water, an appreciable fraction of volatile compounds such as alcohols and volatile acids. These compounds make the condensate too acidic (pH 4–4.5) and with high BOD (>4 g/liter) and COD (>3 g/liter) values making necessary an additional treatment prior to discharge or reuse. The concentrated paste has a high concentration of the polluting organic load, so its combustion induces air pollution (Niaounakis and Halvadakis, 2004; Rozzi and Malpei, 1996). All these processes also have extremely high costs, due to the great energy consumption necessary and the equipment costs that has to be made of materials resistant to corrosion.

Natural evaporation of OMWW in ambient air with the use of solar energy in evaporation ponds or storage lakes (lagoons) has much lower energy costs and it is a simple procedure. It is one of the first processes used and removal of COD ranging from 20–30% to 75–80% has been reported. The waste has a residence time of 7–8 months in the lagoons and large land surface areas are required (about 1 $m^3$ for each 2.5 $m^3$ of OMWW). Several ecological concerns arise including the possibility of groundwater contamination if the bottom of the lagoon is not properly lined against infiltration and leakage and the emissions of methane in the atmosphere due to the anaerobic fermentation of the waste that occurs in the lagoons. These lagoons should be located far enough from residences to avoid the insect and odor nuisances (Azbar et al., 2004; Rozzi and Malpei, 1996).

Combustion and pyrolysis are radical and destructive techniques that eliminate any possibility of further use of OMWW. Both are very expensive methods with high energy requirements, pretreatment of the waste, and posttreatment of the gaseous emissions necessary and expensive equipment needed. For these reasons, they are more suitable for strong wastewaters, concentrated solutions of OMWW, or for olive husk.

In an attempt to minimize the energy costs of thermal processes, several researchers have proposed the combined thermal treatment of OMWW and olive husk. In these processes the required heat for the evaporation of OMWW is produced by the combustion of OMWW concentrated evaporation residue or olive husk or a mixture of these wastes. A critical parameter affecting the feasibility of this disposal approach is the degree of mixing of olive husk and OMWW. As such disposal systems are characterized by a rather high technological level requiring remarkable capital investments and qualified personnel, they are more suitable for centralized treatment plants that serve a large number of mills and gain benefits from the economy of the scale (Caputo et al., 2003; Vitolo et al., 1999).

#### 8.4.1.4 Membrane Processes

Membrane processes also are tested for use in treatment of OMWW because they are effective for separation of oil-water mixtures without adding solvents. Ultrafiltration is the widely considered membrane process for this purpose, while microfiltration and reverse osmosis also have been tested. Two different phases are obtained: the retentate (concentrate) and permeate. Colloidal particles, lipids, and various macromolecules (molecular weights of the order of 10,000 to 100,000 Da) can be prevented from passing through the membrane to the permeate.

With ultrafiltration, only a small amount of retentate (waste) is produced (permeate is 90–95% of the volume of the feed) and very high removal of lipids is achieved. Also, by choosing the appropriate pore size of the membrane used, the composition of the permeate can be controlled. A separation of fats that are rejected by the membrane from salts, sugars, and phenolic substances that pass to the permeate can be achieved, enabling the economic exploitation of these substances.

The capital costs of this operation are extremely high and it is a complicated procedure that needs qualified personnel. The main problem is that severe fouling of the membrane occurs very easily, strongly reducing the membrane efficiency due to gelling substances contained in OMWW. The removal of these substances in a number of pre-treatment steps is therefore absolutely necessary. Also, only a limited concentration factor is achieved and dissolved components such as those determined by the parameter COD are only insufficiently removed and both retentate and permeate still have high COD concentrations and have to be further processed prior to disposal.

For all the reasons stated above, membrane processes are not suitable for the treatment of strong OMWW such as from traditional press systems because of their limited efficiency and their high costs, which make their use just for detoxification purposes economically unprofitable. They can be used as pretreatment steps in processes that aim at the recovery of valuable, expensive components such as polyphenols and flavoring agents from OMWW. Passing through the membrane, the waste becomes concentrated in these substances making their subsequent extraction easier and more economical while the high costs of the membranes are compensated by the high added value of the product. The retentate that has a poor polyphenolic and a high oil content can be used as fertilizer or animal feed after appropriate treatment.

### 8.4.2. Processes That Aim at the Production of Various Products

OMWW may be regarded as an inexpensive source of inorganic and organic compounds to be recovered because of their potential economic interest or their ability to be transformed into products for use in agriculture, biotechnology, and the pharmaceutics industry as well as in thc food industry.

#### 8.4.2.1. Production of Fertilizers (Recycling of the Waste at Land)

*8.4.2.1.a. Biofertilization or Bioremediation.* OMWW should not be directly applied on soil and crops because of its phytotoxic properties. But with certain treatment it could be converted into a useful fertiliser and soil conditioner, due to its high content of water, organic matter, and plant nutrients. $N_2$-fixing bacteria of the genus *Azotobacter* utilize phenols, sugars, organic acids, etc., of OMWW as a carbon source, i.e., they degrade them and convert atmospheric nitrogen into several organic nitrogenous compounds.

So, the process of treating OMWW with an enriched aerobic microbial population of this kind results in a nonphytotoxic thick liquid that could be characterized as an organic soil-conditioner biofertilizer with the following characteristics:

1. Contains exopolysaccharides (microbial metabolites) that favor the formation of stable aggregates.
2. Contains all the major and trace plant nutrients that were originally present in OMWW.
3. It is biologically enriched via the $N_2$-fixation mechanism with organic forms of nitrogen.
4. Contains plant growth-promoting factors such as auxins and cytokinins produced by the bacteria metabolic activity.
5. It is a soil microbial inoculant that allows the establishment of favorable to plants rhizospheric microorganisms and enhances soil suppressiveness.

In this way the waste is fully recycled to land through an environmentally "clean" process while energy and water savings are also achieved at the same time (Chatjipavlidis et al., 1996; Flouri et al., 1994; Ramos-Cormenzana et al., 1995).

*8.4.2.1.b. Composting.* The term "composting" refers to the process of controlled aerobic biological degradation of organic substrates (wastes, residues, etc.). The microorganisms used for this purpose utilize the organic substrate for growth and multiplication in two stages: A thermophilic stage characterized by the quick growth of bacterial populations and subsequent quick degradation of the organic load and a mesophilic stage where growth of fungi takes place and the rate of organic load degradation is much lower. By this process, stabilized, humic substances and mineral salts-containing material are produced that can be used for soil fertility and plant production improvement.

In the case of OMWW composting, OMWW quantities are added to the solid substrate during the thermophilic stage to replace the water evaporated. In this way, composting time is prolonged, the solid substrate becomes enriched in organic matter (that is necessary for the microbial growth), and OMWW is evaporated or consumed. Either fresh OMWW or sludge from evaporation ponds can be used and OMWW sludge shows greater humification rates, stability, and maturity.

Suitable carriers for OMWW are considered those that have heavy loads of N, so as to compensate the relatively low N content of OMWW and neutral or slightly acid pH to minimize N losses. So far, several plant waste carriers have been studied such as cotton waste, maize straw (Cegarra et al., 1996; Paredes et al., 1999), sugar cane bagasse, spent mushroom compost, barley straw, grape marc, and rice hull (Cegarra et al., 1996). Solid wastes from the olive oil industry and olive orchard operations, such as extracted olive press cake and olive tree leaves and branches also have been used as carriers in a pilot plant and a commercial large-scale plant constructed in Crete, Greece (Manios et al., 2004; Parlavantza et al., 1994). Finally, the use of animals' manure (such as poultry manure) and urban wastes that are rich in N have been studied (Paredes et al., 2001). Although use of OMWW as a fertilizer appears to be a viable alternative, it is not sufficient to cope with the enormous quantities of OMWW produced.

#### 8.4.2.2. Recovery of Antioxidants

The olive fruit is rich in simple and complex water-soluble compounds with potential antioxidant properties. The most important of them are polyphenols, flavonoids, anthocyanins, tannins, oleanolic acid, and maslinic acid.

Polyphenols in OMWW have shown to be antioxidant, antibiotic, antimicrobial, and antifungal activity. Formulas of these olive-derived substances can be used as nutrition supplements or skin cosmetics. Due to their antimicrobial properties, they are also used as antimicrobial agents in detergents and rinsing and cleaning agents. Hydroxytyrosol is the most active component of them and it possesses great economic interest due to its remarkable pharmacological and antioxidant properties. It can be used as a food preservative and in pharmacology and cosmetology in topical preparations with anti-aging and anti-inflammatory action. Oleanolic acid regulates cholesterol levels in blood and balances body weight. Maslinic acid has been widely investigated during the last years and it seems to possess anti-inflammatory and antihistaminic activity. It also could be used in pharmacology as an inhibitor of AIDS virus. Finally, anthocyanins are used as natural food colorants (Ramos-Cormenzana and Monteoliva-Sanchez, 2000).

There are scarce data in the literature concerning methods of extraction of antioxidants of high added value from OMWW because most of these processes are patented. The most widely used are solvent extraction (both liquid–liquid and solid–liquid extraction) and adsorption in adsorbent resins. As a pretreatment step, selective concentration of the waste by ultrafiltration or reverse osmosis can be used. For the purification and separation of these substances from the mixture of antioxidants extracted from OMWW, chromatographic methods are used.

Liquid–liquid extraction is a simple and convenient alternative for this purpose. Polar solvents are best for extraction of polyphenols from OMWW and the yield of extraction increases with increasing polarity of the solvent. Among various polar solvents so tested are methyl isobutyl ketone, methyl ethyl ketone, diethylether, and ethyl acetate, the latter of which is considered the most suitable. Apart from the type of solvent used, other factors affecting the process efficiency are the pH of OMWW (better when acidic), the ratio of solvent/waste, and the number of theoretical steps in batch systems (Allouche et al., 2004; Visioli et al., 1999).

Adsorption is the attachment of dissolved compounds (adsorbate) from polluted waters to a solid substance (adsorbent) as a result of attractive interaction of the molecules of the adsorbate with micropores or macropores of the adsorbent having comparable dimensions to that of the molecules. In the case of OMWW, adsorbates are polyphenols and tannins. The most widely used adsorbents for this purpose are activated carbon, activated clay, and superabsorbent polymers.

Activated carbon adsorption installations are associated with extremely high costs coming from both the high initial cost of the material and from subsequently high operational costs. A possible solution to this economic burden could be the use of activated carbons produced by olive stone and solvent-extracted live pulp, inexpensive by-products of the olive oil industry. In this way, the volume of solid waste also could be reduced (Galliatsatou et al., 2002). The use of activated clay is another cheap alternative with maximum removal of polyphenols about 81% and 71% for organic matter (Al-Malah et al., 2000).

#### 8.4.2.3. Production of Biopolymers

The production of biopolymers from OMWW is a very interesting alternative that has been studied during the last years due to the high added value and excellent properties of these substances. Research is oriented toward two main categories of these substances: exopolysaccharides and polyhydroxyalkanoates.

Exopolysaccharides are extracellular polysaccharides that are derived from specific microorganisms under stress conditions. They possess great rheological properties such as great solubility in water, increase of viscosity of a solution even at small concentrations, pseudoplasticity and good stability in solutions. The most important of them, from the economical point of view, are xanthan and pullulan (Ramos-Cormenzana and Monteoliva-Sanchez, 2000).

Pullulan is produced by the fungus *Aureobasidium pullulans*. OMWW is a good quality substrate for its production and production yield of 8 g/liter of a solution of OMWW/water 1:3 has been reported (Ramos-Cormenzana et al., 1995). Xanthan is produced by *Xanthomonas campestris*. Concentrations of OMWW in the substrate of the microorganism greater than 60% have an inhibitory effect. In more diluted substrates, the maximum xanthan production reported is 7.7 g/liter (Lopez et al., 2001; Lopez and Ramos-Cormenzana, 1996).

Polyhydroxyalkanoates (PHA) are reserve polyesters that are accumulated as intracellular granules in a variety of bacteria under unbalanced growth conditions, i.e., in the excess of a carbon source combined with limited concentration of another vital nutrient such as $O_2$, $N_2$, or phosphate. They can be used in the production of completely biodegradable plastics and in the preparation of micro- or nanocapsules that permit the controlled release of an active compound of a medicine so as to prolong its action (Ramos-Cormenzana and Monteoliva-Sanchez, 2000).

The production of these substances is extremely expensive (15–30 $ cost for 1 kg produced) because of the substrates used. The use of OMWW as substrate is a very economical alternative. Strains of *Azotobacter chroococcum* are studied for this purpose. Results report PHA produced up to 50% of the cell dry weight after 24 hours in medium supplemented with OMWW (Gonzalez-Lopez et al., 1996).

#### 8.4.2.4. Production of Biogas

Biogas contains methane and $CO_2$ and can be used as an energy source like natural gas and other calorific gases. It is produced during the anaerobic fermentation of wastewaters. Almost 80% of organic compounds of OMWW are biomethanizable. Theoretically, a yield of 37 $m^3$ of methane per $m^3$ of OMWW could be achieved, but on applying anaerobic fermentation to this wastewater, the high concentration of polyphenols inhibits the normal development of biomethanization. As explained in a previous chapter, by combining unaerobic treatment with an aerobic step where specific microorganisms eliminate the phenolic content, the yield of biogas production can be increased (Fiestas Ros de Ursinos and Borja-Padilla, 1996; Niaounakis and Halvadakis, 2004).

#### 8.4.2.5. Production of Animal Feed

Olive cakes or solid residues of various OMWW processes could be used in animal feeding, as they are rich in oil, carbohydrates, and proteins. Problems arise from OMWW's high concentration of potassium and phenolic compounds which are anti-digestive factors.

This problem could be eliminated and the nutritive value of these wastes improved by special chemical treatments such as with sodium hydroxide and ammonia (Molina Alcaide and Nefzaoui, 1996).

## 8.5. CONCLUSIONS

OMWW treatment and disposal is a problem with great complexity due to the strong nature of the waste and several economical, technical, and organizational constraints involved in the olive oil sector. Practically, all treatment processes developed for domestic and industrial wastewaters have been tested on OMWW but none of them appeared suitable to be generally adopted.

All the approaches so far, although technically successful, lack economic viability. This is because up to now the emphasis has been on treating OMWW like all other wastes, i.e., on reducing polluting loads to legally accepted levels for disposal to the environment. To achieve this goal, sophisticated technical solutions are required that the majority of the small-sized and high geographically scattered olive mills cannot afford.

So, the future olive oil waste management strategy should be toward a combination of detoxifying OMWW and utilizing it, at the same time, for producing valuable by-products. In this way, high costs of detoxification could be compensated. This is the case with the large amounts of phenolic compounds present in OMWW. They constitute the major obstacle in the detoxification of OMWW, while they are products of high added value at the same time. So, the trend now is toward turning this problem to an economic benefit by extracting these compounds. The optimal solution for OMWW treatment will eventually depend on local factors in each and every separate case. However, process approaches that are economically sustainable will constitute the core of the waste treatments.

## 8.6. REFERENCES

Aggelis, G., Iconomou, D., Christou, M., Bokas, D., Kotzailias, S., Christou, G., Tsagou, V., and Papanikolaou, S., 2003, Phenolic removal in a model olive oil mill wastewater using Pleurotus ostreatus in bioreactor cultures and biological evaluation of the process, *Water Res.* **37**:3897.

Aktas, E. S., Imre, S., and Ersoy, L., 2001, Characterization and lime treatment of olive mill wastewater, *Water Res.* **35**:2336.

Alburquerque, J. A, Gonzalvez, J., Garcia, D., and Cegarra, J., 2004, Agrochemical characterization of ''alpeorujo'', a solid by-product of the two-phase centrifugation method for olive oil extraction, *Bioresour. Technol.* **91**:195.

Allouche, N., Fki, I., and Sayadi, S., 2004, Towards a high yield recovery of antioxidants and purified hydroxytyrosol from olive mill wastewaters, *J. Agric. Food Chem.* **52**:267.

Al-Malah, K., Azzam, M., and Abu-Lail, N. I., 2000, Olive mills effluent (OME) wastewater post-treatment using activated clay, *Sep. Purif. Technol.* **20**:225.

Assas, N., Ayed, L., Marouani, L., and Hamdi, M., 2002, Decolorization of fresh and stored-black olive mill wastewaters by Geotrichum candidum, *Process Biochem.* **38**:361.

Azbar, N., Bayram, A., Filibeli, A., Muezzinoglu, A., Sengul, F., and Ozer, A., 2004, A Review of Waste Management Options in Olive Oil Production, *Crit. Rev. Env. Sci. Technol.* **34**:209.

Borja, R., and Gonzαlez, A., 1994, Comparison of anaerobic filter and anaerobic contact process for olive mill wastewater previously fermented with Geotrichum candidum, *Process Biochem.* **29**:139.

Borsani, R., and Ferrando, B., 1996, Ultrafiltration Plant for Olive Vegetation Waters By Polymeric Membrane Batteries, *Desalination* **108**:281.

Caputo, A. C., Scacchia, F., and Pelagagge, P. M., 2003, Disposal of by-products in olive oil industry:waste-to-energy solutions, *Applied Thermal Engineering* **23**:197.

Cegarra, J., Paredes, C., Roig, A., Bernal, M. P., and Garcia, D., 1996, Use of olive Mill wastewater Compost for crop production, *Int. Biodeterioration & Biodegradation* **38**:193.

Cereti, C. F., Rossini, F., Federici, F., Quaratino, D., Vassilev, N., and Fenice, M., 2004, Fermentative decolorization of olive mill wastewater by Lactobacillus plantarum, *Bioresour. Technol.* **91**:135.

Chatjipavlidis, I., Antonakou, M., Demou, D., Flouri, F., and Balis, C., 1996, Bio-fertilization of olive oil mills liquid wastes. The pilot plant in Messinia, Greece, *Int. Biodeterioration & Biodegradation* **38**:183.

Cox, L., Celis, R., Hermosin, M. C., Becker, A., and Cornejo, J., 1997, Porosity and herbicide leaching in soils amended with olive-mill waste water, *Agric. Ecosyst. Environ.* **65**:151.

Dalis, D., Anagnostidis, K., Lopez, A., Letsiou, I., and Hartmann, L., 1996, Anaerobic digestion of total raw olive-oil wastewater in a two-stage pilot-plant (up-flow and fixed-bed bioreactors), *Bioresour. Technol.* **57**:237.

D' Annibale, A., Ricci, M., Quaratino, D., Federici, F., and Fenice, M., 2004, Panus tigrinus efficiently removes phenols, color and organic load from olive-mill wastewater, *Research in Microbiology* **155**:596.

Eroglu, E., Gunduz, U., Yucel, M., Turker, L., and Eroglu, I., 2004, Photobiological hydrogen production by using olive mill wastewater as a sole substrate source, *Int. J. Hydrogen Energy* **29**:163.

Fadil, K., Chahlaoui, A., Ouahbi, A., Zaid, A., and Borja, R., 2003, Aerobic biodegradation and detoxification of wastewaters from the olive oil industry, *Int. Biodeterioration & Biodegradation* **51**:37.

Fiestas Ros de Ursinos, J.A., and Borja-Padilla, R., 1996, Biomethanisation, *Int. Biodeterioration & Biodegradation* **38**:145.

Fiorentino, A., Gentili, A., Isidori, M., Lavorgna, M., Parrella, A., and Temussi, F., 2004, Olive Oil Mill Wastewater Treatment Using A Chemical And Biological Approach, *J. Agric. Food Chem.* **51**:5151.

Flouri, F., Chatjipavlidis, I., and Balis, C., 1994, Agriculture as a receptor of its waste: The case of OMWW'' (In Greek), in: *Proc. of Int.Symposium, Geot.l Ch. of Greece, Sitia, Crete, Greece*, pp. 26–34.

Fountoulakis, M.S., Dokianakis, S.N., Kornaros, M.E., Aggelis, G.G., and Lyberatos, G., 2002, Removal of phenolics in olive mill wastewaters using the white-rot fungus Pleurotus ostreatus, *Water Res.* **36**:4735.

Galiatsatou, P., Metaxas, M., Arapoglou, D., and Kasselouri-Rigopoulou, V., 2002, Treatment of olive mill waste water with activated carbons from agricultural by-products, *Waste Management* **22**:803.

Garcia, G. I., Pepa, J., Venceslada, P. R., Martin, M.A., Santos, M.A., and Ramos Gomez, E., 2000, Removal of phenol compounds from olive mill wastewater using Phanerochaete chrysosporium, Aspergillus niger, Aspergillus terreus and Geotrichum candidum, *Process Biochem.* **35**:751.

Gernjak, W., Maldonado, M. I., Malato, S., Cαceres, J., Krutzler, T., Glaser, A., and Bauer, R., 2004, Pilot-plant treatment of olive mill wastewater (OMW) by solar TiO2 photocatalysis and solar photo-Fenton, *Solar Energy*, In Press.

Gonzalez-Lopez, J., Pozo, C. Martinez-Toledo, V., Rodelas, B., and Salmeron, V., 1996, Production of Polyhydroxyalkanoates by Azotobacter chroococcum H23 in wastewater from olive oil mills (alpechin), *Int. Biodeterioration & Biodegradation* **38**:271.

Improlive 2000: Improvements of treatments and validation of the liquid-solid waste from the two-phase olive oil extraction, *EU Project FAIR CT961420.*

International Olive Oil Council, 2004, Statistics on olive oil, olive-pomace oil and table olives, December 2004, at www.internationaloliveoil.org

Jaouani, A., Sayadi, S., Vanthournhout, M., and Penninckx, M., 2003, Potent fungi for decolourisation of olive oil mill wastewaters, *Enzyme Microb. Technol.* **33**:802.

Javier-Benitez, F., Beltran-Heredia, J., Acero, J. L., and Pinilla, M. L., 1997, Simultaneous Photodegradation and Ozonation plus UV radiation of phenolic acids -major pollutants in agro-industrial wastewaters, *J. Chem. Technol. Biotechnol.* **70**:253.

Lanciotti, R., Gianotti, A., Baldi, D., Angrisani, R., Suzzi, G., Mastrocola, D , and Guerzoni, M. E., 2004, Use of Yarrowia lipolytica strains for the treatment of olive mill wastewater, *Bioresour. Technol.* In Press.

Lesage-Meesen, L., Navarro, D., Maunier, S., Sigoillot, J. C., Lorquin, J., Delattre, M., Simon, J-L., Asther, M., and Labat, M., 2001, Simple phenolic content in olive oil residues as a function of extraction systems, *Food Chem.* **75**:501.

Lopez, M. J., and Ramos-Cormenzana, A., 1996, Xanthan production from olive-mill wastewaters, *Int. Biodeterioration & Biodegradation* **38**:263.

Lopez, M. J., Moreno, J., and Ramos-Cormenzana, A., 2001, Xanthomonas campestris strain selection for xanthan production from olive mill wastewaters, *Water Res.* **35**:1828.

Manios, Maniadaki, Kalogeraki, Mari, Terzakis, Magiatis, Mikros, Agalias, Spanos, and Manios, 2004, Cocomposting olive residuals and green waste on Crete, *Biocycle International* **45**(2):65.

Marques, I.P., 2001, Anaerobic digestion treatment of olive mill wastewater for effluent re-use in irrigation, *Desalination* **137**:233.

Mechichi, T., and Sayadi, S., 2004, Evaluating process imbalance of anaerobic digestion of olive mill wastewaters, *Process Biochem.* **40**:139.

Mitrakas, M., Papageorgiou, G., Docoslis, A., and Sakellaropoulos, G., 1996, Evaluation of various pretreatment methods for olive mill wastewaters, *European Water Pollution Control* **6**:10.

Molina Alcaide, E., and Nefzaoui, A., 1996, Recycling of olive-oil by-products: Possibilities of utilization in animal nutrition, *Int. Biodeterioration & Biodegradation* **38**:227.

Mulinacci, N., Romani, A., Galardi, C., Pinelli, P., Giaccherini, C., and Vincieri, F. F., 2001, Polyphenolic Content in olive Oil waste waters and related olive samples, *J. Agric. Food Chem.* **49**:3509.

Niaounakis, M., and Halvadakis, C. P., 2004, *Olive-Mill Waste Management- Literature Review And Patent Survey*, Typothito-George Dardanos, Athens, Greece.

Oikonomou, D., Koutsaftaki, L., Stefanoudaki, E., Metzidakis, G., Androulakis, I., and Iniotakis, N., 1994, Olive-mill waste water treatment with a membrane process, (In Greek), in: *Proc. of Int. Symposium, Geot.l Ch. of Greece, Sitia, Crete, Greece*, pp. 156–167.

Paredes, C., Bernal, M. P., Roig, A., Cegarra, J., and Sánchez-Monedero, M. A., 1996, Influence of the bulking agent on the degradation of olive-mill wastewater sludge during composting, *Int. Biodeterioration & Biodegradation* **38**:205.

Paredes, C., Cegarra, J., Roig, A., Sanchez-Monedero, M.A., and Bernal, M.P., 1999, Characterization of olive mill wastewater (alpechin) and its sludge for agricultural purposes, *Bioresour. Technol.* **67**:111.

Paredes, C., Bernal, M.P., Roig, A., and Cegarra, J., 2001, Effects of olive mill wastewater addition in composting of agroindustrial and urban wastes, *Biodegradation* **12**:225.

Parlavantza, M., Balis, C., Flouri, F., Chatjipavlidis, I., Belivanis, N., and Blyssides, A., 1994, Composting of olive mill wastes in Sitia, Crete, (In Greek), in: *Proc. Of Int. Symposium, Geot. I.Ch. Of Gr, Sitia, Crete, Greece*, pp.174–184.

Pinto, G., Pollio, A., Previtera, L., Stanzione, M., and Temussi, F., 2003, Removal of low molecular weight phenols from olive oil mill wastewater using microalgae, *Biotechnol. Lett.* **25**:1657.

Piperidou, C. I., Chaidou, C. I., Stalikas, C. D., Soulti, K., Pilidis, G. A., and Balis, C., 2000, Bioremediation of Olive Oil Mill Wastewater: Chemical Alterations Induced by Azotobacter vinelandii, *J. Agric. Food Chem.* **48**:1941.

Ramos-Cormenzana, A., Monteoliva-Sanchez, M., and Lopez, M. J., 1995, Bioremedation of alpechin, *Int. Biodeterioration & Biodegradation* **35**:249.

Ramos-Cormenzana, A., Juarez-Jimenez, B., and Garcia-Pareja, M. P., 1996, Antimicrobial activity of olive-mill waste waters (alpechin) and biotransformed olive oil mill wastewater, *Int. Biodeterioration and Biodegradation* **38**:283.

Ramos-Comerzana, A., and Monteoliva-Sanchez, M., 2000, Potencial biofarmaceutico de los residuos de la industria oleicola ''Potential biofarmaceutical use of the wastes of the olive oil industry'', *Ars Pharmaceutica* **41**(1):129.

Rinaldi, M., Rana, G., and Introna, M., 2003, Olive-mill wastewater spreading in southern Italy: effects on a durum wheat crop, *Field Crops Research* **84**:319.

Rivas, J.F., Beltran, F. J., Gimeno, G., and Frades, J., 2001, Treatment of Olive Oil Mill Wastewater by Fenton's reagent, *J. Agric. Food Chem.* **49**:1873.

Rozzi, A., and Malpei, F., 1996, Treatment and Disposal of Olive Mill Effluents, *Int. Biodeterioration & Biodegradation* **38**:135.

Sabbah, I., Marsook, T., and Basheer, S., 2004, The effect of pretreatment on anaerobic activity of olive mill wastewater using batch and continuous systems, *Process Biochem.* **39**:1947.

Sierra, J., Martν, E., Montserrat, G., Cruaρas, R., and Garau, M. A., 2001, Characterisation and evolution of a soil affected by olive oil mill wastewater disposal, *Sci. Total Environ.* **279**(1–3):207.

Stefanoudaki-Katzouraki, E., and Koutsaftakis, A., 1994, Characteristics of waste of two- and three-phase olive oil mills, in: *Proc. Of Int. Symposium, Geot. I.Ch. Of Greece, Sitia, Crete, Greece*, pp. 19–25.

Visioli, F., Romani, A., Mulinacci, N., Zarini, S., Conte, D., Vincieri, F. F., and Galli, C., 1999, Antioxidant and other biological activities of olive mill waste waters, *J. Agric. Food Chem.* **47**:3397.

Vitolo, S., Petarca, L., and Breski, B., 1999, Treatment of olive oil industry wastes, *Bioresour. Technol.* **67**:129.

Zervakis, G., Yiatras, P., and Balis, C., 1996, Edible mushrooms from olive oil wastes, *Int. Biodeterioration & Biodegradation* **34**:237.

Zouari, N., 1998, Decolorization of olive oil mill effluent by physical and chemical treatment prior to anaerobic digestion, *J. Chem. Technol. Biotechnol.* **73**:297.

Zouari, N., and Ellouz, R., 1996, Toxic effect of coloured olive compounds on the anaerobic digestion of olive oil mill effluent in UASB-like reactors, *J. Chem. Technol. Biotechnol.* **66**:414.

# 9

# Anaerobic Degradation of Animal By-Products

Roland Kirchmayr, Christoph Resch, Martin Mayer, Stephan Prechtl, Martin Faulstich, Rudolf Braun, and Johann Wimmer

## 9.1. INTRODUCTION

As a result of growing meat consumption and production the slaughtering and rendering industry has an increasing output of nonconsumable animal by-products. According to new European legislation the established treatment and disposal technologies are either not allowed anymore or are too costly. Therefore, new ways of treatment for raw slaughterhouse waste products and pretreated materials from rendering plants have been established. As legislation on the recovery of organic materials for animal feed is becoming tighter and more restrictive, anaerobic digestion is a promising alternative for the treatment of these materials, since the process combines material recovery and energy production (Salminen and Rintala, 2002). The careless or rash utilization of animal by-products as substrates for anaerobic digestion subsequently may cause process instability and odor emissions. The conditioning of the digestate, like nitrogen-removal or concentration, may reduce logistic costs for utilization as a fertilizer.

ROLAND KIRCHMAYR, CHRISTOPH RESCH, MARTIN MAYER, RUDOLF BRAUN • BOKU University of Natural Resources and Applied Life Sciences, Institute for Environmental Biotechnology, 3430 Tulln, Austria. STEPHAN PRECHTL AND MARTIN FAULSTICH • ATZ Entwicklungszentrum, 92237 Sulzbach-Rosenberg, Germany. JOHANN WIMMER • Federal Government of Upper Austria, 4021 Linz, Austria.

## 9.2. REQUIREMENTS OF THE EUROPEAN REGULATION EC NO. 1774/2002

### 9.2.1. Hazard Potential of Animal By-products

The inefficient treatment of slaughterhouse waste or improper use of products produced from it led to the pandemic occurrence of animal diseases such as BSE and foot-and-mouth disease. Only a rigorous European Regulation on the treatment and further use of animal by-products could prevent a further spread of diseases. This very extensive Regulation governs the collection, transport, methods, and procedures of treatment, as well as the further disposal, use, or trade in the products. Therefore, the Regulation has an effect on the export to and the import from nonmember countries of the European community.

### 9.2.2. Definitions

***ABP (animal by-products)***: Animal by-products are all bodies or parts of animals and products of animal origin not intended for human consumption, because either they are not fit for human consumption or there is no market for them as foodstuff.
***Biogas plant according to the ABP-Regulation***: This is a plant in which biological degradation of products of animal origin is undertaken under anaerobic conditions for the production and collection of biogas.

### 9.2.3. The EC-Regulation (EC) No. 1774/2002

The Regulation (EC) No. 1774/2002 of the European Parliament and of the Council of October 3, 2002, lays down health rules concerning animal by-products not intended for human consumption (''ABP-Regulation'' or ''Hygiene Regulation'' in committees working on environmental issues) and regulates possible uses and processing rules of animal by-products (ABP). The Regulation was published in the official journal of the European Community of October 10, 2002, L 273, pages 1–95, and has had to be applied directly in all member states of the European Community since May 1, 2003.

Further temporary Regulations and implementing rules were laid down by the European Commission for the cushioning and modification of this Regulation. In this text all amendments that entered into force by February 2006 are included. Specific possibilities of processing and use for animal by-products are listed in this ABP-Regulation. In future, existing and newly developed methods and processes or utilization possibilities may be examined by the Scientific Steering Committee and approved by the European Commission.

### 9.2.4. Animal By-products in Biogas Plants

In the ABP-Regulation animal by-products are divided into three categories:

- Category 1: contains those materials with the highest risk for public health, animals, or the environment (hygienic risk, risk of BSE, etc.).
- Category 2: includes all animal by-products that can be allocated neither to category 1 nor to category 3 (e.g., manure or digestive tract content or animals not fit for human consumption). For manure and catering waste the conditions for approval and for treatment, as well as other criteria for the end product and for the remaining animal by-products, are defined (Figure 9.1 and Table 9.5). Biogas plants that process catering waste or manure can be approved by national rules (pending further EC-legislation).
- Category 3: comprises those animal by-products that would be fit for human consumption, but are (for commercial reasons) not intended for human consumption.

According to Regulation (EC) No. 1774/2002 different substrates for biogas plants require different treatments (no treatment, pasteurization, or sterilization) and the approval of the respective biogas plant according to national regulations or regulation (EC) No. 1774/2002. Figure 9.1 and Table 9.1 give an overview about substrates and the legal force of national or international regulations and a survey of materials designated for treatment in biogas plants.

### 9.2.5. Materials of Category 1

ABP of category 1 represent an increased risk for public health, animals, or the environment. These materials such as specified risk material (SRM), animals suspected of being infected with BSE, ABP with increased concentrations of

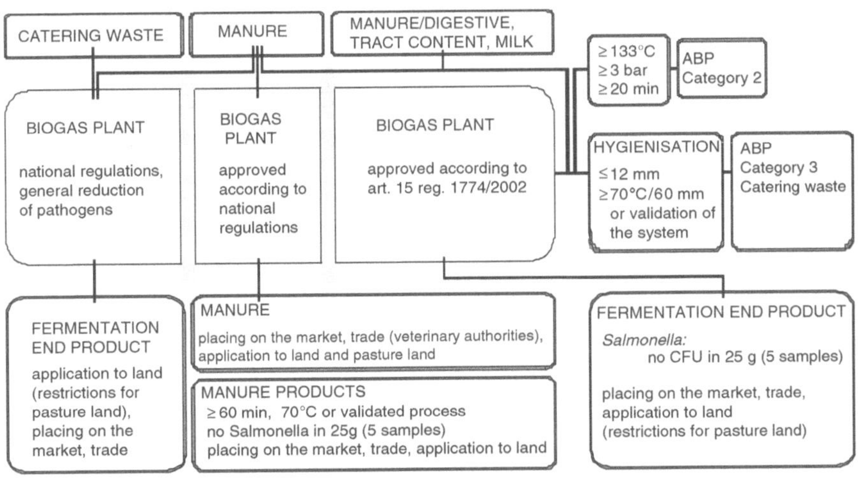


**Figure 9.1.** Substrates and legal force of national and international regulations.

*Table 9.1.* Survey of materials designated for treatment in biogas plants: Three categories of animal by-products

| Category | Material to be treated in biogas plants |
|---|---|
| Category 1 | Not designated |
| Category 2<br>Without preliminary treatment | Manure as well as digestive tract content (separated from the digestive tract; if there is no risk of dispersal of serious-infectious diseases) milk and colostrum |
| Category 2<br>After sterilization with steam pressure and marking (with smell) | All materials classified as category 2 (e.g., perished animals or animals slaughtered, but not intended for human consumption) |
| Category 3<br>In a biogas plant approved in accordance with article 15 of the regulation | All materials classified as category 3 (e.g., meat-containing wastes from the foodstuff-industry, slaughterhouse wastes of animals fit for human consumption) |
| Category 3<br>In biogas plants which are to be approved in accordance with provisions and methods to be adopted or which are approved according to national legislation | Catering waste (except from international means of transport) |

*Table 9.2.* Tissues classified as specific risk material: Regulation (EC) No. 999/2001, amended by Regulation (EC) No. 657/2006

| Animal | Tissue |
|---|---|
| Bovine animals, aged over 12 months | Skull excluding the mandible and including brain and eyes, and the spinal cord. |
| Bovine animals, aged over 24 months | Vertebral column including the dorsal root ganglia, but excluding vertebrae of the tail, the spinous and transverse processes of the cervical, thoracic and lumbar vertebrae and the median sacral crest and wings of the sacrum. |
| Bovine animals, all ages | Tonsils, intestines from duodenum to rectum, mesentery |
| Ovine and caprine animals, aged over 12 months | Skull including brain and eyes, tonsils and spinal cord |
| Ovine and caprine animals, all ages | Spleen as well as ileum |

environmental contaminants, solid materials (> 6 mm) from wastewater treatment in category 1 processing plants and establishments in which SRM is removed (slaughterhouses and cutting plants), and catering waste from international means of transport are not allowed to be processed in a biogas plant.

### 9.2.6. Materials of Category 2

Category 2 comprises all ABP that are neither included in category 1 nor in category 3. These are manure, digestive tract content, milk not fit for human consumption, killed or fallen animals, and solid materials in wastewater streams of slaughterhouses (particle size > 6 mm). ABP of category 2 may be processed in a biogas plant only after sterilization with steam pressure (at least 20 minutes without interruption at a core temperature of more than 133°C and an absolute steam pressure of not less than 3 bar), except manure, digestive tract content, and milk, which need no pretreatment.

#### 9.2.6.1. Manure

The definition of manure according to the ABP-Regulation is excrements and/or urine of farm animals, with or without litter, and guano, either unprocessed, processed, or transformed in biogas plants or composting plants.

Manure, digestive tract content (separated from the digestive tract), milk, and colostrum are materials of category 2. These materials, however, can be fed directly and without any pre-treatment to an approved biogas plant. The fermentation end product of the "transformation" of manure processed in a biogas plant together with other substrates that are not covered by this Regulation (e.g., renewable raw materials or energy crops) may be considered as untreated manure. Conditions for placing "untreated manure" on the market within the boundaries of a member state of the EC, as well as special requirements for transport (marking as "manure," cleaning of containers, etc.) may be laid down by national legislation.

The list of third countries from which EC member states may authorize imports of manure for treatment of the soil is established in a separate regulation (Directive 79/542/EEC). Conditions and special requirements for the placing manure on the market within national boundaries may be laid down by national legislation.

#### 9.2.6.2. Wastewater from Slaughterhouses

For slaughterhouses (or cutting plants removing SRM) and plants processing material of category 1 and 2 (e.g., intermediate and rendering plants) a pretreatment of the wastewater is required that retains all solid materials up to a particle size of 6 mm (e.g., using a screen with a mesh size of 6 mm). Any materials removed from the wastewater by this pretreatment unit (screenings, materials from desanding, grease and oil mixtures, sludge, material removed from drains) are regarded as materials of category 2 or materials of category 1 (for plants processing materials of category 1 or removing SRM), respectively.

Any decomposition or reduction in size of the materials in the wastewater stream prior to the retaining unit is not allowed! Materials removed from the wastewater stream after the pretreatment of the wastewater (flotation sludge, etc.) and the residual wastewater containing no solid particles are not covered by this Regulation and are to be treated in accordance with the relevant wastewater legislation.

### 9.2.7. Materials of Category 3

Category 3 contains all ABP originating from animals fit for slaughter but not intended for human consumption as well as animal by-products from food production and catering waste. Materials of category 3 must be pasteurized before treating in an approved biogas plant. These ABP may be processed in a biogas plant equipped with a hygienization unit which cannot be bypassed. These biogas plants have to be approved according to the approval conditions laid down in article 15 of the ABP-Regulation (see Figure 9.2). The processing standards in the hygienization unit are defined as 70°C during 60 minutes with a maximum particle size of 12 mm. Alternative processes have to be validated in order to demonstrate the achievement of an equal overall risk reduction (Reg (EC) No/208/2006).

#### 9.2.7.1. Catering Waste

Catering waste is grouped into two categories: those from international means of transport (catering from aeroplanes, ships, or railways) are included in category 1 and must be disposed of. All other catering wastes are defined as materials of category 3. Catering waste of category 3 may be processed in biogas plants according to national rules pending the adoption of relevant provisions and approvals of the EC.

Pending adoption of these EC provisions concerning the treatment of catering waste, the application of alternative standards for processing may be authorized for biogas plants processing only catering waste (together with manure as well as energy crops). However, effective reduction of pathogens has to be ensured.

#### 9.2.7.2. Meat and Bone Meal

At present the feeding of animal protein [meat and bone meal (MBM)] to farm animals is prohibited. General exceptions exist for the feeding of animals not intended for human consumption (pets and fur animals). Specific derogations allow the feeding of certain kinds of processed animal protein (i.e., hydrolyzed protein, fish meal) to nonruminants. Furthermore, it is prohibited to feed animals with meat and bone meal produced from bodies or parts of animals of the same species.

#### 9.2.7.3. Fish Waste

Fish waste (category 3) may be fed to a biogas plant passing the hygienization unit or may be processed to microbiologically stable fish silage or compost. The production of fish meal or fish silage for the purpose of feeding farm animals is allowed under certain conditions.

### 9.2.8. Examples of Animal By-products in Biogas Plants

In the following sections two cases are specified, which shall illustrate the application of the "ABP-Regulation" (EC) No. 1774/2002 to biogas plants.

### 9.2.8.1. Biogas Plant at a Pig Slaughtering Facility

Substrates used in the biogas plant for this case study are listed in Table 9.3. This biogas plant is registered by the competent authorities and has to be approved according to article 15 of the ABP-Regulation. A pasteurization unit that cannot be by-passed must be available and a concept of control and monitoring must be followed. The microbiological parameters for digestates (Table 9.4) must be applied. Manure as well as digestive tract content (separated from the digestive tract) may be processed in the bio-gas plant without pretreatment.

All ABP of category 3 such as slaughtering by-products, bones, intestines, as well as blood are to be pasteurized before processing in a biogas plant. Any materials removed from the wastewater stream before the prescribed wastewater pretreatment unit in slaughterhouses are regarded as materials of category 2 and, like these, are to be sterilized by steam pressure prior to processing in a biogas

*Table 9.3.* Example for substrates used in a biogas plant located at a pig slaughterhouse

| Substrate | Category | Required treatment |
|---|---|---|
| Manure from pigs | 2 | No pretreatment required |
| Digestive tract content | 2 | No pretreatment required |
| Digestive tract (fit for human consumption) | 3 | Pasteurization |
| Bones, slaughter by-products | 3 | Pasteurization |
| Blood | 3 | Pasteurization |
| Parts of slaughtered animals[a] (not fit for human consumption) | 3 | Pasteurization |
| Screenings, flotation sludge (> 6 mm) | 2 | Sterilization |
| Content of fat removal devices[b] (< 6 mm) | – | |
| Washings (purely liquid fraction) | – | |

[a] From animals fit for human consumption (ante mortem), identified as not fit for human consumption after postmortem inspection.
[b] Without reduction of particle size and removed from the wastewater stream after the wastewater pretreatment unit.

*Table 9.4.* End product parameters for digestates and fermentation end products

| Product | Product parameter |
|---|---|
| Unprocessed and digested manure<br>Placing on the national market | None |
| Manure products | Heat treatment (> 60 min., > 70°C)<br>*Salmonella:* absence in 5 samples of 25 g each or validated process<br>treatment for the reduction of creators of spores and toxin, storage in special containers |
| Fermentation end products<br>Biogas plant approved in compliance with Article 15 | |
| Fermentation end product<br>Manure and catering waste | According to national regulations (with restrictions) |

*Table 9.5*. Example for substrates used in a biogas plant located at a bovine abattoir

| Substrate | Category | Required treatment |
|---|---|---|
| Manure from bovine animals | 2 | No pretreatment required |
| Rumen content (stomach content) | 2 | No pretreatment required |
| Rumen | 3 | Pasteurization |
| Slaughter by-products, bones | 3 | Pasteurization |
| Parts of slaughtered animals[a] | 3 | Pasteurization |
| Blood | 3 or 1 | Pasteurization or incineration |
| Bones: vertebral column and skull | 1 | Sterilization, incineration |
| Intestines | 1 | Sterilization, incineration |
| Screenings (bigger than 6 mm) | 1 | Sterilization, incineration |
| Content of fat removal devices[b] (particles < 6 mm) | – | |
| Washings (purely liquid fraction) | – | |

[a] From animals fit for human consumption (antemortem), identified as not fit for human consumption after postmortem inspection.
[b] Without reduction of particle size and removed from the wastewater stream after the wastewater pretreatment unit.

plant. Materials of categories 2 and 3 sterilized with steam pressure may be fed to a biogas plant without any further pretreatment. For materials that have passed the wastewater pretreatment unit the ABP-Regulation is not effective.

#### 9.2.8.2. Biogas Plant at a Bovine Slaughtering Facility

Substrates used in the biogas plant for this case study are listed in Table 9.5. The plant is registered by the competent authorities and approved according to article 15 of the ABP-Regulation. Herewith follows the necessity of a pasteurization unit which cannot be bypassed and a concept of control and monitoring for this digestion plant. Additionally, the hygiene parameters for the digestates (requirements on digestates) are to be applied. Manure and rumen content may be fed to the biogas plant without pre-treatment. Rumen, slaughtering by-products, and bones are regarded as ABP material of category 3 fit for human consumption and must be fed to the biogas plant by passing the hygienization unit. If it is guaranteed that no SRM gets into the bloodstream during slaughtering (reliable separation of blood draining unit and SRM removal unit, retention of blood until the submission of a negative BSE test, etc.), bovine blood can be considered as material of category 3. Otherwise, bovine blood is to be ''disposed of'' as a mixture together with SRM.

Materials of category 1 must not be fed to the biogas plant.[1] Content of fat removal devices (achieved after screening with a mesh size of 6 mm) as well as washing (purely liquid fraction) are not subject to any provisions of this Regulation.[2]

[1] At present the biogas technology is not designated for the treatment of material of category 1. It is intended, however, that other methods and processes (other than those mentioned in the ABP regulation) may be assessed by the Scientific Steering Committee (SSC) and approved by the Commission afterward. For that it is required to submit a detailed description of the alternative process method or procedure, including an appropriate risk-assessment, to the Commission of the EC.

[2] They are subject to national and international regulations for wastewater and waste.

### 9.2.9. Requirements of Fermentation End Products (Digestates)

Depending on the used substrates and the final destination of the digestate different end product parameters are defined in Regulation (EC) No. 1774/2002 (Table 9.4).

### 9.2.10. Conditions for the Approval of Biogas Plants and Composting Plants According to Article 15 (EC) 1774/2002

Biogas plants processing animal by-products must be registered by the authorities and approved according to article 15 of the ABP-Regulation. Figure 9.2 illustrates which requirements are to be fulfilled in order to be approved by the (competent) authority.

### 9.2.11. Plants' Own Check / HACCP Concept

The principle of (direct) responsibility of the operator and owner of plants or their representatives is a cornerstone of the demands on the biogas plants on the Regulation (EC) No.1774/2002. By compliance with conditions of acceptance (e.g., nonacceptance of questionable batches) and hygienic operating conditions (e.g., compliance with the "principle of clean and unclean sector," pasteurization

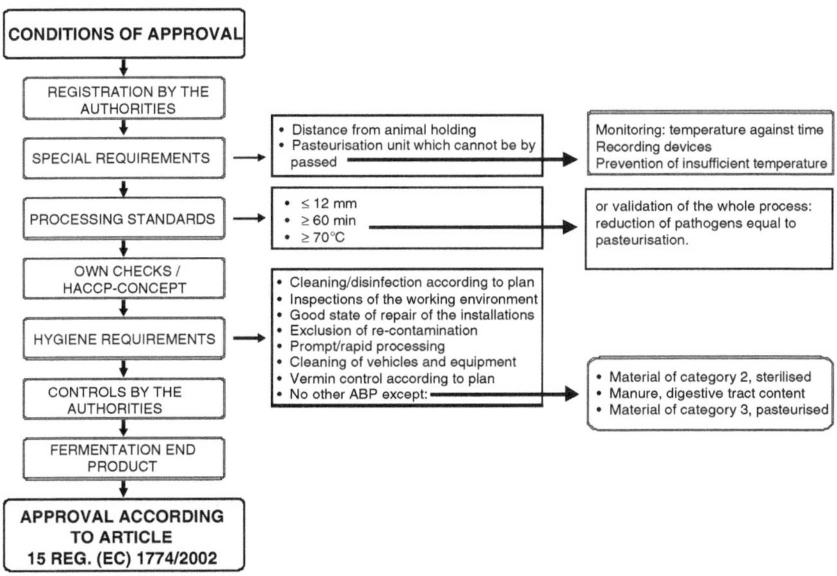


**Figure 9.2.** Survey of the conditions of approval concerning biogas plants according to article 15 of the ABP Regulation (EC) No. 1774/2002.

of the substrate, etc.) the potential health risk that may emanate from a biogas plant or its digestate can be reduced.

In Regulation (EC) 1774/2002 a concept for self-monitoring according to the system of hazard analysis of critical control points (HACCP concept) is demanded of the operators of biogas plants that use animal by-products as substrates. For biogas plants such a (practicable) concept of control and monitoring that follows the HACCP concept has to be developed.

A description of the critical control points (from a hygienic point of view) when operating a biogas plant must include (Kirchmayr et al., 2003):

- description of the place or procedure that is identified as a critical control point,
- description of the actual risk factors and an evaluation of the significance of these,
- preventive activities in order to minimize or delete the risk,
- monitoring: method and frequency,
- definition of critical values, and
- plan for correcting activities when critical values are exceeded.

Operators of biogas plants have to identify, define, and describe the critical control points of their plants. When operating the plant it has to be controlled and documented whether the parameters or standards in the respective control points have been applied. Examples of control points (from a hygienic point of view) are used when operating a biogas plant and for general hygiene requirements are listed in Tables 9.6 and 9.7.

*Table 9.6.* Examples for control points in biogas plants (Kirchmayr et al., 2003)

| Control points (direct) | Possible failures |
|---|---|
| Receipt of raw substrate | |
| Manure | High contamination with pathogens |
| Vegetable substrates | Unintentional pathogen contamination |
| Animal substrate | Insufficient pretreatment |
| Material of category 2 and/or 3; pretreated or not pretreated | |
| Transport | Mixture of untreated raw material and end product (treated) |
| Separation between raw material and end product | |
| Animal by-products | |
| Hygienization | Insufficient hygienization |
| Pasteurization material category 3 | Failure in temperature, time, particle size<br>Exceptionally high pathogen contamination |
| Storage of the end products | Recontamination from environment<br>Growth of remaining pathogens |
| Delivery | Cross-contamination between clean and unclean areas, means of transport and containers |
| Transport of heat-treated material or end product | |

*Table 9.7*. Examples for general hygiene requirements (Kirchmayr et al., 2003)

| General hygiene requirements | Possible failures |
|---|---|
| Maintainance of premises and equipment | Generally dissatisfactory hygiene conditions |
| Intercalibration and check of monitoring and measuring instruments | Insufficient validity of the control |
| Prevention of spreading animal diseases | Insufficient separation and distance from animal holdings and feed storage facilities |
| Pest control | Spreading of pathogens |
| | Cross-contamination between clean and unclean sector |
| Cleaning and disinfection | Internal reinfection (''in-house'') |
| | Continuous contamination of heat-treated substrates |
| | Cross-contamination between clean and unclean sectors (i.e., via aerosols) |
| Education of staff | Uncertainty about aim and methods of the plant's own check system and unclear delegation of responsibility |

## 9.3. CHARACTERIZATION OF WASTE STREAMS

### 9.3.1. Slaughterhouse Wastes

The physical and chemical characteristics of slaughterhouse wastes and pretreated rendering wastes are extensively reported in literature. Analyses of chemical parameters and characteristics have been done for a lot of reasons and different treatment goals. In recent years the focus of waste analyses focused also on methane potential investigations to evaluate the feasibility of these materials for treatment in anaerobic digestion plants. These investigations started on slaughterhouse and rendering wastewaters and were extended to solid organic waste fractions.

Based on these results anaerobic digestion plants have been set up on slaughterhouse and rendering sites, most of them connected to the wastewater treatment facilities. Under the new Regulations (EC 1774/2002) more and different waste fractions may be treated in anaerobic digestion plants. Some of these waste fractions contain high amounts of nitrogen which has to be incorporated in feasibility assessments. In Table 9.8 analysis of waste fraction and raw by-products that accrue directly at the slaughterhouse site are listed. The analysis data are cited from literature or unpublished analysis conducted at own laboratories.

In modern slaughterhouses most waste streams are collected separately. Therefore best substrate mixture for the anaerobic digestion plant may be designed. As this ''best substrate'' cannot be achieved in lot of cases, all listed chemical parameters have to be taken into account when the substrates are designed:

- The COD or VSS contents are used to calculate the derived hydraulic and organic loading rate.
- The TKN load for the anaerobic digester is the result of the nitrogen contents of chosen substrate composition.

*Table 9.8.* Substrate characteristics from waste fractions at slaughterhouses

| Substrate | COD (g/kg) | TKN (g/kg) | TS (%) | VSS (%) | Biogas or methane potential | Reference |
|---|---|---|---|---|---|---|
| *Swine wastes* | | | | | | |
| Gut fill (swine) | | 1.57 | 11.2 | 9.9 | 2,040 $Nm^3_{Gas}/Mg_{VSS}$ | Carpentier et al. (2005) |
| Swine blood | | 17.07 | 14.8 | 14.3 | 1,020 $Nm^3_{Gas}/Mg_{VSS}$ | Carpentier et al. (2005) |
| Swine blood | 340 | 33.13 | 21.48 | 20.44 | | Unpublished results: Authors' own investigations |
| Swine colon | 259 | 1.27 | 6.12 | 5.43 | | |
| Swine gut | 299 | | | 12.08 | | |
| Swine stomach content | 215 | 2.96 | 13.53 | 13.03 | | |
| *Bovine and cattle wastes* | | | | | | |
| Bovine blood | 310 | 21.17 | 19.48 | 18.63 | | Unpublished results: Authors' own investigations |
| Bovine leg bones | 1,181 | 24.89 | 76.13 | 50.01 | 522 $Nm^3_{Gas}/Mg_{COD}$<br>221 $Nm^3_{Gas}/Mg_{VSS}$ | |
| Bovine clutches | 547 | 49.16 | 47.59 | 32.83 | 527 $Nm^3_{Gas}/Mg_{COD}$<br>317 $Nm^3_{Gas}/Mg_{VSS}$ | |
| Bovine shoulder bones | 707 | 36.46 | 69.26 | 40.70 | 330 $Nm^3_{Gas}/Mg_{COD}$<br>191 $Nm^3_{Gas}/Mg_{VSS}$ | |
| Bovine rib bones | 770 | 32.53 | 58.28 | 39.51 | 819 $Nm^3_{Gas}/Mg_{VSS}$<br>422 $Nm^3_{Gas}/Mg_{VSS}$ | |
| Bovine omasum | 646 | 14.01 | 29.59 | 27.48 | | Unpublished results: Authors' own investigations |
| Bovine omasum content | 188 | 6.05 | 11.87 | 10.93 | | |
| Bovine rumen | 119 | 2.17 | 9.55 | 8.43 | | |
| Bovine stomach content | 166 | 11.94 | 10.24 | 10.03 | | |
| *Poultry wastes* | | | | | | |
| Poultry blood | 273 | 10.83 | 10.59 | 9.47 | 878 $Nm^3_{CH_4}/Mg_{VSS}$<br>305 $Nm^3_{CH_4}/Mg_{COD}$ | |
| Poultry blood | | 1.67 | 22.00 | 20.02 | 500 $Nm^3_{CH_4}/Mg_{VSS}$<br>100 $Nm^3_{CH_4}/Mg_{FS}$ | Salminen et al. (2003) |
| Poultry carcass | | | 37.00 | | 200–250 $Nm^3_{CH_4}/Mg_{FS}$ | Chen (1999); Chen and Shyu (1998) |
| Poultry litter | | 1.67–2.99 | 52–81 | 32–53 | 140–220 $Nm^3_{CH_4}/Mg_{VSS}$<br>100–150 $Nm^3_{CH_4}/Mg_{FS}$ | Webb and Hawkes (1985) |
| Poultry manure | | 0.92–3.15 | 20–47 | 12–31 | 200–300 $Nm^3_{CH_4}/Mg_{VSS}$<br>40–60 $Nm^3_{CH_4}/Mg_{FS}$ | Huang and Shih (1981); Safley et al. (1987) |
| Poultry feather | | 3.65 | 24.30 | 23.50 | 200 $Nm^3_{CH_4}/Mg_{VSS}$<br>50 $Nm^3_{CH_4}/Mg_{FS}$ | Salminen et al. (2003) |
| Poultry offal, feet, head | | 2.07 | 39.00 | 37.05 | 700–900 $Nm^3_{CH_4}/Mg_{VSS}$<br>300 $Nm^3_{CH_4}/Mg_{FS}$ | Salminen et al. (2003) |
| Poultry trimmings and bone | | 3.44 | 22.40 | 15.23 | 600–700 $Nm^3_{CH_4}/Mg_{VSS}$<br>150–170 $Nm^3_{CH_4}/Mg_{FS}$ | Salminen et al. (2003) |

The composed substrate has to fulfill both: to provide the organic load and to guarantee a nitrogen concentration within the reactor that is not close to the inhibitory level. The highest nitrogen concentrations have been investigated in the blood of cattle, swine, and poultry.

If bones or carcasses are included in the substrate, the high inorganic part may be the limiting step for the use of these waste fraction. As these bone fragments will sediment in the reactor a process strategy has to be implemented. The gas and methane yield for several investigated bones of cattle and bovine varies widely. Thus the meat that is stuck to the bone depends on the cutting and separation process in the slaughterhouse. The investigated cattle leg bone gave a high gas yield (522 $Nm^3_{gas}/Mg_{COD}$) because of the marrow within the bone.

### 9.3.2. Waste Fractions from Rendering Plants and Food Processing Industries

Twenty-five percent of the total weight of slaughtered animals have to be processed according to the regulations for treatment and use and disposal of these slaughtering by-products (Tritt and Schuchardt, 1992). Lots of these wastes as carcasses and blood are processed in rendering plants to output basically market products, solid organic wastes, and wastewater. Some of the market products are not allowed to be used any longer and also have become waste products that have to be disposed. The currently installed processes that transform slaughterhouse waste and animal by-products into new wastes will not be cost and energy effective in some future cases. Other technologies such as anaerobic digestion may be implemented into existing rendering facilities to dispose animal by-products accomplishing following goals:

- hygienization,
- recovery of valuable products,
- recovery of energy, and
- purification of wastewater.

A series of substrate analysis, laboratory, and pilot-scale experiments (Tritt, 1990; Johns, 1995; Kirchmayr, 1998; Huber, 1998) have been undertaken to assess the feasibility of implementing anaerobic digestion technology into rendering plants. The substrate characteristics and measured biogas yields are listed in Table 9.9. In many cases the available wastes in rendering and slaughterhouses do not have constant chemical characteristics. The parameters vary because of changed or inconstant processes. The waste fractions showing the widest variations are the fat sludges from dissolved air flotation (DAF) and filter-cakes from sieving machines. Peak discharges during washing cycles may overload the flotation or sieving units and cause significant changes in the characteristics of the flotation sludges and sieving residues.

The final process step in a rendering plant will be a wastewater treatment unit that includes an aerobic nitrification and denitrification step. This has to be taken

*Table 9.9.* Substrate characteristics from waste fractions at rendering plants and various wastes from meat processing industry

| Substrate | COD (g/kg) | TKN (g/kg) | TS (%) | VSS (%) | Biogas[a] or methane potential; substrate specification | Reference |
|---|---|---|---|---|---|---|
| *Waste fraction from rendering plant* | | | | | | |
| Homogenate | 975 | 24.40 | 42.70 | 39.40 | 1 140 $Nm^3_{Gas}/Mg_{VSS}$ | Kirchmayr (1998) laboratory and 4 $m^3$ pilot scale |
| Blood | 141 | 15.6 | 9.70 | 9.20 | 630 $Nm^3_{Gas}/Mg_{VSS}$ | |
| Concentrated blood | 21.2 | 3.95 | 1.5 | 1.0 | 940 $Nm^3_{Gas}/Mg_{VSS}$ | |
| Stomach and intestinal content | 205 | 6.41 | 16.5 | 13.7 | 680 $Nm^3_{Gas}/Mg_{VSS}$ | |
| Rumen content | 185 | 3.78 | 14.3 | 12.7 | 370 $Nm^3_{Gas}/Mg_{VSS}$ | |
| Exhaust vapors condensate | 15 | 0.55 | 0.05 | 0.05 | Clear, bright yellow liquid | Huber (1998) laboratory and 4 $m^3$ pilot scale |
| Concentrated blood | 17 | 2.23 | 1.33 | 0.8 | Turbid liquid | |
| Blood and wastewater from the hutches | 54 | 5.34 | 4.99 | 4.06 | Viscous, coagulated blood residual | |
| Homogenate | 705 | 18.18 | 36.25 | 33.19 | Viscous, pasty | |
| Concentrated sludge | 114 | 5.94 | 7.64 | 6.87 | Viscous up to gel-like | |
| Squeezed fat sludge | 1,596 | 15.54 | 88.47 | 81.4 | From fluid to solid, depending on temperature | |
| Substrate mixture[b] | 71 | 2.64 | 3.10 | 2.76 | 500–700 $Nm^3_{Gas}/Mg_{COD}$ $CH_4$: Ø 73% of biogas | |
| *Wastes sludges from sieves and fat flotation* | | | | | | |
| Primary flotation sludge | | | 16.00 | 15.00 | 2,400 $Nm^3/Mg_{VSS}$ | Carpentier et al. 2005, laboratory and 2 $m^3$ pilot scale |
| Filter-cake from sieving | | 9.15 | 31.50 | 30.50 | 1,000 $Nm^3/Mg_{VSS}$ | |
| Floated fat after DAF | 100 | 1.27 | 6.12 | 5.43 | | Unpublished results: Authors' own investigations |
| Fat scraper content (cattle slaughter-house) | 214 | 3.09 | 14.98 | 11.39 | | |

| | | | | | | |
|---|---|---|---|---|---|---|
| *Byproducts from meat processing industry* | | | | | | |
| Squeezed stomach from swine | 347 | 18.89 | 15.40 | 14.03 | 750 $Nm^3_{Gas}/Mg_{VSS}$<br>304 $Nm^3_{Gas}/Mg_{COD}$<br>709 $Nm^3_{CH4}/Mg_{VSS}$<br>287 $Nm^3_{CH4}/Mg_{COD}$ | Unpublished results: Authors' own investigations laboratory scale |
| Concentrated animal fat and proteins | 511 | 2,720 | 30.14 | 26.46 | 607 $Nm^3_{Gas}/Mg_{VSS}$<br>376 $Nm^3_{Gas}/Mg_{COD}$<br>314 $Nm^3_{CH_4}/Mg_{VSS}$<br>195 $Nm^3_{CH_4}7Mg_{COD}$ | |
| Category 3 material (swine) | | 16.29 | 38.5 | 37.8 | 1,670 $Nm^3_{Gas}/Mg_{VSS}$ | Carpentier et al. (2005), laboratory and 2 $m^3$ pilot scale |
| Animal glue leather from tannery industry | 217 | 8.71 | 17.10 | 10.69 | 430 $Nm^3_{CH_4}/Mg_{COD}$ | Grasmug (2004), laboratory scale |
| Carcass meal from rendering plant | 1,101 | 82.70 | 95.77 | 64.58 | | |
| Mixed waste 50% poultry blood; 50% OFMSW | 263 | 8.56 | 11.35 | 8.90 | 60 $Nm^3_{CH_4}/Mg_{FS}$<br>$CH_4$: Ø 68% of biogas | Resch et al. (2005), 4 $m^3$ pilot scale |
| Mixed waste 30% conc. animal fat and proteins 70% OFMSW | 330 | 12.17 | 19.81 | 16.12 | 102 $Nm^3_{CH_4}/Mg_{FS}$<br>$CH_4$: Ø 72% of biogas | |

[a] Best results concerning gas production and stability were achieved with a hydraulic retention time (HRT) of 40 days and a volume load ($B_R$) of 1.5 kg $COD/m^3 \cdot d$, respectively.
[b] Composition of mixture: exhaust vapors condensate 76.1%, concentrated blood 10.0%, homogenate 6.6%, blood and wastewater from the hutches 4.1%, sewage sludge 2.7%, fat sludge 0.5%.

into account when the substrate input for the anaerobic digestion plant is composed of available waste fractions. The denitrification process step in the wastewater treatment unit needs an easy bioavailable carbon source to transform nitrate and nitrite into $N_2$. A waste fraction consisting of easy bioavailable COD and low nitrogen concentration may be used for this purpose. Exhaust vapors condensates from the rendering process achieve these requirements.

The anaerobic digestion technology is widely used in organic waste treatment from households [organic fraction of municipal solid waste (OFMSW)], industrial, and catering facilities (Mata-Alvarez, 2000). Treatment plants that are licensed to accept the mentioned category 3 wastes also may use category 3 animal by-products from slaughterhouses or meat processing industries. As some fractions of animal by products may cause severe problems in the biological treatment process as single substrate the co-digestion with other waste streams may be a favorable option (Grasmug, 2004). The OFMSW is an easily degradable material with low nitrogen contents and therefore is a highly suitable cosubstrate for animal by-products and wastes from meat industry. Results from laboratory and pilot experiments are listed in Table 9.9.

## 9.4. TECHNICAL SCALE EXPERIENCES WITH THE ANAEROBIC DIGESTION OF SLAUGHTERHOUSE WASTES

Since the ABP-Regulation (EC No. 1774/2002) is defining new treatment possibilities and laying down corresponding processing parameters, new pathways for the utilization of slaughterhouse wastes were opened. Some biogas plants in the agroindustrial sector are using pasteurized ABPs as cosubstrates (together with manure, catering waste, and other energy crops). There is only fragmented information available about the limits of the utilization and the behavior of high-protein-containing wastes for anaerobic digestion in laboratory, pilot, and technical scale (Brachtl, 2000; Carpentier et al., 2005; Resch et al., 2005; Kirchmayr, 1998; Grasmug, 2004).

### 9.4.1. Anaerobic Digestion of Slaughterhouse Waste in an Agricultural Biogas Plant

In the following, experiences with a full-scale plant using animal by-products deriving from a pig slaughtering facility processing 16–20 t/d of slaughterhouse wastes will be described. Colon, colon content, blood, and fat scrubber content, stomach content, and process water are used as substrates. The plant originally was designed and constructed like a ''classic'' agricultural biogas plant. Two main fermenter (HRT = 100 d in fermenter 1, HRT = 50 d in fermenter 2) and a post-fermentation tank (without heating, HRT = 40 d) were constructed. The digestate was pasteurized (70°C/60 min) after the anaerobic digestion and subsequently stored in the storage tank. The digestate was used as fertilizer.

*Table 9.10*. Characteristic parameters of the sludge in the fermenter of a biogas plant (monitoring interval: 1 year)

| Sample | | pH | TSS (%) | VSS (mg/L) | TKN (mg/L) | $NH_4$-N (mg/L) | UAN 37°C (mg/L) |
|---|---|---|---|---|---|---|---|
| Fermenter 1 | min | 7.60 | 2.59 | 1.94 | 6,668 | 5,867 | 333 |
| | max | 8.25 | 3.33 | 2.67 | 8,172 | 7,069 | 1,154 |
| Fermenter 2 | min | 7.87 | 2.49 | 1.87 | 7,488 | 6,494 | 518 |
| | max | 8.30 | 2.75 | 2.12 | 8,093 | 7,120 | 1,269 |
| Postfermentation | min | 7.96 | 2.10 | 1.55 | 7,125 | 6,115 | 588 |
| | max | 8.31 | 2.44 | 1.81 | 7,944 | 6,972 | 1,179 |
| Storage tank | min | 7.89 | 1.51 | 1.04 | 6,120 | 5,300 | 581 |
| | max | 8.56 | 4.81 | 3.16 | 7,742 | 7,584 | 1,782 |

Due to a start-up without monitoring, several serious process failures were observed. Hand in hand with process instability, heavy foaming together with the accidental escape of fermentation liquid occurred. The subsequent land application of the (not completely metabolized) digestate contributed to the range of increased odor emissions.

In order to ensure proper operation, the biogas plant concept was redesigned. In section 6 the concept to reduce odor emissions deriving from the processing unit is described.

Previous to anaerobic digestion, the substrate is pasteurized. With this measure the anaerobic digestion will be situated at the ''clean side'' of the process (concerning the requirements of the ABP-Regulation) and the manipulation of the sludge is much easier.

Before feeding into fermenter 1 and fermenter 2 (in parallel) the substrate is further chilled down to 50°C in order to minimize damage of the bacterial biomass in the biogas fermenter. Both fermenters are operated at mesophilic temperatures (35°C). Higher temperatures will increase the amount of unionized ammonia ($NH_3$, UAN) in the system. Table 9.10 shows the calculated UAN concentrations in the specific biogas plant. Levels between 300 and 1,700 mg/liter are close or beyond reported inhibition values (Koster, 1986; Koster and Koomen, 1988; Angelidaki and Ahring, 1994). Calculated on the basis of Eqs. 1 and 2 (Lide, 1993) the increase in temperature will cause an increase in unionized ammonia concentration by 43% (at pH = 8). This in turn could probably lead to a much higher level of inhibition.

According to the high water content of the substrate mixture the total nitrogen content in the fermenter is in the range of 5.8 to 7.5 g/liter. In agricultural cofermentation plants using maize corn as main substrate, ammonia values in the same range have been measured (IFA-Tulln, 2005) apparently without any toxic influence.

$$(1) \qquad \frac{[NH_3]}{[NH_3] + [NH_4^+]} = \left[\frac{1}{1 + 10^{pKa-pH}}\right]$$

$$(2) \qquad pKa = 4 * 10^{-8} * T^3 + 9 * 10^{-5} * T^2 - 0{,}0356 * T + 10{,}072$$

The applied organic loading rate was limited to a maximum of 3.5 kgCOD/$m^{3*}$d. Laboratory-scale experiments to investigate the biodegradability of slaughterhouse wastes under anaerobic conditions showed that an organic loading rate exceeding 2.5 kgCOD/$m^3$.d may cause process stability problems (Kirchmayr et al., 2002).

After reinoculation and operation at low organic loading rate the process is quite stable, although the concentrations of free volatile fatty acids are very high (see Table 9.11). The concentrations of acetic and propionic acid are reduced in the secondary fermenter and storage tank from 5000 to 9000 mg/liter to approx. 1000 to 2000 mg/liter. The organic loading rate may not be considered as very high, but as shown in Section 6.2 the end product of the anaerobic digestion may still cause some annoyances due to odor emissions.

Considering the reasons for the incomplete metabolization only little information is available and different hypotheses about the toxic effect of intermediate metabolites have been described. On the one hand the high concentration of unionized ammonia may cause toxic effects (Koster, 1986; Koster and Koomen, 1988; Angelidaki and Ahring, 1994), on the other hand different intermediate metabolites like phenol, p-cresol, or indole also may cause inhibitory effects (Behmel et al., 1994; Field et al., 1987; Griehl et al., 2002). Higher concentrations of volatile fatty acids also may be identified as the source of inhibitory effects (Kroeker et al., 1979; Duarte et al. 1982; Ahring, 1995).

Further minor process instabilities may be explained as a result of elevated concentrations of cleaning and disinfecting agents contained in the fat scrubber fraction. Also, a partial overload of the flotation device may have caused process instability due to a slight hydraulic overload of the anaerobic reactors.

### 9.4.2. Conclusion

Under the described circumstances the stable operation of a biogas plant using only slaughterhouse waste as substrate is feasible. Strong attention should be paid to the

*Table 9.11.* Volatile fatty acid concentrations in the biogas reactors with slaughterhouse waste as substrates (monitoring interval: 1 year)

| Sample | | Acetic acid (mg/L) | Propionic acid (mg/L) | i-Butyric acid (mg/L) | Butyric acid (mg/L) | Valeric acid (mg/L) | i-Valeric acid (mg/L) | VFA total (g/L) | UVA[a] total (mg/L) |
|---|---|---|---|---|---|---|---|---|---|
| Fermenter 1 | min | 5,196 | 3,539 | 339 | 104 | 893 | 60 | 11.71 | 3.35 |
| | max | 9,451 | 6,235 | 1,213 | 500 | 2,361 | 578 | 19.04 | 23.35 |
| Fermenter 2 | min | 2,784 | 4,240 | 172 | 0 | 331 | 52 | 8.70 | 2.53 |
| | max | 7,832 | 6,425 | 1,006 | 268 | 1,997 | 458 | 16.06 | 9.25 |
| Postfermentation | min | 775 | 1,130 | 39 | 0 | 36 | 0 | 4.34 | 1.54 |
| | max | 4,682 | 6,294 | 529 | 111 | 819 | 162 | 10.36 | 4.59 |
| Storage tank | min | 409 | 140 | 24 | 0 | 17 | 16 | 0.98 | 0.28 |
| | max | 3,370 | 3,141 | 228 | 112 | 923 | 179 | 3.55 | 2.94 |

[a] Calculated as acetic acid.

odor emissions caused by the operation of the biogas plant and by the land application of the digestate. To ensure higher metabolization rates during anaerobic digestion and in order to utilize the microbial activity in the storage tank for further stabilization of the digestate the hygienization unit (pasteurization) should be located before anaerobic digestion and not afterwards.

The anaerobic monodigestion of high nitrogen-containing animal by-products like slaughterhouse wastes causes a very weak equilibrium in the process. Small mistakes may easily cause process failure. Proper training of the plant operator and careful process control (e.g., trough the measurement of volatile fatty acids, ammonia content, and pH) is absolutely necessary in order to ensure long-term digestion operation. Further investigation is necessary to determine which process parameters have to be changed (e.g., nitrogen removal in a bypass stream) in order to increase the reported insufficient degradation rate.

### 9.4.3. ATZ Thermal Pressure Hydrolysis (TPH) of Animal By-products

In this section the application of ATZ thermal pressure hydrolysis (TPH) followed by anaerobic digestion is described (Prechtl et al., 2002, 2003). The TPH process with subsequent anaerobic digestion has been so far successfully explored concerning the recycling of sewage sludge (Hertle and Renner, 1994; Tippmer, 1994; Chwistek et al., 1997) and organic fraction of sorted municipal solid waste (ATZ Entwicklungszentrum, 1997). The sterilization in the TPH reactor (temperature about 230°C, pressure 20–30 bar) results in the splitting of the organic polymers by hydrolysis with the guarantee of a complete hygienization.

After mashing with process water the hydrolysate is converted into biogas in a fixed-bed-loop-reactor. The biogas is used to supply the thermal energy for the TPH process and for $NH_3$-stripping.

Due to good degradation performance in the anaerobic step there is only a small amount of solid residue left, which may easily be dewatered and disposed through incineration. Compared to the existing rendering plants less wastewater is produced due to the process water cycle and the loss of water through the step of $NH_3$-stripping and biogas production. The excess wastewater has a typical composition like a normal wastewater from anaerobic waste fermentation plants, with COD concentrations between 4 to 7 $kg/m^3$ after conditioning.

After pretreatment with the TPH process it is possible to run a stable mono-fermentation of animal by-products with a high organic loading rate of about 10 kg $COD/m^{3*}d$. According to the methane content of up to 77% in the biogas and a biogas yield between 200 to 300 $m^3/mg$ raw material, a combined heat and power station produces up to 780 kWh electric energy from 1,000 kg of raw material.

An anaerobic biofilm reactor with a volume of 85 $m^3$ was used for biogas production from the hydrolysate produced in the TPH pretreatment. With this process a reduction of 80–94% $COD_{dissolved}$ could be achieved. Figure 9.3 shows a flow diagram of the industrial-scale application.

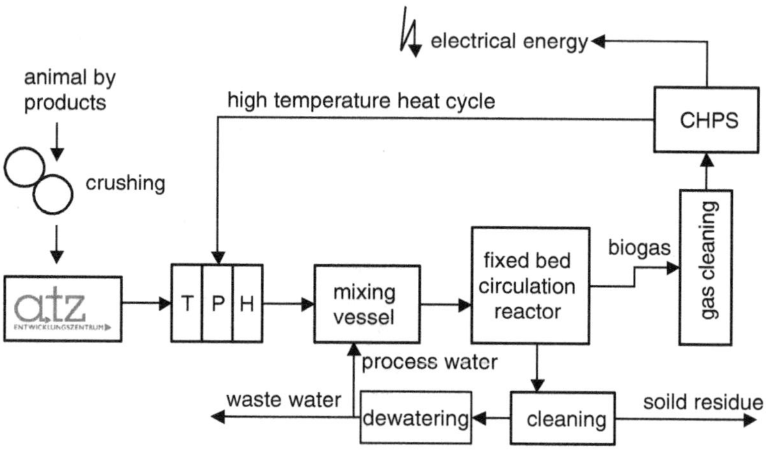


**Figure 9.3.** Flow diagram of the process combination ATZ-TPH and digestion of animal by-products.

Contrary to the present production of meat and bone meal, which consumes fossil energy, the process combination of TPH and anaerobic digestion is able to produce renewable energy from animal by-products. The biogas generated from the organic share of the animal by-products by fermentation can be used to produce an excess of thermal and electric energy for the whole process.

Since May 2005 a full-scale TPH plant, with a capacity of about 40,000 mg/year, for the treatment of manure and renewable substrates is in operation in Germany. The plant was constructed by the R. Scheuchl company, a licensee of ATZ Entwicklungszentrum.

## 9.5. NITROGEN REMOVAL IN THE ANAEROBIC DIGESTION

Anaerobic digestion effluents and its subsequent handling now are gaining more interest not only because plant sizes are growing significantly, but also because of rising environmental interest and politics, and thus implementation of stricter limits throughout the European Union. But effluent purification also is interesting in terms of on-site liquid recycling. Many industrial wastes have to be diluted, suspended, pulped, etc., in order to condition the waste stream for anaerobic digestion. As some slaughterhouse waste streams have extremely high nitrogen contents dilution is a proper way to lower the nitrogen concentration in the reactor (Siegrist et al., 2005). Actually, the following treatment strategies are presently implemented, whereas agricultural use is a final depose, the other technologies provide regenerated water applicable for further use on-site:

- Agricultural use as fertilizer
- Biological treatment

- Magnesium ammonium phosphate precipitation (MAP)
- Ammonia stripping

In the following section the treatment concepts will be described briefly and the effluent quality after treatment will be outlined.

### 9.5.1. Agricultural Use as Fertilizer

Agricultural use of digestion effluent is the only treatment approach where the whole stream, including the liquid and the solid fraction, is disposed. In some European countries this is a common approach, regulated by the amount of nitrogen applied to the field. Thus the council Directive 91/676/EEC is limiting the use of nitrogen fertilizer to 170 kg N/ha/a. In terms of costs this solution can be a very cheap one, but disadvantages are that huge storage vessels have to be built and the use is laborious. For larger anaerobic digestion plants or for industrial plants handling large volumes, agricultural use will not pose a solution. Particularly for larger amounts of digester effluent, fertilizer use is not appropriate as the corresponding demand on the agricultural area will be enormous. Along with the rising demand for agricultural area, the transportation costs will rise dramatically.

### 9.5.2. Biological Nutrient Removal

Biological nutrient removal by means of carbon, nitrogen, and phosphor removal (Obaja et al., 2003) is a well know and common practice at least when municipal wastewater is treated (Verstraete and Philips, 1998). Treatment of ammonium-rich wastewater in industrial application on the other hand is still under way. One challenge is the potentially inhibitory characteristics of ammonium-rich wastewater on the biological process (Brond and Sund, 1994), another is the poor carbon to nitrogen ratio (Ra et al., 2000).

As far as the inhibitory effects (Anthonisen et al., 1976) are concerned a smart treatment concept will overcome these difficulties. Only the initial break in phase attention has to focus on the special wastewater property. With a smooth start-up high loaded industrial wastewater also can be treated, even more so as activated sludge systems are well adaptable.

For denitrification an organic carbon source is mandatory. A poor carbon to nitrogen ratio therefore will result in reduced denitrification (Elefsiniotis et al., 2004). With optimal nitrification an accumulation of nitrate or nitrite will take place. If a lack in carbon source is identified, treatment concepts are proposed using external carbon source. Such systems are realized in advanced biological treatment mainly following the sequencing batch reactor (SBR) concept (Vallés-Morales et al., 2004).

All together these processes comprise two steps of treatment (Carrera et al., 2003). The first step is oxidation of the ammonium to nitrite or nitrate followed by denitrification.

The two steps of biological nitrification (Pambrun et al., 2004):

(3) Nitritation: $NH_4^+ + 1.5\ O_2 \rightarrow NO_2^- + 2H^+ + H_2O$

ammonia-oxidizing bacteria

(4) Nitratation: $NO_2^- + 0.5\ O_2 \rightarrow NO_3^-$

nitrite-oxidizing bacteria

Subsequent to nitrification, denitrification takes place. During an anoxic period, nitrite and nitrate are used as an oxygen source. The $N_2$ builds up and is released to the environment. Thus the total nitrogen concentration in the anaerobic digester effluent is reduced, making the liquid ready for recycling.

Recently, apart form classic denitrification anammox plants (Fux et al., 2002), anaerobic ammonium oxidation also is applied. Thereby ammonium is converted to nitrogen gas under anoxic conditions with nitrite as the electron acceptor (Mulder et al., 1995).

### 9.5.3. Magnesium Ammonium Phosphate Precipitation (MAP)

Crystallization of ammonium and phosphate through MAP from anaerobic digestion effluents is implemented not only for removing these nutrients, but also for recovering them (Doyle and Parsons, 2002; Uludag-Demirer et al., 2005; Battistoni et al., 2000). Hence, this not only contributes to minimizing the environmental impacts but also contributes to the increasing global demand for nitrogenous fertilizer (Mulder, 2003) and the limited phosphorus rock reserves (Steen, 1998). Struvite precipitation has already been applied to various wastewaters such as swine waste (Burns et al., 2001), tanning factories (Tunay et al., 1997), and anaerobic supernatant (Yoshino et al., 2003) and follows the general and simplified reaction outlined in Eq. 5.

(5) $$Mg^{2+} + PO_4^{3-} + NH_4^+ \Leftrightarrow MgNH_4PO_4 * 6H_2O$$

For $Mg^{2+}$ various sources can be used, like $Mg(OH)_2$, MgO, $MgCl_2 * 6H_2O$, and others. Depending on the initial molar ratio of Mg:N:P (Altinbas et al., 2002) an addition of a phosphorus source also may be necessary. For optimal results the pH has to be adjusted (Nelson et al., 2003), but this results in a rise in salinity. This elevated salinity has to be considered when opting for an internal recycling of the liquid. As far as final discharge is concerned, a subsequent treatment is obligatory in order to meet the discharge quality as the total carbon load needs to be considered.

### 9.5.4. Ammonium Stripping

Prior to stripping of ammonium the suspended solid concentration should be lowered and the pH of the supernatant has to be increased. The higher the pH, the more the equilibrium of ammonium to ammonia is shifted toward $NH_3$. That is necessary for economical operation of the stripper. After this pretreatment step, free ammonia ($NH_3$) is stripped off by air or steam (Siegrist, 1996).

Stripping of volatile compounds is a standard industrial application, but stripping anaerobic digester effluent needs further care. The high suspended solids concentration especially has to be considered. Like MAP precipitation, ammonium stripping barely affects the total COD load, and therefore not sufficient for direct wastewater discharge. But in terms of internal recycling of the liquid, this technology is highly suitable and furthermore a valuable product, ammonium concentrate, can be obtained and sold. Similarly to MAP precipitation, ammonium stripping is only possible with the usage of chemicals resulting in a significant rise in salinity.

### 9.5.5. Purification Grade

The purification grade not only depends on the chosen technology but also on the quality of the substrate used for anaerobic digestion (Figure 9.4).

As the quality as well as the quantity of digester effluent varies tremendously from plant to plant (Graja and Wilderer, 2001) there is no first-rate treatment technology. Depending on the desired quality, different technologies have to be chosen and in some cases even a combination of them will be beneficial. In Figure 9.4 the average effluent qualities are outlined. COD is hardly affected by MAP precipitation and ammonium stripping, total nitrogen removal is in the range of 80 to 90%. Biological treatment without the use of an external carbon source will

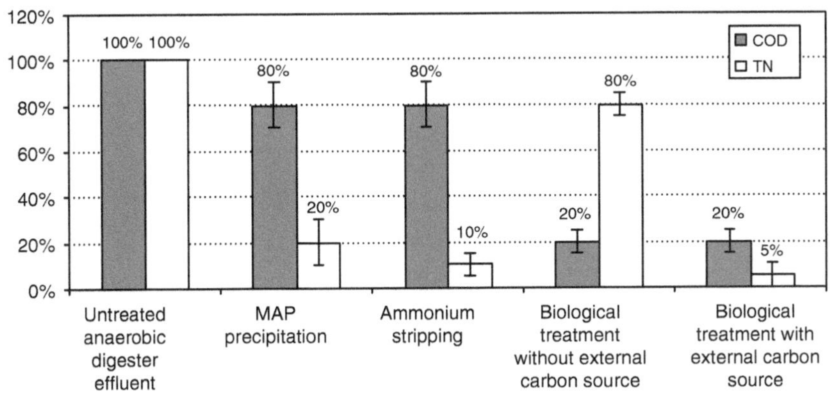


**Figure 9.4.** Effluent qualities after different treatment technologies by means of COD and TN, calculated on the basis of untreated anaerobic digester effluent. The different treatment options account for different purification levels. The figure shows that COD is nearly unaffected by MAP precipitation and ammonium stripping. Best result can be achieved with biological technologies (Kabdasli et al., 2000; Graja and Wilderer, 2001).

largely eliminate the COD, but the nitrogen content is not significantly affected (Sharma and Ahlert, 1977). Biological treatment with external carbon source results in up to 100% removal on the condition that the activated sludge is totally removed.

### 9.5.6. Conclusion

On a case-by-case basis the optimal solution will be either a single-treatment process or a combination of several ones. For internal recycling of the liquid for dilution, suspending, pulping, etc., chemical–physical but even more so biological treatment approaches are suitable. For final discharge only a biological treatment meets the legal limits.

Moreover, the ''end of the pipe'' approach will not always be the best, especially when dealing with high nitrogen-loaded waste streams like slaughterhouse wastes. Particularly unionized ammonium nitrogen (UAN) will create problems during anaerobic fermentation, as UAN is inhibits the biological processes. Thus ammonium removal already in the input material will have some beneficial effects: higher loading rates of the anaerobic digester, more stable biological processes, and effluent with a dramatically reduced nitrogen load, simplifying further treatment steps.

The implementation of the stripping procedure in full scale prior to the anaerobic digestion is not reported, but research work is ongoing. Figure 9.5

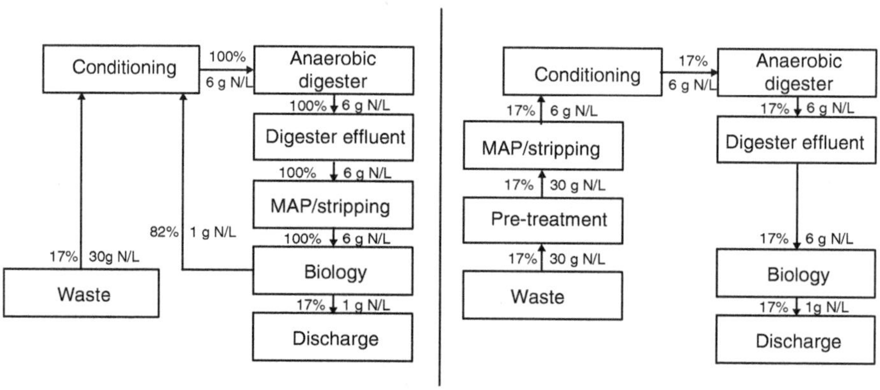


**Figure 9.5.** The left flow schema shows a treatment concept with postnitrogen removal. The right flow schema shows a treatment concept with prenitrogen removal. The % values indicate the mass flow, with the appropriate nitrogen concentration in grams nitrogen per liter. The difference between post- and pretreatment is that with pretreatment only 17% with high nitrogen content has to be treated, whereas with posttreatment 100% with a relatively low concentration need treatment. Thus the installed treatment capacity has to be accordingly larger. In both systems the nitrogen concentration in the anaerobic digester is 6 g N/L, in the left schema by 5.8-fold dilution with cleaned effluent, in the right schema by nitrogen removal by MAP/stripping.

shows the different flow schemas. In the input material the greater part of the nitrogen is organically bound to the substrate matrix. Hence prior to the nitrogen removal process an adequate pretreatment process has to be installed, decomposing the organic matrix and releasing the nitrogen as ammonium. A further benefit when removing the nitrogen load from the input material is that the high loaded waste streams can be treated. Hence the removal will work more efficiently as the concentration difference is higher, resulting in a reduction of costs.

## 9.6. ODOR EMISSION

Several biogas plants processing animal by-products such as slaughterhouse wastes and catering waste are facing opposition in residential areas due to massive odor emissions. In the following, the concept to prevent odor emissions deriving from the processing unit is presented. Odor emissions do not only derive from the processing unit, but also the digestate used as fertilizer can cause annoyances.

### 9.6.1. Odor Emissions from the Processing Unit

The following explanations are focused on the description of the odor emissions from a biogas plant processing slaughterhouse wastes such as intestines, blood, and fat scrubber content. Beside extensive problems concerning the technical process (in the first stage stable fermentation could not be achieved), massive and extensive odor problems occurred already in the start-up phase. The first redevelopment (reduction of the organic loading rate, treatment of the waste air deriving from the central storage tank, and an outdoor storage tank with container biofilters) did not result in a reduction of the complaints about odor emissions.

Olfactometric emission measurements showed extremely high concentrations of odorous substances and loads especially in the central storage tank. The concentration of $H_2S$ (up to approx. 160 ppm) and organic sulfur compounds like dimethyl sulfide, dimethyl disulfide, etc. (approx.50 $mg/m^3$) in the raw waste air of the pump storage and the central storage tank lead to distinctive acidification of the biofilter materials and further to insufficient deodorization (Figure 9.6). Similarly also the diffuse odor emissions deriving from the gas engine cooling air caused by unsuitable sealing were noticeable. The emissions coming from the open storage tank could not be quantified. Estimation on laboratory-scale (adjustment of equation) showed a high potential for odor emissions.

A reclamation concept developed on the basis of the described measurements (Figure 9.7) specified the separated collection of all highly concentrated exhaust air streams and their usage as combustion air for the gas engines after alkaline washing. Reduced charged waste air streams will be collected and treated in available biofilter systems. To avoid further emissions of the open storage tank the liquid surface was covered with shredded straw.

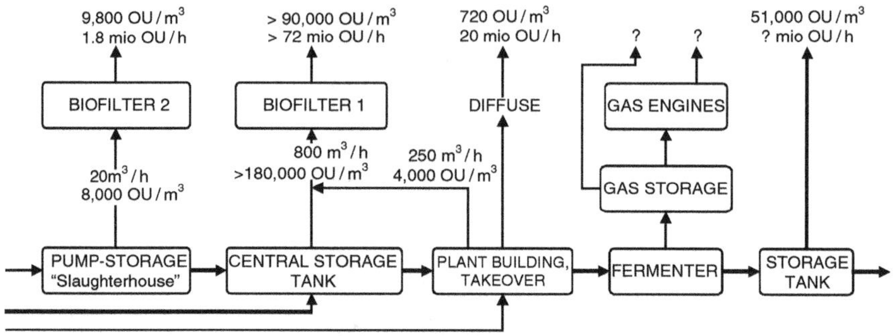


**Figure 9.6.** Olfactory emissions after the first reclamation trial.

Subsequent checkups confirmed the success of the reclamation strategy. High odorous substance loads coming from the central storage tank and hygienization tank can be effectively reduced by using the gas engines for thermal treatment of the exhaust air. Nitrogen oxide aromas dominate the olfactory odor tone of the exhaust air coming from the gas engines. Raw gas smells were not detectable any more.

The application of heavily loaded waste air streams is not assumed to essentially influence the odor emissions of the gas engines. The estimated odor emissions of the gas motors are remarkably high, probably due to high $NO_X$ emissions of more than 2000 mg/m$^3$ (remark: Austrian emission threshold value: 400 mg/m$^3$$NO_2$). At the time of recording the data both biofilters were operating with approx. 35% of their capacity (in order to generate good degradation performance).

The main causes of annoyance should now be the emissions from the open storage tank and the manipulation of the digestate. Thus subsequently covering the storage tank (reclamation step 2) seems necessary.

To avoid widespread annoyance in a residential area the operation of biogas plants using slaughterhouse wastes as substrates can only be achieved by:

- Storing the substrate and digestate in closed containers, which are connected to the gas system (storage tank) or ventilated sufficiently with subsequent proper treatment of the waste air.
- The emerging waste air including the waste air from hygienization and pasteurization facilities should not be treated in biofilter systems because of the high content of inorganic and organic sulfur compounds. The best solution seems to be to utilize the waste air as combustion air in the existing gas engines (after pretreatment in alkaline scrubbers). As experience shows the engine operation is not noticeably affected as long as the waste air fraction does not exceed 50% of the combustion air.
- Performing all other takeover and manipulation activities (including manipulation of the digestate) in closed, sufficiently ventilated buildings (target

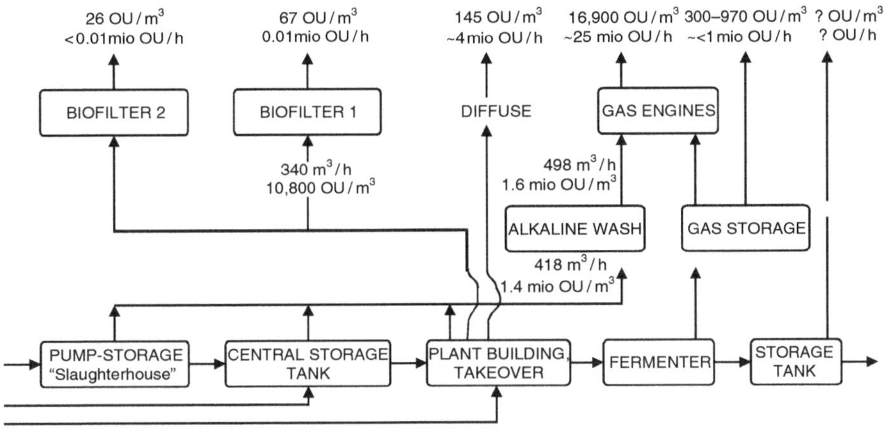


**Figure 9.7.** Olfactory emissions after implementation of reclamation step 1.

air change number 2–5 to prevent diffuse olfactory emissions). The waste air from the manipulation hall is less odorous, and thus a treatment in biofilter systems seems to be adequate.

- In general all activities that lead to surface contamination with malodorous substances should be minimized (immediate removing of puddles, frequent cleaning, etc.).

### 9.6.2. Odor Emission during Land Application of Slaughterhouse Digestates

Odor emissions are not only reported from the processing unit, the land application also may result in noticeable odor annoyances. In order to understand land application annoyances two aspects have to be considered: the odor emission concentrations and the hedonic odor tone, i.e., its position on the ''pleasant–unpleasant scale.''

Figure 9.8 gives an overview of the odor emissions observed during test applications of manure and some biogas digestates. The odor emissions of digested slaughterhouse wastes (in this particular case good mineralized, VFA concentrations below 1 g/liter) are comparable with the odor emission deriving from pig and cattle manure (Table 9.12 and Figure 9.8). In this test series only the emissions from incomplete digested catering waste (VFA concentration approx. 13 g/liter) are worse.

Comparing the hedonic odor tone of the different effluents of biogas plants (Table 9.13) it appears that the smell of the digestate of slaughterhouse waste is characterized as very disagreeable (remark: in this case: 7.9 g/liter VFA). Three out of four persons tested characterized the smell as galling and one as very disagreeable. The same result also may be obtained with pig manure. Although

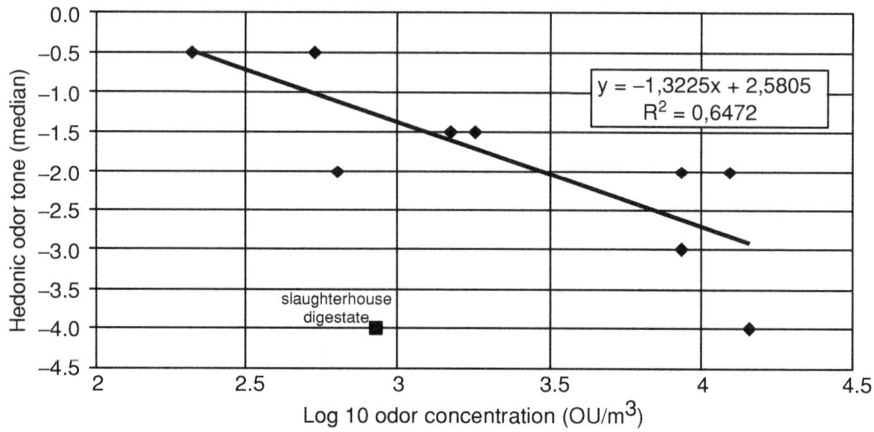


**Figure 9.8.** Dependence of hedonic odor tone and dilution of different manures and biogas digestates (Weber–Fechner law).

*Table 9.12.* Soil application of odorous effluents of biogas plants (gas collection: 16.2 $m^3/m^2 * h$)

| | Time after application (soil incorporation 4 h after application) | | | | | |
|---|---|---|---|---|---|---|
| | 2 | 6 | 24 | 72 | 168 | Applied digestate[a] |
| Substrate | Odor concentration air over soil ($OU/m^3$) | | | | | ($m^3$/ha) |
| Cattle manure | 94 | 30 | 59 | 34 | 34 | 29.4 |
| Pig manure | 164 | 38 | 33 | 20 | 43 | 30.3 |
| Effluent slaughterhouse waste | 118 | 16 | 25 | 22 | 22 | 16.4 |
| Effluent catering waste | 750 | 64 | 31 | 31 | 25 | 14.3 |
| Effluent energy crops | 49 | 26 | 29 | 22 | 19 | 32.3 |

[a] According to a single application of approx. 100 kg N per ha.

the digestate (not well digested) of catering waste is emitting high amounts of odor units (Table 9.12) the smell is not characterized as that offensive.

Generally the experiments show a correlation between odor concentration and hedonic odor tone following Weber–Fechner law except the slaughterhouse digestate produced a very unpleasant odor characteristic at comparably low odor concentrations.

Odor emissions deriving from good mineralized slaughterhouse waste should not exceed odor emissions of pig manure during land application. With a weak reactor performance (corresponding with high VFA concentrations) a much higher level of annoyance deriving from the digestate cannot be excluded.

*Table 9.13.* Hedonic characterization of biogas plant effluents. Four persons nominating the corresponding hedonic characterization

| | Hedonic characterization | | | | | | | | | |
|---|---|---|---|---|---|---|---|---|---|---|
| | galling | | not pleasant and not unpleasant | | | | | | delightful | Odor conc. |
| | −4 | −3 | −2 | −1 | 0 | 1 | 2 | 3 | 4 | (OU/m³) |
| Pig manure | 3 | 1 | | | | | | | | 14,500 |
| Pig manure II | | | 3 | 1 | | | | | | 12,400 |
| Cattle manure | | | | 2 | 1 | 1 | | | | 210 |
| Effluent slaughterhouse waste | 3 | 1 | | | | | | | | 850 |
| Cofermentation I | | 1 | 1 | 1 | 1 | | | | | 1,800 |
| Cofermentation II | | 2 | | 2 | | | | | | 8,600 |
| Energy crops + manure | | | | 2 | 2 | | | | | 540 |
| Cofermentation III | | 1 | 3 | | | | | | | 640 |
| Cofermentation IV | | 3 | 1 | | | | | | | 8,600 |
| Catering waste | | 2 | 2 | | | | | | | 1,500 |

## 9.7. ABBREVIATIONS

| | |
|---|---|
| ABP | Animal by-products |
| BSE | Bovine spongiform encephalopathy (mad cow disease) |
| Cat. | Category |
| CFU | Colony forming units (number of bacteria) |
| COD | Chemical oxygen demand |
| conc. | Concentration |
| DAF | Dissolved air flotation |
| EC | European Community |
| FS | Fresh substance |
| HRT | Hydraulic retention time |
| MAP | Magnesium ammonium phosphate |
| OFSMW | Organic fraction of municipal solid waste |
| TKN | Total Kjeldahl nitrogen |
| TSE | Transmissible spongiform encephalopathy (e.g. BSE) |
| TSS | Total suspended solids |
| VFA | Volatile fatty acids |
| VSS | Volatile suspended solids |

## 9.8. REFERENCES

Ahring, B. K., Sandberg, M., Angelidaki, I., 1995. Volatile fatty acids as indicators of process imbalance in anaerobic reactors, Applied Microbiology and Biotechnology, **43**, 559–565.

Altinbas, M., Yangin, C., and Ozturk, I., 2002, Struvite precipitation from anaerobically treated municipal and landfill wastewaters, *Water Sci Technol* **46**:271.

Angelidaki I., and Ahring B. K., 1994, Anaerobic thermophilic digestion of manure at different ammonia loads: effect of temperature, *Water Res* **28**(3):727–731.

Anthonisen, A. C., Loehr, R. C., Prakasam, T. B. S., and Srinath, E. G., 1976, Inhibition of nitrification by ammonia and nitrous acid, *Journal WPCF* **48**(5):835.

ATZ Entwicklungszentrum, 1997, Neue Möglichkeiten der Verwertung von Biogenen, *Blick durch Wissenschaft und Umwelt* **11**:36.

Battistoni, P., Pavan, P., Prisciandaro, M., and Cecchi, F., 2000, Struvite crystallization: a feasible and reliable way to fix phosphorus in anaerobic supernatants, *Water Res* **34**:3033.

Behmel, U., Leupold, G., and Soenargo, F., 1994, Production of biogas from vegetable wastes. Part 2. Phenols and phenylcarboxylic acids in the process water from spent grain, *Chem Mikrobiol Technol Lebensm* **16**(1/2):49–61.

Brachtl E., 2000, Pilotversuch zur Cofermentation pharmazeutischer Proteinabfälle mit Rinderjauche, diploma thesis, IFA-Tulln, BOKU-University, Vienna.

Brond, S., and Sund, C., 1994, Biological removal of nitrogen in toxic industrial effluents, high in ammonia, *Water Sci Technol* **29**:231.

Burns, R., Moody, L., Walker, F. and Raman, D., 2001, Laboratory and in situ reductions of soluble phosphorus in swine waste slurries, *Env Technol.* **22**:1273.

Carpentier J., Platteau W., Vanwalleghem J., Steenhoudt D., and Verstraete W., 2005, Anaerobic digestion of solid slaughterhouse waste: potential of renewable energy for Belgium, in: *Proceedings of the 4th Symposium on Anaerobic Digestion of Solid Waste*, Copenhagen, pp. 649–655.

Carrera, J., Baeza, J. A., Vicent, T. and Lafuente, J., 2003, Biological nitrogen removal of high-strength ammonium industrial wastewater with two-sludge system, *Water Res* **37**(17):4211.

Chen, T.-H., 1999, Anaerobic treatment of poultry mortalities, in: *Proceedings of the 2nd International Symposium on Anaerobic Digestion of Solid Wastes*, J. Mata-Alvarez, A. Tilche, F. Cecchi, ed., Barcelona, pp. 69–72.

Chen, T.-H., and Shyu, W.-H., 1998, Chemical characterization of anaerobic digestion treatment of poultry mortalities, *Bioressource Technol* **63**:37–48.

Chwistek, M., Jung, R., and Bischof, F., 1997, Minimierung der Klärschlammmenge - Das Verfahren der temperatur-aktivierten Druckhydrolyse, *GWA Gewässerschutz Wasser Abwasser* **3**:168.

*Commission Regulation (EC) No 208/2006 of 7 February 2006 amending Annexes VI and VIII to Regulation (EC) No 1774/2002 of the European Parliament and of the Council as regards processing standards for biogas and composting plants and requirements for manure* (OJ L 36, 02/02/2006 P. 0025–0036).

*Commission Regulation (EC) No 657/2006 of 10 April 2006 amending Regulation (EC) No 999/2001 of the European Parliament and of the Council as regards the United Kingdom and repealing Council Decision 98/256/EC and Decisions 98/351/EC and 1999/514/EC (OJ L 116, 29.4.2006, P. 9–13)*

*Commission Regulation (EC) No 808/2003* of May 12, 2003 amending Regulation (EC) No 1774/2002 of the European Parliament and of the Council laying down health rules concerning animal by-products not intended for human consumption (OJ L 117/1,13/05/2003 P.0001–0009).

*Commission Regulation (EC) No 811/2003* of May 12, 2003 implementing Regulation (EC) No 1774/2002 of the European Parliament and of the Council as regards the intra-species recycling ban for fish, the burial and burning of animal by-products and certain transitional measures (OJ L 117/14,13/05/2003 P.0014–0018).

*Commission Regulation (EC) No 197/2006 of 3 February 2006 on transitional measures under Regulation (EC) No 1774/2002 as regards the collection, transport, treatment, use and disposal of former foodstuffs (OJ L 32, 4.2.2006, P. 13–14).*

*Commission Regulation (EC) No 92/2005 of 19 January 2005 implementing Regulation (EC) No 1774/2002 of the European Parliament and of the Council as regards means of disposal or uses of animal by-products and amending its Annex VI as regards biogas transformation and processing of rendered fats (OJ L 19, 21.1.2005, P. 27–33).*

*Commission Regulation (EC) No 181/2006 of 1 February 2006 implementing Regulation (EC) No 1774/ 2002 as regards organic fertilisers and soil improvers other than manure and amending that Regulation (OJ L 29, 2.2.2006, P. 31–34)*

Council Directive *79/542/EEC* of December 21, 1976 drawing up a list of third countries from which the Member State authorize imports of bovine animals, swine and fresh meat (OJ L 146,14/06/1979 P.0015–0017).

Doyle, J. D. and Parsons, S. A., 2002, Struvite formation, control and recovery, *Water Res* **36**(16):3925.

Drescher, H. P., 1998, Neue Verfahrenskombination Hydrolyse - Verbrennung zur Klärschlammaufbereitung, *GWA Gewässerschutz Wasser Abwasser* **165**:47/1.

Duarte, A. C. and Anderson, G. K., 1982, Inhibition modelling in anaerobic digestion. *Water Sci Technol* **14**:749–763.

Elefsiniotis, P., Wareham, D. G., and Smith, M. O., 2004, Use of volatile fatty acids from an acid-phase digester for denitrification, *J Biotechnol* **114**:289.

Field, J. A., Lettinga, G., and Geurts, M., 1987, The methanogenic toxicity and anaerobic degradability of potato starch wastewater phenolic amino acids, *Biol Wastes* **21**:37–54.

Fux, C., Boehler, M., Huber, P., Brunner, I. and Siegrist, H., 2002, Biological treatment of ammonium-rich wastewater by partial nitritation and subsequent anaerobic ammonium oxidation (anammox) in a pilot plant, *J Biotechnol* **99**:295.

Graja, S. and Wilderer, P. A., (2001) Characterization and treatment of the liquid effluents from the anaerobic digestion of biogenic solid waste. Water Sci Technol. **43**(3), 265–274.

Grasmug M., 2004, Untersuchungen zur Kofermentation in der kommunalen und industriellen Anaerobtechnik, PhD thesis, IFA-Tulln, BOKU-University, Vienna.

Griehl C., Junghannß U., Bieler S., and Vollmer R., 2002, Mikrobiologische und analytische Untersuchungen zur Vergärbarkeit proteinreicher Substrate, *Chem Ing Tech* **5**:695–696.

Hertle, A., and Renner, G., 1994, Kostengünstige Klärschlammentsorgung mittels Flüssigphasenhydrolyse. *AWT Abwassertechnik* **2**:38.

Huang, J. J. H., Shih, J. C. H., 1981, The potential of biological methane generation from chicken manure, *Biotechnol Bioeng* **23**:2307–2314.

Huber, J., 1998, Pilotversuche zur Implementierung der Anaerobtechnik in die industrielle Tierkörperverwertung, diploma thesis, IFA-Tulln, BOKU-University, Vienna.

IFA-Tulln, 2005, Monitoring of agricultural biogas plants, data not published.

Johns M. R., 1995, Developments in wastewater treatment in the meat processing industry: a review, *Biores Technol* **54**:203–216.

Kabdasli, I., Tünay, O., Öztürk, I., Yilmaz S. and Arikan O., 2000, Ammonia removal from young landfill leachate by magnesium ammonium phosphate precipitation and air stripping, **41**(1):237.

Kirchmayr, R., 1998, Anaerobe Vergärung von Schlachtabfällen. Scale up vom Labor in den Pilotmaβstab, diploma thesis, IFA-Tulln, BOKU-University, Vienna.

Kirchmayr, R, Baumann, F., Braun, R., 2002, Perspectives of Anaerobic Digestion in the Treatment of Animal By-Products - Possibilities and Limits of AD-Technology – Is a TSE-Post Treatment Monitoring possible?, in: *Proceedings of Impacts of Waste Management Legislation on Biogas Technology*, Tulln, Austria.

Kirchmayr, R., Scherzer, R., Baggesen, D., Braun R. and Wellinger, A., 2003, Animal By-Products and Anaerobic Digestion - Requirements of the European Regulation (EC) No 1774/2002, IEA, Task 37 Energy from Biogas and Landfill Gas.

Koster, I. W., 1986, Characteristics of the pH-influenced adaptation of methanogenic sludge to ammonium toxicity, *J Chem Tech Biotechnol* **36**:445–495.

Koster, I. W., and Koomen, E., 1988, Ammonium inhibition of the maximum growth rate of hydrogenotrophic methanogens at various pH-levels and temperatures, *Microbiol Biotechnol* **28**:500–505.

Kroeker, E. J., Schulte, D. D., Sparling, A.B. and Lapp, H.M., 1979, Anaerobic treatment process stability, *J Water Pollu Control Fed* **51**(4), 718–727.

Lide, D., 1993, *Handbook of Chemistry and Physics. A Ready-reference Book of Chemical and Physical Data*, CRC Press, Boca Raton.

Mata-Alvarez, J., Macé, S., and Llabrés, P., 2000, Anaerobic digestion of organic waste. An overview of research achievements and perspectives, *Biores Technol* **74**:3–16.

Mulder, A., 2003, The quest for sustainable nitrogen removal technologies, *Water Sci Technol* **48**:67.

Mulder, A., van de Graaf, A.A., Robertson, L.A., and Kuenen, J.G., 1995, Anaerobic ammonium oxidation discovered in a denitrifying fluidized bed reactor, *FEMS Microbiol Ecol* **16**:177.

Nelson, N., Mikkelsen, R. and Hesterberg, D., 2003, Struvite precipitation in anaerobic swine lagoon liquid: effect of pH and Mg:P ratio and determination of rate constant, *Biores Technol* **89**:229.

Obaja, D., Mace, S., Costa, J., Sans, C. and Mata-Alvarez J., 2003, Nitrification, denitrification and biological phosphorus removal in piggery wastewater using a sequencing batch reactor, *Biores Technol* **87**:103.

Pambrun, V., Paul, E. and Spérandio, M., 2004, Treatment of nitrogen and phosphorus in highly concentrated effluent in SBR and SBBR processes, *Water Sci Technol* **50**(6):269.

Prechtl, S., Schneider, R., and Faulstich, M., 2002, Biogas production from animal by-products – Experiences with a technical plant, in: *Proceedings of the 3rd International Symposium Anaerobic Digestion of Solid Wastes*, Munich, Germany, pp. 200–204.

Prechtl, S., Schneider, R., and Faulstich, M., 2003, *Mono fermentation of animal by-products - Experiences with a technical plant*, in: *Proceedings of the International IWA Workshop Anaerobic digestion of slaughterhouse wastes*, Narbonne, France.

Ra, C.S., Lo, K.V., Shin, J.S., Oh, J.S., and Hong, B.J., 2000, Biological nutrient removal with an internal organic carbon source in piggery wastewater treatment, *Water Res* **34**:965.

*Regulation (EC) No 1774/2002* of the European Parliament and of the Council of October 3, 2002 laying down health rules concerning animal by-products not intended for human consumption (Official Jounal L 273 10/10/2002 P.0001–0095).

*Regulation (EC) No 999/2001* of the European Parliament and of the Council of May 22, 2001 laying down rules for the prevention, control and eradication of certain transmissible spongiform encephalopathies (OJ L147,31/05/2001 P.0001–0040).

Resch, C., Kirchmayr, R., Grasmug, M., Smeets, W., and Braun, R., 2005, Optimised anaerobic treatment of house-sorted biodegradabile waste and slaughterhouse waste in a high loaded half technical scale digester, in: *Proceedings of the 4th Symposium on Anaerobic Digestion of Solid Waste*, Copenhagen, pp. 615–622.

Safley, L. M., Vetter, R. L., Smith, D., 1986. Operating a full-scale poultry manure anaerobic digester, Biological Wastes, **19**(2):79–90.

Salminen, E., and Rintala, J., 2002, Anaerobic digestion of organic solid poultry waste – a review, *Biores Technol* **83**:13–26.

Salminen, E., Einola, and Rintala, J., 2003, The methane production of poultry slaughterhouse waste and effect of pre-treatment on the methane production of poultry feather, *Envir Technol* **24**(9): 1079–1086.

Sayed Jr., L.M., Vetter, R.L., and Smith, D., 1987. Operating a full-scale poultry manure anaerobic digester, *Biol Wastes* **19**:79–90.

Sharma, B., and Ahlert, R.C., 1977, Nitrification and Nitrogen Removal, *Water Res* **11**(10):897.

Siegrist, H., 1996, Nitrogen removal from digester supernatant – comparison of chemical and biological methods, *Water Sci Technol* **34**(1–2):399.

Siegrist, H., Hunziker, W. and Hofer, H., 2005, Anaerobic digestion of slaughterhouse waste with UF-membrane separation and recycling of permeate after free ammonia stripping, *Water Sci Technol* **52**(1–2):531.

Steen, I., 1998, Phosphorus availability in the 21st century: management of a non-renewable resource. Phosphorus Potassium, September–October:25–31.

Tippmer, K., 1994, *Verfahren der Flüssigphasenhydrolyse zur Verwertung von biogenen Restmüllfraktionen mit Biogasgewinnung*, Thyssen Still Otto Company, Bochum, Germany.

Tritt, W.P. and Schuchardt F., 1992, Materials flow and possibilities of treating liquid and solid waste from slaughterhouses in germany – a review, *Biores Technol* **41**:235–245.

Tritt, W.P., 1990, The application of anaerobic technology in slaughterhouses and carcasses removal plants, *Die Fleischmehlindustrie* **9**:162–170.

Tunay, O., Kabdasli, I., Orhon, D. and Kolcak, S., 1997, Ammonia removal by magnesium ammonium phosphate precipitation in industrial wastewaters, *Water Sci Technol* **36**:225.

Uludag-Demirer, S., Demirer, G.N. and Chen, S., 2005, Ammonia removal from anaerobically digested dairy manure by struvite precipitation, *Process Biochem*, in press.

Vallés-Morales, M.J., Mendoza-Roca, J.A., Bes-Piá, A. and Iborra-Clar, A., 2004, Nitrogen removal from sludge water with SBR process: start-up of a full-scale plant in the municipal wastewater treatment plant at Ingolstadt, Germany, *Water Sci Technol* **50**(10):51.

Verstraete, W. and Philips, S., 1998, Nitrification-denitrification processes and technologies in new contexts, *Environmental Pollution* **102**:717.

Webb, A.R., Hawkes, F.R., 1985, Laboratory scale anaerobic digestion of poultry litter: gas yield-loading rate relationship, *Agric Wastes* **13**:31–49.

Yoshino, M., Yao, M., Tsuno, H. and Somiya, I., 2003, Removal and recovery of phosphate and ammonium as struvite from supernatant in anaerobic digestion, *Water Sci Technol* **48**:171.

# 10

# Utilization of Whey

Gunther Pesta, Roland Meyer-Pittroff, and Winfried Russ

## 10.1. INTRODUCTION

Large amounts of wastewaters emerge from milk processing in dairies, which are one of the largest sources of industrial effluents. Thus, wastewater and organic residues produced from dairy production have the potential to be converted into economic gain if the proper processing technology is employed.

Among others anaerobic digestion (AD) is a promising technology to gain economic benefits. The main advantage of AD is the production of biogas, a renewable energy source. A valuable effluent also is obtained, which mainly can be used as an excellent soil fertilizer.

Most of wastewater in dairies results from cleaning and rinsing processes of transport lines, technical equipment, or tanks. Carbohydrates, fats, and proteins from milk are the main contributors of organic load. Modern companies lowered their use of fresh water and reduced the amount of wastewater. Separating organic residues of production processes from wastewater reduces the organic freight additionally. Due to this measure the average organic load ranges between 2.5 and 2.3 g COD/liter (COD = chemical oxygen demand). Nevertheless the average composition of dairy wastewater has a wide range concerning characteristic parameters, as shown in Table 10.1.

Predominantly aerobic wastewater treatment processes are implemented and run in dairies. Their effectiveness in order to reduce the organic freight is out of discussion. Nevertheless a high amount of sludge surplus is generated by an aerobic process, causing additional costs for sludge treatment.

Instead of metabolizing the organic load by aerobic processes, producing worthless carbon dioxide ($CO_2$) and biomass, respectively, sludge, anaerobic digestion processes should be implemented. Preferable two-phase systems digest not only organically high loaded wastewater but also solid organic residues with a high water content.

GUNTHER PESTA, ROLAND MEYER-PITTROFF, AND WINFRIED RUSS • Technische Universität München, Chair for Energy and Environmental Technologies of the Food Industry, Weihenstephaner Steig 22, 85350 Freising-Weihenstephan, Germany.

*Table 10.1.* Average composition of dairy waste water (adapted from Rosenwinkel, 2000)

| Parameter | Unit | Daily average | Variation limit |
|---|---|---|---|
| Amount of waste water | $m^3/t_{milk}$ | 1–2 | 0.5–0.4 |
| $BOD_5$ | mg/L | 500–2,000 | 1–160,000 |
| COD concentration | mg/L | 700–3,000 | 1–150,000 |
| $COD/BOD_5$ | – | 1.45 | 1.1–2.8 |
| TKN | mg/L | 30–150 | – |
| $NO_3$-N | mg/L | 20–130 | – |
| Phosphor, total | mg/L | 20–100 | – |
| Oil and lipids | ml/L | 50–250 | – |
| Sedimentable substances | ml/L | 1–3 | 0–250 |
| pH | – | 9–10.5 | 1–13 |

AD produces biogas which substitutes fossil energy sources. Electrical and thermal energy is gained by utilizing biogas with combined heat and power units (CHP). A recent review on anaerobic treatment of dairy waste waters is given by Demirel et al. (2005).

All liquid and pasty effluents from milk processing are utilizable for anaerobic digestion. Nowadays separated streams of by-products are used as feedstock for various products, such as food or beverage ingredients. Whey, a by-product in cheese processing dairies, and its alternative utilizations are introduced and thoroughly described. Due to its composition whey offers the possibility to obtain an energetic surplus of electricity and thermal energy by burning biogas, gained out of anaerobic digestion.

## 10.2. CHARACTERISTICS OF WHEY

Whey is a by-product in milk processing, the world production amounts to about 82 million metric tons, and especially the acid whey is seen as a waste product. In general three types of whey have to be differed: sweet, acid, and casein whey. Table 10.2 shows the average composition of these types of whey. Without regard to its origin the main component of whey is the carbohydrate lactose. The amount of lactose exceeds 70% of total dry matter (DM). Proteins and inorganic substances (ash) follow with differing weight proportions.

## 10.3. STATUS QUO OF WHEY TREATMENT AND FURTHER UTILIZATION

Whey may be separated by means of ultrafiltration or reversed osmosis into its main components protein, lactose, and delactosed permeate (DLP). Whey with an average dry matter of 6% is first deproteinized and whey protein concentrate (WPC) is

*Table 10.2.* Composition of different types of whey in percent by mass fresh material (Demmler, 1968; Kessler, 1996)

| Parameter | Unit | Sweet whey | Acid whey | Casein whey |
|---|---|---|---|---|
| Dry matter (DM) | % | 6.20 | 5.70 | 6.10 |
| Lactose | % | 4.80 | 4.60 | 4.70 |
| Protein | % | 0.75 | 0.30 | 0.50 |
| Fat | % | 0.05 | < 0.01 | < 0.01 |
| Ash | % | 0.60 | 0.80 | 0.90 |
| pH | – | 6.10 | 4.60 | 4.40 |

*Table 10.3.* Characteristics of delactosed permeate (DLP)

| Parameter | Unit | Amount |
|---|---|---|
| pH | | 5.5 |
| Dry matter | % by mass | 23.5–25.0 |
| Volatile solids | % by mass | 84 |
| Lactose | % by mass | 14.0–16.0 |
| Protein | % by mass | 2.6–3.7 |
| Ash | % by mass | 3.7–4.0 |
| Salt | % by mass | 2.4 |
| Sodium (Na) | mg/L | 4,286 |
| Potassium (K) | mg/L | 12,000 |
| Magnesium (Mg) | mg/L | 643 |
| Calcium (Ca) | mg/L | 2,000 |
| Manganese (Mn) | mg/L | 1.0 |
| Phosphorus (P) | mg/L | 3,143 |
| Chloride (Cl) | mg/L | 7,720 |
| Sulfate (SO4) | mg/L | 2,450 |
| Copper (Cu) | mg/L | 0.75 |
| Lead (Pb) | mg/L | < 0.01 |
| Zinc (Zn) | mg/L | 0.94 |
| Arsenic (As) | mg/L | < 0.01 |

sold. Furthermore, the deproteinized permeate is concentrated to a dry matter of about 60%, followed by a crystallization of the lactose. The purified lactose is sold and DLP disposed. It is not feasible to dry DLP-like whey because of the high content of salts.

Overall the DLP remains with a pH of 5.2–5.5 and a dry matter of about 25–30%, consisting of approximately 15% protein, 15–20% ash, and 60–70% lactose. The characteristics of the effluent DLP are shown in Table 10.3. This remaining residue is costly in disposal due to its high COD. Aggravated legislations increased the costs for disposal, new approaches have to be developed for further investigations to produce new products from DLP, or to realize a disposal with additional benefits, like the production of biogas, a renewable energy source.

### 10.3.1. Ethanol Production from Whey

The production of ethanol is considered to be a feasible technique to enhance the profit from the by-products of whey treatment and is used for both, producing spirits or liquid fuels. Ethanol production can be performed either as a fermentation with or without enzymatic pretreatment. Many of the microorganisms are not capable for direct conversion of lactose to ethanol. Thus enzymatic pretreatment with ß-galactosidase is required. The fermentation is feasible with *Aspergillus niger* at a pH between 3.5 and 4.5 or *Streptococcus lactis* at a higher pH level ranging from 6 to 7 (Fischbach, 1987). The problem of fermentation with enzymatic pretreatment is that microorganisms will metabolize first glucose, easy to decompose, instead of galactose, which is more difficult to metabolize. Consequently the fermentation yield will decrease.

Ethanol production from whey is possible with lactose-fermenting yeast strains as well. According to Bothast et al. (1986) and Sanderson and Reed (1985) yeast strains with good ethanol yields are:

- *Kluyveromyces marxianus (Kluyveromyces fragilis, Kluyveromyces lactis)*; with an ethanol yield of approximately 0.5 g ethanol/g lactose
- *Candida pseudotropicalis (Candida kefyr)*

Growth limitations due to an inhibiting effect of ethanol on the growth of *Kluyveromyces marxianus* occurs at a concentration of ethanol of about 45–95 g ethanol/$dm^3$ depending on exposing time (Vienne and von Stokar, 1985). Another alternative has been intensively studied by using recombinant DNA techniques for achieving the expression of genes that code for ß-galactosidase and lactose permease system of *Kluyveromyces lactis* in *Saccharomyces cerevisiae* (Gonzales Siso, 1996).

The most common substrate for lactose-fermenting yeast strains is whey permeate from ultrafiltration. Different systems to produce ethanol from whey are described as follows:

- Dansk Gaerings Industry process: The Dansk Gaerings Industry process is a continuous process with *Streptococcus fragilis*. From 42 $dm^3$ whey permeate containing 4.4% lactose microorganisms produce 1 $dm^3$ of ethanol. The efficiency of the fermentation is about $\eta = 0.8$ (Fischbach, 1987).
- Carbery process: The Carbery process is characterized by a batch fermentation and a continuous distillation. A lactose stream produced during whey ultrafiltration is pumped into one of the eight fermentation vessels in the fermentation room. Yeast is added to each fermentation vessel at the beginning of vessel filling. Conditions in each fermentation vessel are set to enable the fast and efficient conversion of lactose to ethanol. The fermentation process is complete in each vessel within 24 hours. The yeast used for fermentation is recovered and reused a number of times before it is discarded. The yeast is recovered at the end of the fermentation by settling

and mechanical clarification known from the beer filtration. At the end of fermentation the ethanol level is between 3.5% by volume and 4.2% by volume (Carbey Group, 2005). The produced ethanol has food-grade quality and its distillate (96.5% by volume ethanol) is further processed to spirits (Fischbach, 1987; Moebus and Teuber, 1986; Delaney, 1981). The efficiency of the fermentation is about $\eta = 0.86$ (Kessler, 1996). Figure 10.1 shows a process diagram and summary of milk ingredients and products produced at Carbery.

- Milbrew process: The aim of the Milbrew process is the production of single-cell protein and ethanol. It is a continuous process with *Kluyveromyces fragilis* NURL Y1109 (Everson, 1978; Delaney, 1981).

A mix of whey permeate, whey, and DLP may be appropriate as raw material for ethanol fermentation. The main target in the fermentation of the mix is to start with high concentrations of lactose and at the same time with an acceptable amount of minerals (Fischbach, 1987). The effluents have to be treated with the total dairy effluents.

Appropriate plants for ethanol production from whey are available from APV International, West Sussex, Great Britain, and Alfa Laval, Hamburg, Germany (Fischbach, 1987). Production costs were estimated in 1986 at 0.61 €/$dm^3$ ethanol (Fischbach, 1987) and are listed in Table 10.4. In 1987, market price of ethanol was approximately 0.57 €/$dm^3$.

Coton (1980) tried to evaluate the production of potable ethanol from whey permeate in 1980 and calculated for Great Britain a profit of 107 pounds/1,000 kg lactose. He remarks that due to the strictly limited ethanol market for use in vinegar manufacturing and in drinks there usually exists severe competition between

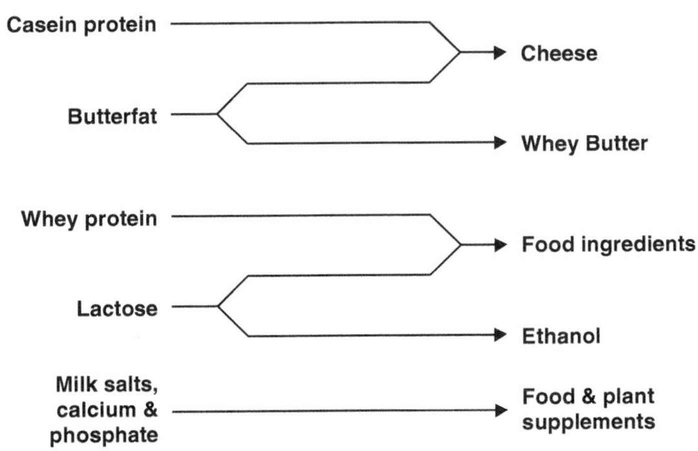


**Figure 10.1.** Process diagram and summary of milk ingredients and products produced at Carbery (Carbery, 2005).

*Table 10.4.* Estimated costs for ethanol production (95.5% by volume) from whey (Fischbach, 1987)

| Costs | €/dm$^3$ |
|---|---|
| Investment | 0.21 |
| Personnel | 0.11 |
| Auxiliaries | 0.08 |
| Energy | 0.21 |
| Total | 0.61 |

ethanol produced from molasses versus other sources. Thus ethanol production from whey permeate for drinks is only in certain cases economically attractive.

The production of ethanol in New Zealand is described by Maddox and Gutierrez (1996). Further specific problems in the production of ethanol from whey were discussed in Compagno et al. (1995), El-Samragy and Zall, (1988), Banat et al. (1996), Szczodrack et al. (1997), Kleine et al. (1995), Kamini and Gunasekaran (1987), Maiorella and Castillo (1984), Zayed and Foley (1987), Marth (1970), Moulin et al. (1980), Moulin and Galzy (1981), Yoo (1975), Vienne and von Stockar (1983), and Walker et al. (1985).

Today production costs for bioethanol vary between 0.15 and 1.07 €/dm$^3$, depending on the feedstock for ethanol production. The available price at the market today for industrial ethanol is about 0,50 €/dm$^3$ (Schmitz, 2005). Economic feasibility of ethanol production from whey permeate has to be evaluated critically in each single case by concerned local, subjective factors.

### 10.3.2. Demineralization

High salt concentrations in the fermenting solution can inhibit microorganism growth (Fischbach, 1987; El-Samragy and Zall, 1988; Mahmoud, 1980; El-Nimr et al., 1982). Therefore, demineralization seems to be obligate for further biological treatment of DLP. Besides different physical effects like precipitation, a high salt content requires halophilic microorganisms which are able to produce ethanol of adequate concentration. The unspecified mineral fraction of demineralization could be processed to licking stones for animals, though the demand probably is very low (Fischbach, 1987; Coton, 1980).

Interesting possibilities for the utilization of specific salty fractions lie in the recovery of melting salts or table milk salts (salt substitutes), production of defined supplements for beverages, the feed or cosmetic industries, etc. (Herrmann, 1998). High concentrations of sulfate in the salty fraction of DLP could be handicapping for some applications. Horton (1995) mentioned one United States whey processor who supplied mother liquor from permeate lactose manufacture to Japan for use in nutritional beverages. The countries most active in use of the milk salts according to Horton (1995) are Japan and Finland.

### 10.3.3. Precipitation of Calcium Phosphate

Calcium and phosphate rank approximately 16% of the total salt content. A removal of both would improve the "degradability" of whey significantly.

Calcium phosphate precipitation is enhanced by raising pH, temperature, and concentration. These conditions are found with evaporator-fouling processes. According to Durham (1997), calcium phosphate precipitation is slow and involves a number of stages. Precipitation of calcium phosphate is regarded as an issue which requires more investigation. Table 10.5 shows an overview of hardly soluble calcium phosphates.

Calcium phosphate milk salts are currently marketed by DMV International, NL, and Valio, SF, as nutritional supplement to combat osteoporosis (Durham et al., 1997; DMV, 2005). The structure of the crystals as well as the purity of milk salts are of interest.

Further investigations in a German dairy resulted in an opportunity to induce precipitation by increasing the pH in the effluent of an electrodialysis plant with CaO and NaOH. Overnight a settlement of crystalline calcium phosphate was observed. The characteristics of the sediment and feasible techniques for solid/fluid fluid separation have been investigated by Schwab (1995).

### 10.3.4. Precipitation of Phosphate and Sulfate in DLP

The precipitation of phosphate was investigated by Behmel et al. (1998). The pH value of DLP was increased with 1, 2, and 3% of CaO. After 1 hour reaction the samples were centrifuged (2,000 U/min, 10 min). The results of pH, dry matter, and phosphate in the liquid phase are shown in Table 10.6.

The best results were achieved with a CaO concentration of 1% by mass. The phosphate concentration could be reduced about 87%. Further investigations have to been done to minimize the CaO concentration but at the same time increasing the precipitation.

### 10.3.5. Ion Exchange

By ion exchange a removal up to 95% of the minerals from whey may be feasible. The main disadvantages are high consumption of regenerated chemicals and of fresh water (Durham et al., 1997).

*Table 10.5.* Hardly soluble calcium phosphates

| Name | Empirical formula | Solubility product at 20°C in (mol/L)ν |
|---|---|---|
| Calciumhydrogenphosphatdihydrate | $CaHPO_4 * 2\ H_2O$ | $2{,}570 * 10^{-7}$ |
| Calciumhydrogenphosphatanhydride | $CaHPO_4$ | $9{,}120 * 10^{-8}$ (at 37°C) |
| Tetracalciummonohydrogentriphosphate | $Ca_4H(PO_4)_3 * 2.5\ H_2O$ | $1{,}268 * 10^{-47}$ |
| β-Tricalciumphosphate | $Ca_3(PO_4)_2$ | $1{,}574 * 10^{-30}$ |
| Hydroxylapatite | $Ca_5(PO_4)_3OH$ | $3{,}192 * 10^{-59}$ |

*Table 10.6.* Characteristics of treated DLP

| | DLP | DLP + CaO (1% by mass) | DLP + CaO (2% by mass)[a] | DLP + CaO (3% by mass)[a] |
|---|---|---|---|---|
| pH | 5.2 | 7.1 | 11.4 | 11.8 |
| Dry matter (precipitation) | 37.0% | 38.2% | 37.5% | 36.1% |
| Dry matter (liquid phase) | 27.9% | 28.2% | 30.1% | 29.3% |
| Ash (precipitation) | 7.1% | 10.5% | 14.1% | 15.3% |
| Ash (liquid phase) | 6.3% | 6.0% | 7.5% | 8.0% |
| Volatile solids (precipitation) | 80.7% | 72.5% | 62.5% | 57.8% |
| Volatile solids (liquid phase) | 77.4% | 78.6% | 75.2% | 72.8% |
| Phosphate (liquid phase) | $14.9\,g/dm^3$ | $2.0\,g/dm^3$ | $1.6\,g/dm^3$ | $2.3\,g/dm^3$ |

[a] Poor separation between liquid phase and precipitation.

### 10.3.6. Electrodialysis

Electrodialysis is a rather effective method for the removal of salts and bases upon the movement of ions through a membrane system forced by electric fields. Using electrodialysis 50% demineralization is feasible; higher levels require a disproportionate increase in recirculating time and electrical power (Durham et al., 1997). The main disadvantage is the high consumption of electrical energy in a range from 10 to 28 kWh/kg of demineralized whey powder. The effluent may be enriched with ions, according to the applied membranes, flow rate, and electrical field. It can be treated more easily with respect to precipitation.

### 10.3.7. Nanofiltration

Nanofiltration ($p = 10$–$30$ bar) has replaced electrodialysis as the preferred demineralization process (Durham et al., 1997). Preferentially monovalent ions like $Na^+$, $K^+$ or $Cl^-$ are removed from the whey, whereas the removal of divalent ions like calcium phosphate is limited. Approximately 94% are retained because of their size, especially in a hydrated state.

The permeate contains about 0.3–0.5% of solids. A further treatment with reverse osmosis leads to clean water and salt concentrate. The monovalent salt mixture could be used as natural low sodium table salt substitute or as supplement for sports and health beverages.

## 10.4. ALTERNATIVES OF WHEY UTILIZATION

In the literature there exist many approaches for the utilization of whey. The suitability for utilization of DLP instead of whey as well as economical aspects of the applications have to be investigated.

### 10.4.1. Ethanolic Cofermentation with Cereals or Molasses

The method is based on replacing water with whey in preparation of cereals fermenting media. In the first stage ethanolic fermentation in whey is performed with *Kluyveromyces marxianus subsp. marxianus*. Such prefermented whey is then used for dilution of molasses instead of water and the resulting medium subjected to the second-stage fermentation performed with *Saccharomyces cerevisiae* (Fischbach, 1987; Zakrzewski and Zmarlicki, 1988; Friend and Shahani, 1979).

### 10.4.2. Production of Lactic Acid

The anaerobic fermentation of whey with lactic acid bacteria for the production of lactic acid is feasible but probably not profitable (Fischbach, 1987). The worldwide need is only approx. 20,000,000–25,000,000 kg/a.

Lactic acid is produced by homofermentative lactic acid bacteria, such as *Lactobacillus bulgaricus* or *Lactobacillus plantarum* (Sanderson and Reed, 1985). Lactic acid is separated from the fermenter solution with electrodialysis or $H_2SO_4$ after neutralization with $Ca(OH)_2$ or $CaCO_3$ (Moebus and Teuber, 1986; Sanderson and Reed, 1985).

### 10.4.3. Production of Ammonium Lactate

Experimental feeding of ruminants has shown that 27% of total nitrogen of the feed can be replaced by nitrogen of fermented whey containing ammonium lactate (Moebus and Teuber, 1986). For the production of ammonium lactate whey or permeate are fermented under anaerobic conditions with *Lactobacillus bulgaricus*. The reaction broth is continuously neutralized with ammonia, resulting in fermented ammoniated condensed whey, so-called "FACW" (Fischbach, 1987; Coton, 1980; Zadow, 1988). Delaney (1981) estimated a sales price of 175–180 €/1,000 kg.

### 10.4.4. Production of Single-Cell Protein (SCP)

SCP is used primarily for feeding but also for human nutrition (Gonzalez Siso, 1996; Lembke and Meinhold, 1975; Sanderson and Reed, 1985; Maiorella and Castillo, 1984). The problem for a feasible economic solution of this application is that salable amounts are restricted (Fischbach, 1987). The Danish fermentation industry calculated in 1980 that biomass production is likely to result in a financial loss (Coton, 1980).

Industrial microbial biomass production from cheese whey for use as food started in France at Fromagerie Le Bel, Vendome (Gonzalez Siso, 1996). The capacity in 1986 was about 250,000–300,000 kg whey per day (Moebus and Teuber, 1986).

### 10.4.5. Production of Bakery Yeast

An industrial process for producing bakery yeast (*Saccharomyces cerevisiae*) was adopted by the Nutrisearch Company, Cincinnati, Ohio. After enzymatic treatment with immobilized ß-galactosidase, an inoculation with yeast strains that are able to metabolize galactose is required (Anonymous, 1984; Gonzalez Siso, 1996; Zadow, 1988).

### 10.4.6. Production of Biogas

The production of biogas from whey can be based on different products of the whey treatment (Fischbach, 1987; Moebus and Teuber, 1986; Schindler, 1986; Zayed and Foley, 1987):

- whey
- permeate (deproteinized whey)
- mother liquor from lactose production
- residues of ethanol production

Demirel et al. (2005) gives a detailed review on anaerobic treatment of dairy wastewaters.

### 10.4.7. Whey Drinks

A well sold lactic acid fermented whey drink is Rivella (Switzerland). The supplement of ß-galactosidase can increase the sweetness of such products (Moebus, 1986; Marth, 1970).

The maximum concentration of sodium ions in whey drinks is about 45 mg/100 g (Fischbach, 1987), so demineralized whey is usually used for preparation of these products (Sanderson and Reed, 1985). Further abstracts about whey drinks are given by Schwab (1995).

## 10.5. FURTHER SCIENTIFIC APPROACHES OF WHEY UTILIZATION

Literature provides a wide range of fermentation possibilities using lactose as feedstock or supplement:

- proteolytic enzymes (Castro and Cantera, 1996)
- ß-galactosidase (Gonzalez Siso, 1996; Maiorella and Castillo, 1984)
- antibiotics (Coton, 1985)
- organic acids (Gonzalez Siso, 1996), for example, gluconic acid (El-Sayed et al., 1996), acetic acid, citric acid (*Aspergillus niger*; *Candida pseudotropicalis*) (Sanderson and Reed, 1985; Friend and Shahani, 1979)

- propionate-containing whey to incorporate in bread formulations (fermentation with *Propionibacterium sp.*) (Champagne et al., 1989; Sanderson and Reed, 1985)
- vitamins (Coton, 1985; Marth, 1970)
- amino acids (Coton, 1985; Sanderson and Reed, 1985)
- 2,3-butanediol, basic material for the chemical industry (Sienkiewicz and Riedel, 1990)
- food gums (Gonzalez Siso, 1996; Coton, 1985)
- starter cultures (Moebus and Teuber, 1986)
- wine, after demineralization (Sienkiewicz and Riedel, 1979; Delaney, 1981; Friend and Shahani, 1979)
- supplement to bakery products (Seibel et al., 1984)

Few of these possibilities have been developed on a commercial scale, mainly because other substrates can be used alternatively and produced in a more cheaper way. Molasses is usually a cheap commodity that can be assembled on any site in very large quantities.

## 10.6. CONCLUSION AND PROSPECTS

Several alternatives for the utilization of whey have been discussed in the literature. But only a few are commercialized. The high lactose concentration of the DLP ethanol fermentation seems to be the most promising technique for further investigations.

High salt concentrations in the fermenting solution can inhibit microorganism growth (Fischbach, 1987; El-Samragy and Zall, 1988; Mahmoud, 1980; El-Nimr et al., 1982). Thus, screened halophilic microorganisms with good ethanol yields or practicable solutions for an economic demineralization have to be found. The optimized precipitation of calcium phosphate could reduce the salt concentration in the DLP and at the same time produce a new salable by-product. Further, if it is necessary for good ethanol yields, the high potassium, sodium, and chloride concentrations have to be reduced with suitable demineralization methods. The possibility of selling milk salts should be taken into account.

Therefore, the research about ethanol production from DLP should include:

- production of ethanol from delactosed permeate with different microorganisms,
- selection of the most halophilic microorganisms compared to ethanol yields after demineralization,
- if demineralization is necessary: optimization of the method; investigations about using the minerals: create tailor-made products, isotonic drinks,
- optimization of the fermentation with a ''mixed strategy'' of different carbon sources (Figure 10.2), and
- economic optimization of the process (cost).

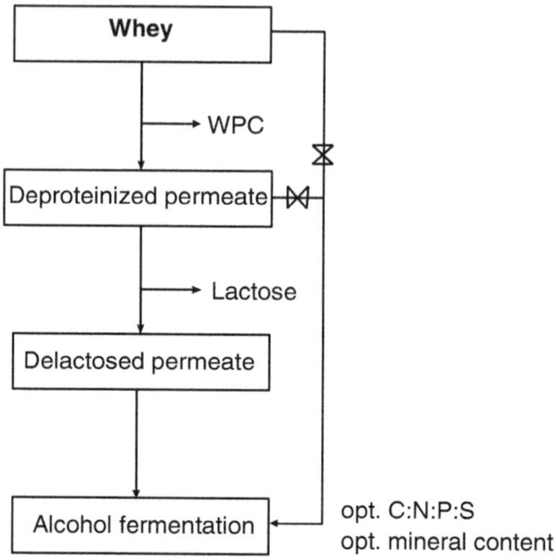


**Figure 10.2.** ''Mixed strategy'' of different carbon sources (whey, whey permeate, delactosed whey permeate) for an optimized ethanol production from whey.

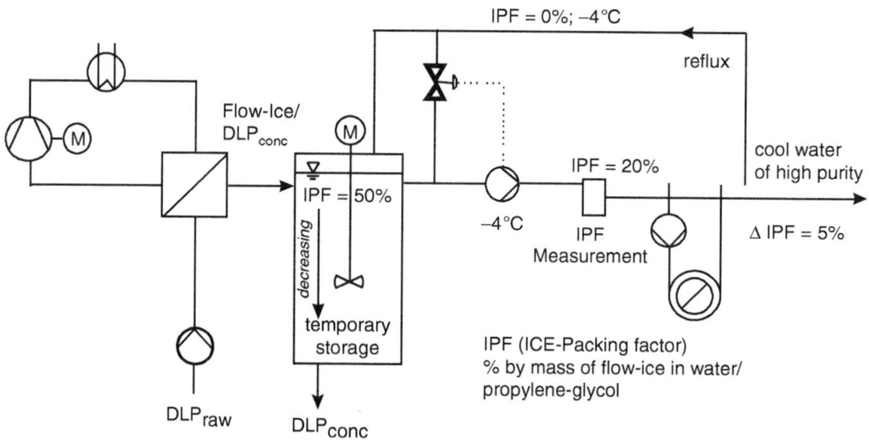


**Figure 10.3.** Production of flow ice from DLP (Behmel et al., 1998).

After fermentation an effluent is left with a high content of minerals and nearly without any carbohydrates. The further utilization as fertilizer has to be analyzed. If a water treatment concept for the entire dairy is planned, the fermentation of DLP to biogas instead of ethanol should be considered.

Another possible investigation is to concentrate the effluent first by producing flow ice for cooling processes (Figure 10.3). The melted ice does not contain any solubles. Utilizing the process up to 50% of the solution are frozen out as flow ice, which can serve as a cooling agent and can be used after melting as pure, cold water. This concept has been investigated successfully in Japan for utilization of wastewater from the chemical industry and different sources (Behmel et al., 1998).

In Conclusion, it has to be put on record that only a combination of various measures will lead to a cost-effective optimization of waste (water) and by-product treatment. For example, the implementation of AD plants in dairies shall be accompanied by restructuring the energy and wastewater management. Especially the utilization of biogas surplus heat needs an intensive consideration of the production process.

## 10.7. REFERENCES

Anonymous 1984, Fa. Nutrisearch, Cincinnati, Immobilized enzyme and fermentation technologies combine to produce bakers yeast from whey, *Food Technol* **6**:26–27.

Banat, I. M., Singh, D., and Marchant, R., 1996, The use of a thermotolerant fermentative kluyveromyces marxianus IMB3 yeast strain for ethanol production, *Acta Biotechnol* **16**(2-3):215–223.

Behmel, U., Kunda, C., and Meyer-Pittroff, R., 1998, Internal report, Technische Universität München.

Bothast, R. J. et al., 1986, Ethanol production by 107 Strains of Yeasts on 5, 10 and 20% Lactose, in: *Biotechnol Lett* **8**(8):593–596.

Carbery Group, 2005, Ballineen (October 10, 2005); http://www.carbery.com/products/ethanol/process.html

Castro, S., and Cantera, A. M. B., 1996, Aprovechamiento de desechos industriales por métodos bioquímicos, *Biotecnologı́a Aplicada* **13**(3):213.

Champagne, C. P., Baillargeon-Côte, C., and Goulet, J., 1989, Whey fermentation by immobilized cells of propionibacterium shermanii, *J Appl Bacteriol* **66**:175–184.

Compagno, C., Porro, D., Smeraldi, C., and Ranzi B. M., 1995, Fermentation of whey and starch by transformed saccharomyces cerevisiae cells, *Appl Microbiol Biotechno.* **43**:822–825.

Coton, S. G., 1980, Whey technology-the utilization of permeates from the ultrafiltration of whey and skim milk, *J Soc Dairy Technol* **33**(3):89–94.

Coton, S. G., 1985, Whey resources and utilization, *J Soc Dairy Technol* **38**(4):97–100.

Delaney, R. A. M., 1981, Recent developments in the utilization of whey, *Cult Dairy Prod J* **16**:11–22.

Demirel, B., Yenigun, O., and Onay, T. T., 2005, Anaerobic treatment of dairy wastewaters: a review, *Process Biochem* **40**:2583–2595.

Demmler, G., 1968, *Milcheiweiβ und Molkeprodukte, Handbuch der Lebensmittelchemie III.*, pt. 1, Tierische Lebensmittel, Springer, Berlin.

DMV International, 2005, Veghel (October 10, 2005); http://www.dmv-international.com/default.asp

Durham, R. J., Hourigan, J. A. et al., 1997, Whey fractionation: wheying up the consequences, *Food Aust* **49**(10):460–465.

El-Nimr, A. A., Badr Eldin, S. M., Ghali, Y., and Yousef, Y. B., 1982, Salted whey utilization (I) Salt tolerance of yeasts, *Egyptian J Dairy Sci* **10**:1–5.

El-Samragy, Y. A., and Zall, R. R., 1988, The influence of sodium chloride on the activity of yeast in the production of single cell protein in whey permeate, *J Dairy Sci* **71**:1135–1140.

El-Sayed, M. M., El-Deeb, S. A., Abdel-Rehim, N. Z., Mostafa, H. E., and Khorshid, M. A., 1996, Utilization of cheese whey for producing gluconic acid as a mean for pollution control, *Milchwissenschaft* **51**(5):266–268.

Everson, T. C., 1978, *Whey-derived Gasohol - a Reality? Proceedings of Whey Products Conference*, Minneapolis/Minnesota (October 9/10, 1978), pp. 62–80.

Fischbach, G., 1987, Molke als Rohstoff, dissertation, Christian-Albrechts-Universität, Kiel.

Friend, B. A., and Shahani, K. M., 1979, Whey fermentation, *N Z J Dairy Sci Technol* **14**:143–155.

González Siso, M. I., 1996, Cheese whey as a resource: whey utilization, *Bioresource-Technol* **57**(1): 1–11.

Herrmann, M., 1998, Wirtschaftliche Verwertung von Lab- und Sauermolke mittels der Elektrodialyse-Entsalzung, *Deutsche Milchwirtschaft* **42**:1420–1422.

Horton, B. S., 1995, Commercial utilization of minor milk components in the health and food industries, *J Dairy Sci* **78**(11):2584–2589.

Kamini, N. R., and Gunasekaran, P., 1987, Simultaneous production of lactose by *Kluyveromyces fragilis* and zymomonas mobilis, *Curr Microbiol* **16**:153–157.

Kessler, H. G., 1996, *Lebensmittel- und Bioverfahrenstechnik,* Kessler, Munich, p. 525 ff.

Kleine, R., Achenbach, S., and Thoss, S., 1995, Whey disposal by deproteinization and fermentation, *Acta Biotechnol* **15**(2):139–148.

Lembke, A., and Meinhold, K., 1975, Mikrobielles Protein aus Nebenprodukten von Molkereien und die Behandlung von Abfallprodukten in der Milchviehhaltung, *Berichte über Landwirtschaft* **192**:539–606.

Maddox, I. S., and Gutierrez, N. A., 1996, Biotechnological developments in New Zealand, *Crit Rev Biotechnol* **16**(2):119–143.

Mahmoud, M. M., 1980, Ultrafiltration in the manufacture of soft pickled cheese and production of ethanol and single cell protein from whey, *Diss Abstr Int* **41**(3):878.

Maiorella, B. L., and Castillo, F. J., 1984, Ethanol, biomass, and enzyme production for whey waste abatement, *Process Biochem*:157–161.

Marth, E. H., 1970, Fermention products from whey, in: *Byproducts From Milk*, B.H. Webb, ed., Avi Publishing, Westport, pp. 43–82.

Moebus, O., and Teuber, M., 1986, Verfahren zur Vergärung von Molke und Ultrafiltrationspermeaten, *Kieler Milchw Forsch* **38**(2):119–130.

Moulin, G., Guillaume, M., Galzy, P., 1980, Ethanol production by yeast in whey ultrafiltrate, *Biotechnol Bioeng* **22**:1277–1281.

Moulin, G., and Galzy, P., 1981, Ethanol production from whey, in: *Advances In Biotechnology*, Vol. II Fuels, Chemicals, Foods and Waste Treatment, M. Moo-Young, ed., Pergamon Press, Toronto, pp. 181–187.

Rosenwinkel, K.-H., 2000, Industrieabwasser, ATV-Handbuch, Verlag Ernst, Berlin.

Sanderson, G. W., and Reed, G., 1985, Fermented products from whey and whey permeate, in: *New Dairy Products Via New Technology,* IDF-Seminar Atlanta, October 1985, pp. 141–157.

Schindler, M., 1986, Verfahren zur Reinigung und energetischen Nutzung von Käsereiabwasser und Molke, dissertation, Eidgenoessische Technische Hochschule, Zürich.

Schmitz, N., 2005, Stand und Perspektiven von Bioethanol in Deutschland, Europäische Biomassekonferenz 2005, Graz, Austria.

Schwab, C., 1995, Die Lösung des Molkenproblems - Vom nutzlosen Nebenprodukt zum Erfrischungsgetränk, *Lebensmitteltechnik* **6**:46–49.

Szczodrak, S., Szewczuk, D., Rogalski, D., and Fiedurek, J., 1997, Selection of yeast strain and fermentation conditions for high-yield ethanol production from lactose and concentrated whey, *Acta Biotechnol* **17**(1):51–61.

Seibel, W., Brümmer, J.-M., and Stephan, H., 1984, Studien über die Verwendungsmöglichkeiten von Milchpulver- und Molkenerzeugnissen bei der Herstellung von Backwaren, *dmz* **3**:75–83.

Sienkiewicz, T., and Riedel, C.-L., 1979, Aspekte der Molkenverwertung, *Lebensmittelindustrie* **26**(7):306–310.

Sienkiewicz, T., and Riedel, C.-L., 1990, *Whey and Whey Utilization*, Th. Mann, Gelsenkirchen-Buer.

Vienne, P., and von Stockar, U., 1983, Ethanol from whey permeate: strain selection, temperature, and medium optimization, *Biotechnol Bioeng Symp* **13**:421–435.

Vienne, P., and von Stockar, U., 1985, An investigation of ethanol and other limitations occurring during the fermentation of concentrated whey permeate, *Biotechnol Lett* **7**(7):521–526.

Walker, L. P., Pellerin, R. A., Rao, A. M., Hang, Y. D., Kalter, R. J., Boisvert, R. N., and Gabler, E. C., 1985, Feasibility assessment of whey-based ethanol facilities, *Energy in Agriculture* **4**:93–107.

Wiegland, S., 1998, Die anaerobe Behandlung von Abwasser aus der Elektrodialyse von Molke, diploma thesis, Technische Universität München.

Yoo, B. W., 1975, Effect of certain factors on cell mass and ethanol production by lactose fermenting yeast, *Diss Abstr Int* **36**(2):641.

Zadow, J. G., 1988, Fermentation of whey and permeate, *Bull IDF* **233**:53–60.

Zakrzewski, E., and Zmarlicki, S., 1988, Ethanolic fermentation in whey and whey-molasses mixtures, II, Two-stage fermentation process of ethanol production from whey and beet molasses, *Milchwissenschaft* **43**(8):492–496.

Zayed, G. Z. A., and Foley, J., 1987, The influence of fermentation conditions on ethanol yields from whey permeate, *Irish J Food Sci Technol* **11**:109–118.

# 11

# Utilization of Plant By-Products for the Recovery of Proteins, Dietary Fibers, Antioxidants, and Colorants

Vasso Oreopoulou and Constantina Tzia

## 11.1. INTRODUCTION

Processing of fruits, vegetables, and oilseeds results in high amounts of waste materials such as peels, seeds, stones, and oilseed meals. Disposal of these materials usually represents a problem that is further aggravated by legal restrictions. Plant waste is prone to microbial spoilage; therefore, drying is necessary before further exploitation. The cost of drying, storage, and transport poses additional economical limitations to waste utilization. Therefore, agroindustrial waste often is utilized as feed or fertilizer. However, demand for feed or fertilizer varies and depends on agricultural production. Moreover, valuable nutrients contained in agroindustrial wastes are lost. Thus new aspects concerning the use of these wastes as by-products for further exploitation on the production of food additives or supplements with high nutritional value have gained increasing interest because these are high-value products and their recovery may be economically attractive.

It is well known that agroindustrial by-products are rich in dietary fibers, some of which contain appreciable amounts of colorants, antioxidant compounds or other substances with positive health effects, while some of them, like the oilseed meals, are rich in proteins. Dietary fibers consist of polysaccharides and lignin that are neither digested nor absorbed in the human small intestine. The beneficial role of dietary fibers in reducing risk of coronary heart disease and certain types of cancer has been pointed out by several epidemiological studies. The recommended daily intake is 30–45 g per day and person, while the estimated consumption in the West

VASSO OREOPOULOU AND CONSTANTINA TZIA • National Technical University of Athens, School of Chemical Engineering, Laboratory of Food Chemistry and Technology, Athens, GR 15780, Greece.

is about 20 g per day and person (Thebaudin et al., 1997). Cereals make up to 35–60% of the actual consumption while fruits and vegetables contribute only 7–15% (Grigelmo-Miguel and Martin-Belloso, 1999b). Cereals are rich in insoluble dietary fibers, which are mainly related to intestinal regulation, while fruits and vegetables contain appreciable amounts of soluble dietary fibers, which help in lowering blood cholesterol and glucose intestinal absorption. Soluble dietary fibers should account for 30–50% of the total dietary fibers intake to improve dietary benefits. Apart from health effects dietary fibers show some functional properties, such as water-holding capacity, swelling capacity, increasing viscosity or gel formation which are essential in formulating certain food products. Formulated food products with high dietary fibers content are now commercially available. Dietary fibers incorporated in these products are obtained mainly by cereals. Citrus and apple pomace are also raw materials for the production of dietary fiber powders. However, the use of by-products from other fruits and vegetables seems promising, since all these materials are rich in soluble dietary fibers.

Several other organic micronutrients, such as carotenoids and polyphenols, impart health benefits. Carotenoids are mostly used as natural food colorants, but also some of them (β-carotene, apocarotenal) have a vitamin A activity as well as antioxidant activity in vitro and in vivo (Kiokias and Gordon 2004). Polyphenols also act as antioxidants; they scavenge free radicals which are responsible for serious diseases and for the oxidation of lipids, proteins, and DNA. Several studies have revealed their antimicrobial, antithrombotic, antimutagenic, and anticarcinogenic activities (Kandaswami and Middleton, 1997; Sahu and Green, 1997). In addition to health benefits, the supplementation of food products with antioxidants delays the formation of off-flavors and rancidity and extends the shelf life of the product. Carotenoids and polyphenols are extensively distributed in several plant by-products. Thus most fruit and vegetable by-products could serve as raw materials for their recovery. Moreover, these antioxidants or colorants would have a natural origin, which is in accordance with the demands of consumers for ''all natural'' because of the possible toxicity of synthetic additives.

The role for food proteins in human nutrition is substantial, while according to modern nutrition recommendations the supply of proteins in human diet should rely mostly on vegetable proteins, i.e., cereals, legumes, oilseeds, etc., than on those from animal sources, i.e., meat. Because the amounts of proteins required to cover the nutritional needs of the world population are continuously increasing, the investigation and finding of new protein sources are of interest. In addition to their nutritional value, proteins offer great potential as functional food ingredients providing useful properties when incorporated into foods. In order to utilize a by-product as a protein source it should both present high protein content and protein value (quality) based on well-balanced essential amino acids. An additional requirement to utilize a material for food purposes is the absence of allergic or toxic substances or the application of a proper pretreatment for their efficient removal. Several protein products (flours, concentrates, or isolates), depending on their protein content, can be introduced to food products in order to improve their nutritional value as well as their sensory and functional properties. Such protein products already have been used as meat extenders in processed meat products or as protein-enriching agents in nutritional beverages. Particularly, soy protein products are used for the production of several foods such as: imitation cheese, whipped toppings, soy milk, and baked products.

Today consumers are highly aware of the close relationship between nutrition and health and they want to include health-promoting ingredients in their diets. Foods fortified with nutraceuticals, or functional foods in other terms, are expected to be the new food category with an expanding demand by future generations. Therefore, food scientists focus their effort on the development of new products with improved nutritional profiles. Natural ingredients recovered from agroindustrial by-products have specific dietary and functional properties and can be utilized effectively to develop this new food category.

## 11.2. BY-PRODUCTS FROM PLANT PROCESSING

Fruit, vegetable, and oilseed processing result in various amounts of by-products depending on the raw material. Some typical yields of by-products, expressed as percent of the processed raw material weight are presented in Table 11.1, together with the accordingly estimated annual worldwide produced wastes. Laufenberg et al. (2003) presented several utilization concepts for vegetable by-products and commented that the raw materials mostly used are wastes from citrus, carrot, grape, apple, sugar beet, sugarcane, potato, and cereals. In addition to uses based on

*Table 11.1.* Annual worldwide processed quantities and resulting wastes of selected fruits and vegetables

| Fruit or vegetable | Annual processed (million Mt) | By-product/ processed fruit (%, wet basis) | Estimated annual waste (million Mt) |
|---|---|---|---|
| Orange and other citrus fruits | 31.2[a] | 50 | 15.6 |
| Apple | 12.0[a] | 25–35 | 3.0–4.2 |
| Pear | 1.7[a] | NA[f] | |
| Peach (canned) | 1.0[a] | NA | |
| Grape | 50[b] | 15–20 | 5–9 |
| Cooking banana | 30[b] | 30 | 9 |
| Kiwifruit | 1.0[c] | 30 | <0.3 |
| Tomato | 30.0[d] | 3–7 | 0.9–2.1 |
| Potato | 150[e] | 15–45 | 22–67 |
| (frozen potatoes) | 6.5[a] | | |
| Carrot | 23.6[c] | 30–40 | <10 |

[a] Processed in 2003/4 according to ffas.usda.gov/psd/complete_tables.
[b] Schieber et al. (2001).
[c] Produced in 2003/4 according to faostat.fao.org/faosta/servlet.
[d] Commission of the European Communities (2000) report.
[e] Estimated as approximately 50% of the world production.
[f] NA: nonavailable

ingredients that are covered in the present chapter, he presented potential exploitation for the production of natural flavors by solid-state fermentation and as bioadsorbents for wastewater treatment.

### 11.2.1. Citrus Fruits

Citrus is the largest fruit crop worldwide, with approximately 100 million metric tons produced annually (FAO, 2005). Oranges account for 60% of the total and the major utilization is processing for juice recovery. Brazil and the United States are the major producers; they grow around 60% of the world oranges and process 85% of their production (Braddock, 1995). The Mediterranean countries also present significant production and processing. The EU countries more specifically produce annually approximately 10 million metric tons (Commission of the European Communities, 2002).

The large quantities of processed fruits result in large amounts of by-products. Industrial utilization of these by-products had been encountered before 1920. Citrus peel, remaining after juice extraction, is the primary waste fraction, amounting to almost 50% of the fruit mass (Braddock, 1995). It is processed to dried pulp cattle feed and molasses, the latter being incorporated into the cattle feed or fermented for the production of alcohol or other products. Pectin is also produced from the peel by acid extraction, dietary fibers by mechanical processing, while the recovery of flavonoids and carotenoids are new potential applications (Kimball, 1999). Juice pulp from the finishing process and essences recovered from the juice and the peel press liquor, amounting to approximately 5% of the fruit mass, also are by-products that find industrial utilization. Citrus seeds amount to 0.1–5% of the fruit mass, depending on the variety. They can be used for oil extraction and recovery of terpenoids, while the meal remaining from the extraction is a good source of proteins. However, the cost of collecting sufficient quantities of seeds in one place limits commercial exploitation (Braddock, 1995).

### 11.2.2. Apple

World apple production is estimated to nearly 27 million metric tons annually (FAO, 2005). Around 9 million tons are produced in EU (Commission of the European Communities, 2002). Processed apples worldwide amounted to upto 12.0 million metric tons (Table 11.1); China is the greater processor (3.8 million Mt), followed by the United States (1.44 million Mt). Apple pomace is the by-product of processing for the recovery of apple juice and amounts to upto 25–35% of the processed fruit (Schieber et al., 2003; Wank and Thomas, 1989). Around 1.3 and 0.25 million metric tons of apple pomace are produced annually in the United States and Germany, respectively (Wank and Thomas, 1989). Although it is conventionally used as an animal feed, production of pectin is considered to be the most reasonable utilization approach from both economical and ecological points of view. Pectin consists of 10–15% of apple pomace, on a dry weight basis (Schieber et al., 2003) and is recovered by acid extraction and precipitation. Pectin currently produced from apple pomace has more superior gelling properties than citrus pectin, but presents a brown hue that may limit its incorporation into light-colored foods. The color of apple pomace and extracted pectin is formed by the oxidation of phenolic constituents that are coextracted with pectin and partially precipitated. Recently a new process has been proposed (Schieber et al., 2003) for the removal of phenolic compounds from the pomace extract through absorption by a hydrophobic styrene–divinylbenzene copolymerizate, followed by elution with ethanol for the recovery of polyphenols. The produced pectin by the aforementioned method is light-colored, while the recovered polyphenols may be used as antioxidants or as ingredients for functional foods. Apple kernel is a promising protein source due to high protein content (37%) (Karson et al., 1994).

### 11.2.3. Peach and Apricot

Around 0.7 million metric tons of peaches are processed to natural juice and/or canned in syrup in the EU (Commission of the European Communities, 2002). The pomace remaining after the production of juice can be used for the recovery of pectin. Pectin recovered from fresh peach pomace is highly methoxylated and has good gelling properties (Pagan et al., 1999).

Kernels constitute another main by-product of peaches or apricots processing. Apart from the use of kernel oil in cosmetics, peeled kernels serve as a raw material for the production of persipan. This requires debittering by hydrolysis of the cyanogenic glycosides. Grinding, soaking, and cooking reduce the cyanogenic glycosides content, but complete hydrolysis may be achieved by the use of β-glucosidase (Schieber et al, 2001). Apricot and peach kernels also are good sources of protein containing 21–45% and 28% protein, respectively; the main problem for their use in foods is the presence of amygdalin (Abd El-Aal et al., 1986; Rahma and Abd El-Aal, 1988).

### 11.2.4. Grape

Grapes are the world's second greatest fruit crop with an annual production of more than 60 million metric tons. Around 80% of the total crop is used in wine making, leaving 15–20% of the fruit weight as grape pomace. According to these data, 5–9 million metric tons of pomace are produced annually (Louli et al., 2004; Schieber et al., 2001). Many compounds, including tartrates, citric acid, grape seed oil, hydrocolloids, and dietary fibers, can be recovered from grape pomace. However, a great deal of research studies on the health effects of grape and wine phenolics appeared during the last years and established grape phenols as the most valuable by-products of the grape pomace. These phenols, recovered by extraction, include anthocyanins, catechins, flavonoids, and phenolic acids and can be used as antioxidants or functional food ingredients. The residue remaining after the extraction of phenols can be used as dietary fiber, while an antioxidant dietary fiber product also has been developed (Saura-Calixto, 1998).

### 11.2.5. Banana

Annual world production of bananas is about 70 million metric tons (FAO, 2005). The EU is a minor producer (0.7 million Mt), but imports up to 3.0 million Mt (Commission of the European Communities, 2002). Cooking bananas (plantains) production approximates 30 million metric tons worldwide, while the banana peels comprise 30% of the ripe fruit; several attempts for their utilization have been reported (Schieber et al., 2001). They include the production of bioprotein, ethanol, and several enzymes ($\alpha$-amylase, hemicellulases, cellulases). The recovery of anthocyanins and carotenoids seems promising, according to recent investigations (Pazmino-Duran et al., 2001).

### 11.2.6. Kiwifruit

By-products result from the rejected kiwifruits, which comprise up to 30% of the total crop [1 million metric tons worldwide, according to FAO (2005)] and from kiwifruit pomace after juice production (Schieber et al., 2001). Apart from the biotechnological production of several compounds, kiwifruit waste could be utilized for the recovery of polyphenols and of dietary fibers that account for up to 25% of the dry material (Martin-Cabrejas et al., 1995).

### 11.2.7. Tomato

The world production of tomatoes for processing totaled around 29.6 million metric tons in 1999/2000. The United States is the leading producer country (11.6 million Mt), followed by the EU (9.1 million Mt) and Turkey (1.86 million Mt) (Commission of the European Communities, 2002). About 3–7% of the raw material is lost as waste during processing for tomato juice or similar products (Schieber et al., 2001). The pomace consists of the crushed skins and seeds of the fruit. It is rich in protein (20–23% on dry basis) and fat (12–18%, mostly located in the seeds), while crude fibers comprise the third main component (12–30%) (Liadakis, 1999). Tomato seed oil has a high content in unsaturated fatty acids, especially in linoleic acid; therefore, it is extracted, refined, and used as edible oil or as an ingredient in cosmetics, etc. The defatted material (tomato seed meal) is exploited as animal feed, though protein isolation attracted interest, since protein amounts to 40–55% of the dry meal and presents a high nutritional value (Liadakis, 1999; Liadakis et al., 1995). Other valuable components are carotenoids, especially lycopene, which are principally located in the skins (Baysal et al., 2000).

### 11.2.8. Potato

Potato is the largest vegetable crop worldwide, amounting to approximately 320 million metric tons annually (FAO, 2005). Processing of potatoes (mainly for the production of chips, French fries, and dehydrated products) has presented a steady increase during the last decades, exceeding considerably the amount of the vegetable consumed as fresh (Kadam et al., 1991; Schieber et al., 2001). Solid waste generated during processing consists mostly of potato peels but also contains green, immature, and cull potatoes and amounts to 15–45% depending on the procedure applied (Schieber et al., 2001). It is used as animal feed, though fermentation for the production of single cell or alcohol has been considered (Natu et al., 1991). Water from potato processing is used for the recovery of proteins by heat coagulation (Natu et al., 1991). Recent investigations suggested the use of water extracts from potato processing waste for the recovery of antioxidants (Rodriguez de Sotillo et al., 1994, 1998).

### 11.2.9. Carrot

Carrot juice and blends thereof are among the most popular nonalcoholic beverages. A steady increase in carrot juice consumption has been observed in Western countries during the last years, leading to a rise in juice production (Chen and Tang, 1998). Despite the considerable improvements in processing technology to increase yield, a considerable amount of valuable components is retained in the pomace, which accounts for 30–40% of the vegetable (Sims et al., 1993). Carotenes (up to 2 g/kg dry matter) are considered to be the most valuable components to be recovered from the pomace, while other major constituents are uronic acids, neutral sugars, and fibers (Schieber et al., 2001).

### 11.2.10. Oilseeds

Thirty to thirty-five million metric tons of oilseeds are processed in the EU. They yield two products: oil, chiefly for human consumption, and meal (cake), which is actually used as animal feed. Soybeans are the major processed crop (around 50% of the total), followed by rapeseeds (around 33%) and sunflower seeds (around 18%) (Commission of the European Communities, 2002). Oilseeds contain available amounts of oil (17–47%). Oilseed meals come from oil processing after the recovery of the oil by mechanical pressing or solvent extraction. Oilseed meals deriving after desolventizing are significant protein sources due to high protein content, low price, and high availability (Visser and Thomas, 1987; Friedman, 1996). The nutritional properties of the oilseed meal proteins are determined by their amino acid composition and they can be converted into edible-grade products. Their utilization in human nutrition is limited because of the presence of phenolic substances, reduced sugars, or even toxic factors. The cottonseed kernel makes up about 55% of the seed's weight and contains 39% protein. After processing, cottonseed yields about 45% meal by weight. A major problem in the use of cottonseed as an animal feed is the presence of gossypol, a yellow polyphenolic pigment (Liadakis et al., 1993). Sunflower seed contains protein between 15 and 20% by weight. The presence of chlorogenic, quinic, and caffeic acid in the seed imparts dark color to the protein products resulting from the meals. The soybean has the highest protein content of all oilseeds; soybean contains 43% protein and the defatted meal 52% protein on a dry basis. Even though the soybean oil is the dominant vegetable oil worldwide, the major economic value of soybean is in the protein. Soybean contains undesirable constituents including trypsin inhibitors, hemagglutinins, phytic acid, saponins, and enzymes. Defatted soybean meal is used mostly as an animal feed; however, some is also processed into protein products, which are incorporated in foods.

## 11.3. RECOVERY OF FUNCTIONAL COMPOUNDS

A general flow diagram for the recovery of various dietary components from fruit peel is presented in Figure 11.1. Depending on the fruit some of the indicated steps may be omitted. Some aspects that have to be considered in the processing steps are the following:

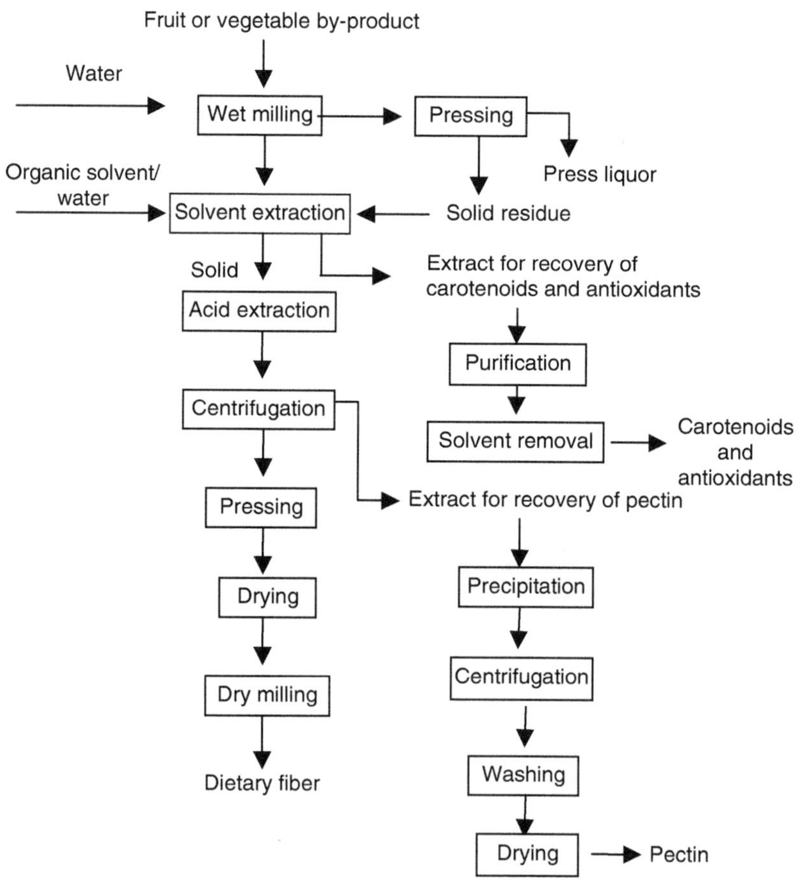


**Figure 11.1.** Flow diagram for the utilization of fruit or vegetable by-products.

1. *Wet milling* is necessary to facilitate and improve the yield of the following extraction steps. Too small particle size of the raw material, however, will result in retention of high amounts of water and will cause difficulties in pressing. Hammer mills are preferred to colloidal mills, in order to obtain a good control of particle size.

2. *Solvent extraction* is used for the recovery of colorants and antioxidants. Several organic solvents or mixtures with water can be used, and their choice depends on the raw material and the polarity of the compounds to be extracted. The process may be accomplished in a packed bed reactor operating under reflux of the solvent or in an agitated vessel followed by filtration to separate the extract from the residue. Repeated extraction steps increase the extract yield. The extraction conditions, such as temperature, time, solvent-to-material ratio, and pH, have to be optimized with respect to quantitative recovery and quality attributes of the extracted substances (for example, some of them may be degraded and decrease in activity at higher temperatures).

3. *Drying* methods and conditions vary for different products. Vacuum ovens or spray dryers, operating at low temperatures, are suitable for high-quality pectin (Kimball, 1999; Schieber et al., 2001). Drum dryers can be used for drying dietary fibers, though milder drying conditions result in higher quality, especially if the product contains bioactive substances (Larrauri, 1999). Drum drying has little effect on total dietary fibers consistency of the products, but it causes heat degradation of pectin and reduction of total sugar content (Wank and Thomas, 1989).

## 11.3.1. Extraction of Carotenoids, Other Colorants, and Antioxidants

Antioxidants and colorants have been extensively incorporated as additives into foods to improve quality. In the past decades synthetic or natural-identical ingredients have been used, but the toxicity of certain compounds as well as the recent consumer demand for ''all natural'' has increased the interest of the food industry to use compounds of natural origin. Natural colorants and antioxidants are found in many agroindustrial by-products and their recovery and purification might be a beneficial solution. Moreover, all these compounds impart health benefits, as pointed out by several researchers; therefore, their incorporation into food may upgrade the latter to functional food, increasing consumer acceptance.

### 11.3.1.1. Carotenoids and Other Colorants

Carotenoids can be extracted mainly from carrot pomace, orange peel, and tomato pomace. Organic solvents can be used for the extraction (Kew and Berry, 1970; Ting and Hendrickson, 1969) and acetone results in the highest yield compared to ethanol, petroleum ether, and hexane (Aravantinos-Zafiris et al., 1992). If fresh, wet raw material is used, initial washing with water removes most free sugars and other soluble compounds, like flavonoid glycosides, and increases the purity of the extracts. A sequential washing with acetone removes water and facilitates carotenoid extraction in the following steps. Successive extractions are needed for quantitative recovery, e.g., two extractions with acetone, at a solvent-to-peel ratio of 2:1 and ambient temperature, removes approximately 90% of the total carotenoids from orange peel (Aravantinos-Zafiris et al., 1992). Maceration of the raw material by enzymic preparations increases yield (Stoll et al., 2003) but may affect negatively the quality of pectin; thus, it should be carefully applied when pectin is recovered in a following step.

The extracted carotenoids may be obtained as a crude pigment, after solvent evaporation at low temperature, if preceding washings were accomplished. Alternatively, a purification step is needed by solvent–solvent transfer to hexane (Aravantinos-Zafiris et al., 1992).

Pigments of carrot pulp waste have been produced by spray drying or freeze-drying. Freeze-dried powders showed higher pigment stability than spray-dried ones, though the stability can be greatly enhanced by employing appropriate packaging methods and storage conditions (Chen and Tang, 1998; Tang and Chen, 2000). Other colorants that can be recovered from plant by-products are anthocyanins from grape pomace (Mazza, 1995) or banana bracts (Pazmino-Duran et al., 2001).

#### 11.3.1.2. Antioxidants

The common plant phenolic antioxidants are tocopherols; flavonoids and related compounds like coumarins, cinnamic acid derivatives, and chalcones; phenolic diterpenes; and phenolic acids. They act as free-radical scavengers, retard oxidative rancidity, and thus protect oils, fats, and fat-soluble components and delay the development of unpleasant flavors and odors resulting from oxidation. By-products like grape skins and seeds (Castillo et al., 2000; Gamez-Meza et al., 1999; Katalinic, 1999; Louli et al., 2004), citrus peel and pulp (Benavente-Garcia et al., 1997; Miyake et al., 1997; Ting and Newhall, 1965), apple pomace (Lu and Foo, 2000; Schieber et al., 2003), potato peel waste (Rodrigues de Sotillo et al., 1994, 1998), onion by-products (Chu et al., 2000; Suh et al., 1999), bean hulls (Dur et al., 1997; Onyenebo and Hettriarachchy, 1991), etc., are rich in compounds with antioxidant activity. Oilseed meals remaining after oil extraction, wastes of olive oil production, and by-products from cereal processing are other potential sources of antioxidants (Moure et al., 2001; Oreopoulou, 2003). The antioxidant contents of some typical by-products are presented in Table 11.2.

The antioxidants are isolated by solvent extraction and both extraction yield and antioxidant activity of the extracts are strongly dependent on the solvent, due to the variant antioxidant potential of compounds with different polarity. Nonpolar solvents (hexane, petroleum ether) can be used for the recovery of tocopherols and certain phenolic terpenes. Ethyl ether and ethyl acetate are very efficient for the recovery of flavonoid aglycons, low-molecular-weight phenols, and phenolic acids. Solvents of higher polarity (ethanol or ethanol–water mixtures) additionally can extract flavonoid glycosides and higher molecular weight phenols, resulting in

*Table 11.2.* Antioxidant content in various by-products

| By-product | Phenol content (g/kg dry matter) | Reference |
|---|---|---|
| Apple pomace | 2.4 | Schieber et al. (2003) |
| Orange peel | 1.8[a] | Fernandez-Lopez et al. (2004) |
| | 2.4[b] | Braddock (1995) |
| Lemon peel | 13.3[c] | Coll et al. (1998) |
| | 1.9[a] | Fernandez-Lopez et al. (2004) |
| Grapefruit peel | 1.6[a] | Fernandez-Lopez et al. (2004) |
| | 2.4[d] | Braddock (1995) |
| Grape pomace | 13.8[e] | Louli et al. (2004) |
| | 43[f] | Larrauri et al. (1997) |
| Potato peel | 7.8[g] | Velioglu et al. (1998) |
| | 0.48[e] | Rodriguez de Sotillo et al. (1994) |
| Red onion scale | 105.5[g] | Velioglu et al. (1998) |
| Sunflower hulls | 97.5[g] | Velioglu et al. (1998) |
| Buckwheat hulls | 39.0[g] | Velioglu et al. (1998) |
| Durum wheat bran | 27.7[h] | Moure et al. (2001) |
| Oat hulls | 0.6 | Xing and White (1997) |

[a] In fresh peel, expressed as g/kg of fresh fruit.
[b] Hesperidin (g/kg) in the fresh fruit.
[c] Flavonoids (eriocitrin and hesperidin).
[d] Naringin (g/kg) in the fresh fruit.
[e] Expressed as gallic acid.
[f] Expressed as tannic acid.
[g] Expressed as ferulic acid.
[h] Phenolic acids analyzed by HPLC.

higher yields of total extracted polyphenols. However, many undesirable substances are frequently coextracted in the latter case and purification is necessary to isolate the antioxidant fraction.

In addition to the solvent used, other extraction parameters that affect yield and antioxidant activity of the extract are the temperature and time of the extraction and the pH in case of aqueous ethanol solutions (Moure et al., 2001; Oreopoulou, 2003). According to Wettasinghe and Shahidi (1999) the extraction of antioxidants from an oilseed meal with aqueous ethanol presents the highest yield at ethanol concentration in the range of 50–60%, while the antioxidant activity of the extract increases with temperature and time up to 70°C and 60 min, respectively, and declines afterward. In general, temperatures lower than 60–70°C are suggested for both extraction and drying of the raw material because they induce less damage to the antioxidant compounds. Acidic conditions favor the extraction with aqueous or alcoholic medium for some raw materials, like grape by-products (pH 3–3.5) (Saint-Cricq de Gaulejac et al., 1999), olive rape (pH 4), or wheat bran (Moure et al., 2001), while alkaline pH results in maximum antioxidant activity of extracts from oat fiber (Lehtinen and Laasko, 1998). The published results suggest that suitable solvents and optimum conditions have to be determined for each raw material. Recently the extraction of natural antioxidants with supercritical $CO_2$ has been investigated; however, the results show that the process is more suitable for the purification of the extracts (Louli et al., 2004; Oreopoulou, 2003).

### 11.3.2. Protein Recovery

Many methods and processes have been developed for the production of protein products by purifying or isolating the proteins from proteinaceous food materials (Lusas et al., 1982). More particularly, these processes comprise either protein purification by removal of nonprotein components or protein recovery by extraction and precipitation. Products deriving from these processes are protein concentrates or protein isolates, respectively; such products were first referred for soybeans. Afterwards a similar technology was also applied to other oilseeds or meals after oil processing, as well as to legumes and cereals and to some food processing wastes or by-products such as wastes from tomato or fruit juices processing. The complete utilization of oilseeds as protein sources providing various protein products, which is industrially applicable, is presented in Figure 11.2 (Considine and Considine, 1982; Visser and Thomas, 1987).

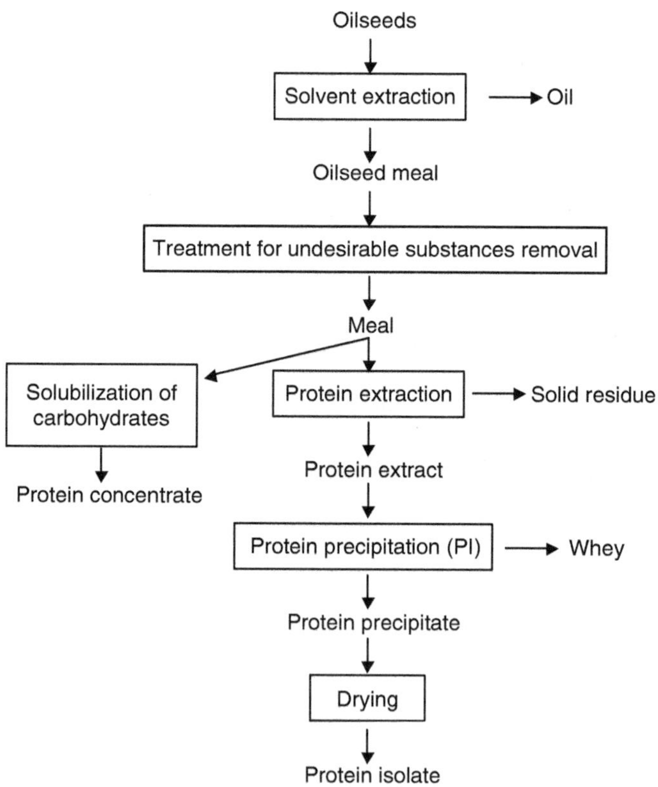


**Figure 11.2.** Complete utilization of oilseeds as protein sources.

The protein products are categorized based on their protein content as protein flours, concentrates, and isolates with about 50%, 70%, and higher than 95% protein content, respectively, and appropriate technology is applied for their production. These products are commonly derived from oilseeds and more often soybeans (Berk, 1992). Soy protein products of the above three types are commercially available and their production exceeds one million tons in total.

#### 11.3.2.1. Protein Flour

Protein flours and grits are simplified products directly resulting from typical (industrial) oil processes. The oil is conventionally solvent extracted by means of hexane, i.e., in soybean, oil, or alternatively mechanical pressing or propressing and solvent extraction may be used, i.e., in cottonseed oil. When protein meals and flours are intended for human nutrition several parameters should be considered, such as the quality of raw oilseeds, thorough dehulling, the extent of heat treatment, and the application of hygienic conditions during processing. So, in the case of flour production for edible use, before oil extraction, the oilseeds must be purified from foreign materials, impurities, and hulls by passing through scalpers and aspirators, then dried and tempered. Then, in order to facilitate extraction, the seeds are cracked in roll mills and after dehulling by aspiration and screening they are conditioned, commonly by heating at 65–70°C until their moisture content reaches 10.5–11% by weight of the seed, and finally flaked. After solvent extraction the defatted flakes are desolventized depending on their intended use either by steam, followed by toasting at temperatures above 93°C for the deactivation of undesirable enzymes inhibitors, or under vacuum (flash) to avoid protein denaturation. Toasted flakes are used as animal feeds, while the resulting white flakes from flash processing are suitable for edible uses (Berk, 1992; Considine and Considine, 1982; Milligan, et al., 1981; Tandy, 1991). The dried flours are ground in two steps using a hammer mill to certain particle sizes.

Full fat flours may be produced following a similar process with the one used for defatted flours without the solvent extraction step for oil removal. Also, refatted and lecithinated flours are produced with the addition of various amounts of fat or lecithin in a defatted flour, respectively. A typical composition of defatted soy flour is 53% protein and less than 1% fat, while that of full-fat flour is 41.5% protein and 18–21% fat, respectively.

#### 11.3.2.2. Protein Concentrate

Protein concentrate process is based on the extraction of carbohydrates, salts, and other low-molecular-weight substances from the flakes, leaving behind the insoluble proteins. The process usually starts with ground defatted flakes and the products contain at least 70% protein. The most common extraction methods are with (1) water–alcohol, (2) dilute acid solution, and (3) hot water after protein denaturation. By using these extracting media the dissolution of significant amounts of protein should be minimized. The resulting products have different functional properties depending on the extracting medium (method).

In the first method, methanol, ethanol, or isopropanol aqueous solutions (60–70%, w/w) are referred, as they are able to extract the fraction of soluble carbohydrates without significantly affecting the dissolution of proteins. Simultaneously, the removal of undesirable phenolic components or the deactivation of trypsin inhibitors present in seeds, i.e., in soybeans, and consequently the improvement of the product quality may be obtained. The extraction may be repeated several times, taking care to ensure that most proteins remain in the insoluble material. The protein concentrate is obtained after alcohol removal by flash desolventising of the extracted flakes and grinding (Berk, 1992; Bernardini, 1983; Mustakas et al., 1962).

In the second method defatted flakes or flour are mixed with a diluted acid solution at low pH, i.e., for soy at 4.5, and the insoluble material is separated by centrifugation from the whey containing the impurities. The method is based on the fact that the solubility of proteins at the isoelectric point is minimal, while the carbohydrates remain under solution. The insoluble material called ''isoelectric concentrate'' is dried and may be resolved after grinding, neutralized, and spray-dried (Berk 1992; Considine and Considine, 1982; Waggle and Kolar, 1979).

In the third method the flakes are toasted to denature the proteins and then are extracted with hot water at 66–93°C, at appropriately adjusted pH, i.e., for soybean in the range of 5.0–7.5. The method is based on the decrease of proteins solubility after their denaturation while maximizing carbohydrate dissolution (Berk, 1992; Waggle and Kolar, 1979).

In addition to the previous methods membrane technology has been recently developed for protein concentrates production. The extract resulting from flakes extraction with dilute acid is subjected to ultrafiltration using membranes of 100,000 MWCO. Thus, protein separation from low, molecular-weight substances (salts, carbohydrates, peptides, colorants, odorous materials, etc.) is achieved and the protein concentrate is obtained by spray-drying of the retentate (Kosseoglu and Rhee, 1994; Lawhon and Lusas, 1984).

The protein concentrates produced by applying all the above methods present a protein content of 70–72%, and improved sensory properties. They have 2–2.5 times higher value or they cost 80% greater on protein basis than flours. Usually from 1 tn flakes about 750 kg of protein concentrate are yielded.

#### 11.3.2.3. Protein Isolate

Protein isolate process has been developed based on the extraction, purification, and recovery of proteins from flakes. The protein isolation technology was for the first time applied to soy proteins. Contrary to protein concentration processes, the isolation approach is protein dissolution while most of the impurities remain in the solids (Berk, 1992). Although isolates can be produced from full-fat flakes, ground defatted flakes are usually used as in the case of protein concentrates facilitating the subsequent extraction. Protein isolates are usually products with more than 90% protein content, and having characteristic properties they can find certain useful food applications.

A typical procedure for protein isolate production commonly consists of three main steps: (1) protein extraction, (2) protein precipitation, and (3) drying of the protein precipitate. Especially the first two steps (extraction, precipitation), which are directly linked to protein solubility, are significant for the efficient isolation of proteins. For this reason the controlling factors that affect protein solubility have been studied in processes using various protein sources. The economic feasibility of protein isolation process depends on the optimization of each one of these steps (Liadakis et al., 1995; Park and Bean, 2003; Salcedo-Chávez et al., 2002; Tzia, 2003).

Following, the factors affecting protein extraction and precipitation steps are described and the conditions depending on the oilseed or protein source used are suggested. Prior to processing of oilseeds or meals for protein isolation, some pretreatments are required for their preparation or the removal of toxic or other undesirable substances (González-Pérez et al., 2002). Cottonseed and sunflower seed meals are treated with acid butanol to lower the gossypol and the chlorogenic acid content to permitted levels (0.00104% and 0.33%, respectively) (Liadakis et al., 1993; Tzia et al., 1998). In soybean meal the inactivation of antinutritional factors is achieved by heat treatment applied for solvent removal.

*11.3.2.3.a. Extraction.* The factors that influence extraction and therefore should be adjusted during the process are the extracting medium, temperature, pH, agitation, liquid-to-solids ratio, particle size of ground solid material, flakes formation, and extraction time. The extraction conditions should be selected so as to achieve the maximal protein dissolution while keeping them unaffected. Aqueous and some nonaqueous organic solvents have been examined in protein extraction, based on the solubility of oilseeds storage proteins: albumins are soluble in water (pH 6–8), globulins in dilute salt solutions, prolamines in ethanol, and glutelins in dilute acid or alkali. The extraction is usually carried out with water by adjusting the pH at 8–10 (by NaOH, $Ca(OH)_2$, or $NH_3$) as most of oilseeds proteins, i.e., soy proteins, are highly soluble in strong alkali and at elevated temperatures of 50–55°C. These conditions do not practically cause protein denaturation, and consequently alteration of functional properties; proteins may deteriorate at high pH values and high temperatures or through Maillard reactions with carbohydrates. In addition to water-alkali, various salt solutions also have been used in the protein extraction of oilseed meals such as $Na_2SO_3$ and NaCl in sunflowerseed and cottonseed (Mieth and Kroll, 1984; Yazicioglu et al., 1981), $Na_2CO_3$ and $NaHCO_3$ in sunflower seed (Taha et al., 1981), $CaCl_2$, $MgSO_4$, and $Na_2SO_4$ in cottonseed (El-Tinay et al., 1980), hexametaphosphate in flaxseed (Wanasundara and Shahidi, 1996), and NaCl and $Na_2CO_3$ in tomato seed or meal (Eggers, 1974; Liadakis et al., 1998; Tchorbanov et al., 1986). As far as the operating conditions are concerned, the particle size of ground solid material should be suitable in order to facilitate the extraction process and the subsequent protein solution separation. Agitation and extraction time

(45–60 min) as well should be adequate to maximize the protein yield. Agitation should be adequately adjusted so as to enhance protein extraction without destroying the flakes. For the liquid-to-solids ratio selection, the water holding in the flakes and the water waste due to separating whey after extraction should be taken into account, while the most common ratios used are 1:10–1:20.

*11.3.2.3.b. Precipitation.* After extraction, the aqueous extract is separated from the solids by screening, centrifugation or decanting, and polish-filtration (Berk, 1992; Bernardini and Bernardini, 1974). The solid material containing mainly dietary fibers after drying may be used as animal feed or for other non-food purposes.

The solubilized proteins contained in the extract may be recovered by precipitation. Precipitation is affected by factors that decrease protein solubility and should be adjusted in order to optimize the entire protein isolation process. Various agents including organic solvents, divalent cations, heat, acid/bases (for pH adjustment), salts, nonionic polymers, and polyelectrolytes have been used in protein precipitation (Belter et al., 1988; Glatz and Fisher, 1986; Hamel and Hunter, 1990; Wheelwright, 1991). During precipitation the formation of submicron particles takes place, followed by their aggregation; insoluble protein particles are formed in either native form or in a readily reversible form. A simple method usually applied is acid precipitation at the isoelectric point (pI) of proteins. A food-grade acid may be used for extract acidification, i.e., acetic, hydrochloric, phosphoric, or sulfuric acid (Liadakis, 1999; Liadakis et al., 1998). The pI of proteins should be determined; for most common oilseeds it is between pH 4 and 5. By isoelectric precipitation a high purity concentrated undernatured product is produced, while the highest protein recovery is accomplished (Bell et al., 1983; Wheelwright, 1991). The protein curd is collected by centrifugation or filtration, washed, and dried. Spray-drying is the most preferable method as it induces less protein deterioration and has lower cost compared to other drying methods, i.e., by drums or freezing. As in the case of protein concentrates membrane isolation process (MIP) also has been developed for protein isolates. Membranes of 10,000–100,000 MWCO are used for the concentration of the protein extract followed by drying (Lawhon et al., 1981). A typical isolation process yields about one third of the starting mass as a protein isolate, while the insoluble residue as well as the whey which occur after drying are used as animal feeds. Because the isoelectric isolates have low functionality, they may be redissolved in water (proteinates) and neutralized with alkali (sodium, potassium, or calcium hydroxide or ammonia) or modified by treatment with proteolytic enzymes; these products have improved functional properties.

Protein isolates have greater than 90% protein content and due to the isolation process they should not contain undesirable taste, odor, and color substances and should be practically free from toxic constituents or antinutritive factors. Protein isolates 5–7 have times higher value or 3 times greater cost on protein basis than flours. Usually from 1 tn flakes about 300–350 kg of protein isolate are yielded.

### 11.3.3. Recovery of Dietary Fibers

Dietary fibers are a combination of chemically heterogeneous substances, such as cellulose, hemicelluloses, pectin, gums, and lignin. Commonly used definitions are based on their physiological role: ''dietary fibers are oligosaccharides, polysaccharides, and the (hydrophilic) derivatives which cannot be digested by the human digestive enzymes to absorbable components in the upper alimentary tract. This includes lignins'' (Thebaudin et al., 1997). The constituents of dietary fibers are divided into soluble (pectin, gums) and insoluble (cellulose, most hemicelluloses, lignin). According to their consistency dietary fibers (DF) are classified as soluble (SDF) or insoluble (IDF). Upon hydration soluble fibers form a gel or a thickened network, while insoluble fibers swell and absorb up to 20 times their weight in water. Both SDF and IDF have been claimed to provide several health benefits (Figuerola et al., 2005; Thebaudin et al., 1997). However, regardless the nutritional benefits, dietary fibers can be incorporated into food only if the final product maintains appealing sensory characteristics. The addition of dietary fibers affects mainly the texture of the product and these ingredients are used for textural modification or stabilization. Especially the use of SDF contributes to the stabilization of the foods structure (dispersions, emulsions, etc.), through gel formation or thickening of the continuous phase. IDF on the other hand increases firmness of the products and provides higher fat absorption capacity. Therefore, the selection of preparations with proper SDF/IDF ratio is critical to achieve the desired functional properties.

Isolation of dietary fibers from plant by-products can follow the recovery of other constituents, like carotenoids, antioxidants, or proteins. Alternatively, dietary fibers may be recovered by omitting one or all of the aforementioned processing steps. Depending on the process followed the dietary fiber product has different composition, and therefore different functional and nutritional properties. Pectin belongs to dietary fibers and has excellent gelling properties. For this reason pectin extraction and purification, separately from the other fibers, is usually performed and will be examined in more details below.

#### 11.3.3.1. Pectin Extraction

Extraction of pectin is accomplished by the use of mineral acids, usually hydrochloric or nitric acid. The extract is separated from the solid residue and pectin is precipitated by the addition of ethanol or $AlCl_3$ solution. Purification of the precipitated pectin involves washing with acidified, alkaline, and finally neutral alcohol. The product is dried at low temperature (around 50°C) in vacuum or ventilated ovens or in spray-dryers. The extraction conditions affect both the yield and quality of pectin. Yield is increased by decreasing pH and increasing temperature or time of extraction. However, the gelling quality of pectin is adversely affected by the decrease in pH and especially by the increase of temperature, a fact mainly attributed to the hydrolysis of pectin. To limit extended hydrolysis mild extraction conditions may be used, i.e., temperature around 80°C and pH 1.2–1.6, depending on the raw material. (Aravantinos-Zafiris and Oreopoulou, 1992; Pagan et al., 1999).

The raw materials currently used for pectin production are citrus peel and apple pomace. However, peach pomace also yields good quality pectin (Pagan et al., 1999) and other fruit by-products could be used as well. Pectin extraction may follow a carotenoid and antioxidant recovery step or may be applied to the raw material just after juice extraction. In the current practice pectin is produced from the dried coarsely ground raw material, which remains after juice extraction. Drying and storage conditions of fruit peel or pomace affect the yield and the gelling properties of the produced pectin. Storage of dry, powdered peach pomace for several months, even at low temperature (3°C) and 95% humidity affects negatively the quality of pectin, even though it improves yield (Pagan et al., 2001). Therefore, storage conditions have to be carefully selected, in case production from fresh wet peel or pomace is not feasible.

On the other hand, the presence of polyphenols in the raw material, like in apple pomace, causes a dark hue in the produced pectin. If drying is accomplished under controlled conditions so that the temperature of the pomace does not exceed 50–60°C, the polyphenols remain unaffected and can be recovered by absorption with an appropriate resin from the solution of pectin extraction (Schieber et al., 2003).

#### 11.3.3.2. Isolation of Dietary Fibers

Mild processing so as to obtain dietary fibers rich in biologically active substances (carotenoids, antioxidants, etc.) includes the following steps:

1. Grinding to coarse particles when necessary, as in the case of fresh citrus peel or grape pomace.
2. Washing the pomace or ground peel with warm water (around 30°C) to remove free sugars while at the same time minimizing loss of valuable components. Usually two successive washings at a water-to-residue ratio of 2:1 remove approximately 90% of the free sugars (Aravantinos-Zafiris et al., 1994). Alternatively, Larrauri (1999) proposed the use of hot water to remove most of the free sugars. He reported that water-holding capacity increases with washing, while total polyphenols and SDF content decrease. Loss of the latter constituents, as well as removal of free sugars, are also affected by the particle size of the raw material and increase as particle size decreases. Also, washing facilitates drying and contributes to a lighter color of the final product.
3. Pressing of the washed material to remove excessive water.
4. Drying at low temperature (60°C or lower) to avoid changes in functional properties and bioactive components. More specifically a significant reduction of polyphenols content is observed at temperatures above 60°C, while

loss of carotenoids occurs above 40°C (Larrauri, 1999; Larrauri et al., 1997). High drying rates at 110°C also are found to decrease the water-holding capacity of the produced fibres (Larrauri, 1999).

5. Grinding to a powder of the desired size. The particle size of the final product affects its functional properties like water-and oil-holding capacity, suspension in water, etc. Grinding to fine powder may adversely affect the functional properties, since it not only causes particle size reduction but also alters the fiber matrix structure (Figuerola et al., 2005).

The dietary fibers products, which are obtained simply by mechanically processing the fruit and vegetable residues, have the typical flavor and color of the raw material. Apart from biologically active substances they are rich in soluble dietary fibers, especially pectin. Potential by-products for their recovery are citrus peel [with total dietary fiber content (TDF) 30–65%, on dry basis] (Grigelmo-Miguel and Martin-Belloso, 1999a), apple pomace (TDF: 60–90%), pear pomace (TDF: 35–45%), peach pomace (TDF: 35–55%) (Figuerola at al., 2005; Fernandez-Lopez at al., 2004; Grigelmo-Miguel and Martin-Belloso, 1999a,b; Pagan et al., 2001). The ratios of IDF to SDF of some plant wastes and DF concentrates are shown in Table 11.3. In most fruit and vegetable by-products this ratio approximates 2/1, though there are variations depending on the variety and maturity of the crop; especially some apple varieties have a high IDF/SDF ratio.

If an additional extraction step is introduced before the mechanical processing, i.e., solvent extraction for the recovery of carotenoids and antioxidants or acid extraction for the recovery of pectin, the resulting dietary fibers product is altered respectively, i.e., it may present a low content of carotenoids and antioxidants or a low content of SDF. The removal of carotenoids and antioxidants is accompanied by the removal of most flavor and color constituents of the raw material, and therefore the production of light-colored, odorless DF. In the case of pectin removal, as it is performed with acid, the residue requires washing to remove anions or perhaps residual acidity. For example, successive washing with 95% ethanol, and water can sufficiently remove nitrates remaining after pectin extraction with nitric acid from orange peels (Aravantinos-Zafiris et al., 1994). The obtained product amounts to up to 42% of the original peel (on dry basis) and contains 84% TDF, 25% of which was SDF, while its protein, sugar, and ash content are very low.

The consistency and particle size of a DF product plays a major role in the effectiveness of its incorporation into a food. Good or poor incorporation also depends on the temperature and addition method, as well as ionic strength and consistency of the food material. The DF provides volume, thickening, and texture to a food and can be used to replace fat, and thus reduce calories. Effective incorporation of the appropriate DF products can result in the development of foods with potentially reduced calories, enriched in DF (and biologically active micronutrients, if desired), with favorable sensory characteristics.

*Table 11.3.* Dietary fibers conent of some residues and DF concentrates obtained by washing, drying, and grinding of fruit and vegetable by-products (g/100 g dry matter)

| By-product | TDF | IDF | SDF | IDF/SDF | Reference |
|---|---|---|---|---|---|
| Wheat bran | 44 | 41 | 3 | 14.2 | Grigelmo-Miguel & Martin-Belloso (1999b) |
| Oat bran | 24 | 20 | 4 | 5.6 | Grigelmo-Miguel & Martin-Belloso (1999b) |
| Soybean shells | 65 | 57 | 8 | 7.4 | Laufenberg et al. (2003) |
| Grape pomace | 59 | 56 | 3 | 24.4 | Laufenberg et al. (2003) |
| Apple DF | 60 | 46 | 14 | 3.4 | Grigelmo-Miguel & Martin-Belloso (1999b) |
| Granny smith | 61 | 57 | 4 | 12.9 | Figuerola et al. (2005) |
| Liberty | 90 | 82 | 8 | 9.9 | Figuerola et al. (2005) |
| Apple pomace | 62 | 48 | 14 | 3.4 | Laufenberg et al. (2003) |
| Orange DF | 38 | 24 | 14 | 1.8 | Grigelmo-Miguel & Martin-Belloso (1999a) |
| Valencia | 64 | 54 | 10 | 5.3 | Figuerola et al. (2005) |
| Lemon DF | | | | | |
| Eureka | 60 | 51 | 9 | 9.9 | Figuerola et al. (2005) |
| Fino | 68 | 62 | 6 | 5.3 | Figuerola et al. (2005) |
| Lemon pulp | 51 | 28 | 23 | 1.2 | Laufenberg et al. (2003) |
| Grapefruit DF | | | | | |
| Ruby | 62 | 56 | 5 | 12.7 | Figuerola et al. (2005) |
| Marsh | 44 | 38 | 6 | 5.9 | Figuerola et al. (2005) |
| Pear DF | 36 | 22 | 14 | 1.6 | Grigelmo-Miguel & Martin-Belloso (1999b) |
| Pear pomace | 44 | 36 | 8 | 4.8 | Laufenberg et al. (2003) |
| Peach DF | 36 | 26 | 10 | 2.7 | Grigelmo-Miguel & Martin-Belloso (1999b) |
| Peach pomace | 54 | 35 | 19 | 1.9 | Pagan et al. (2001) |
| Kiwi pomace | 26 | 19 | 7 | 2.6 | Laufenberg et al. (2003) |
| Carrot pomace | 30 | 19 | 11 | 1.8 | Laufenberg et al. (2003) |
| Potato pulp | 16 | 9 | 7 | 1.5 | Laufenberg et al. (2003) |
| Sugar beet pulp | 75 | 50 | 25 | 2,0 | Laufenberg et al. (2003) |
| Cauliflower DF | 43 | 35 | 8 | 4.4 | Femenia et al. (1997) |
| Artichoke DF | 59 | 45 | 14 | 3.2 | Grigelmo-Miguel & Martin-Belloso (1999b) |
| Asparagus DF | 49 | 39 | 10 | 3.7 | Grigelmo-Miguel & Martin-Belloso (1999b) |

## 11.4. REFERENCES

Abd El-Aal, M. H., Hamza, M. A., and Rahma, E. H., 1986, In vitro digestibility, physicochemical and functional properties of apricot kernels, *Food Chem.* **19**:197.

Aravantinos-Zafiris, G., and Oreopoulou, V., 1992, The effect of nitric acid extraction variables on orange pectin, *J. Sci. Food Agric.* **60**:127.

Aravantinos-Zafiris, G., Oreopoulou, V., Tzia, C., and Thomopoulos C. D., 1992, Utilisation of orange by-products-Orange peel carotenoids, *J. Sci. Food Agric.* **59**:77.

Aravantinos-Zafiris, G., Oreopoulou, V., Tzia, C., and Thomopoulos C. D., 1994, Fiber fraction from orange peel residues after pectin extraction, *Lebensm. Wiss. Technol.* **27**:468.

Baysal, T., Ersus, S., and Starmans, D. A. G., 2000, Supercritical $CO_2$ extraction of β-carotene and lycopene from tomato paste waste, *J. Agric. Food. Chem.* **48**:5507.

Bell, D. J., Hoare, M., and Dunnill, P., 1983, The formation of protein precipitates and their centrifugal recovery, *Adv. Biochem. Eng./Biotechnol.* **26**:1.

Belter, P. A., Cussler, E. L., and Hu, W. S., 1988, *Bioseparations. Downstream Processing for Biotechnology*, Wiley, New York, pp. 221–236.

Benavente-Garcia, O., Castillo, J., Marin, F. R., Ortuno, A., and Del Rio, J. A., 1997, Uses and properties of citrus flavonoids, *J. Agric. Food. Chem.* **45**:4505.

Berk, Z., 1992, *Technology of Production of Edible Flours and Protein Products from Soybeans*, FAO Agricultural Services Bulletin 97, FAO, Rome.

Bernardini, E., 1983, *Oilseeeds, Oils and Fats*, B.E. Oil Publishing House, Rome, pp. 361–363, 483–493.

Bernardini, E., and Bernardini, M., 1974, Industrial apparatus for the production of protein superconcentrates from seed meals, *Trop. Sci.* **16**:23.

Braddock, R. J., 1995, By-products of citrus fruit, *Food Technol.* **49**:74.

Castillo, J., Benavente-Garcia, O., Lorente, J., Alcaraz, M., Redondo, A., Ortuno, A., and Del Rio, J. A., 2000, Antioxidant activity and radioprotective effects against chromosomal damage induced in vivo by X-rays of flavan-3-ols (procyanidins) from grape seeds (*vitis vinifera*): comparative study versus other phenolic organic compounds, *J. Agric. Food. Chem.* **48**:1738.

Chen, B. H., and Tang, Y. C., 1998, Processing and stability of carotenoid powder from carrot pulp waste, *J. Agric. Food. Chem.* **46**:2312.

Chu, Y. H., Chang, C. L., and Hsu, H. F., 2000, Flavonoid content of several vegetables and their antioxidant activity, *J. Sci. Food Agric.* **80**:561.

Coll, M. D., Coll, L., Laencina, J., and Tomas-Barberan, F. A. 1998, Recovery of flavanones from wastes of industrially processed lemons, *Z. Lebenm. Unters. F. A.* **206**:404.

Commission of the European Communities, 2002, *The Agricultural situation in the European Union*, 2000 Report, Brussels, 11.02.2002 COM (2002) 67 final.

Considine, D. M., and Considine, G. D., 1982, *Foods and Food Production Encyclopedia*, Van Nostrand Reinhold Co. Inc., New York, pp. 1610–1622, 1856–1883, 2029–2047.

Dur, P. D., Yen, W. Y., Du, P. C., and Yen, G. C., 1997, Antioxidant activity of Mung bean hulls, *JAOCS* **74**:1059.

Eggers, L. K., 1974, *Some Biochemical and Electron Microscopic Studies of the Protein Present in Seeds Recovered from Tomato Cannery Waste*, Ph.D. Dissertation, The Ohio State Univ., Columbus, OH.

El-Tinay, A., Chandrasekhar, H., and Ramanatham, G., 1980, Protein and gossypol extractability from cottonseed flour, *J. Sci. Food Agric.* **31**:38.

FAO Production Yearbook, 2005, hendry.ifas.ufl.edu.

Femenia, A., Lefebvre, A. -C., Thebaudin, J. -Y., Robertson, J. A., and Bourgeois, C. -M., 1997, Physical and sensory properties of model foods supplemented with cauliflower fiber, *J. Food Sci.* **62**:635.

Fernandez-Lopez, J., Fernandez-Guines, J. M., Aleson-Carbonell, L., Sendra, E., Sayas-Barbera, E., and Perez-Alvarez, J. A., 2004, Application of functional citrus by-products to meat products, *Trends Food Sci. Tech.* **15**:176.

Figuerola, F., Hurtado, M. L., Estevez, A. M., Chiffelle, I., and Asenjo, F., 2005, Fiber concentrates from apple pomace and citrus peel as potential fiber sources for food enrichment, *Food Chem.* **91**:395.

Friedman, M., 1996, Nutritional value of proteins from different food sources. A Review, *J. Agric. Food. Chem.* **44**:6.

Gamez-Meza, N., Noriega-Rodriges, J. A., Medina-Juarez, L. A., Ortega-Garcia, J., Cazarez-Casanova, R., and Angulo-Guerrero, O., 1999, Antioxidant activity in soybean oil of extracts from Thomson grape bagasse, *JAOCS* **76**:1445.

Glatz, C. E., and Fisher, R. R., 1986, Modelling of precipitation phenomena in protein recovery, *ACS Symposium Series*, 314, pp. 109–120.

González-Pérez, S., Merk, K. B., Vereijken, J. M., Van Koningsveld, G. A., Gruppen, H., and Voragen, A. G. J., 2002, Isolation and characterization of undernatured chlorogenic acid free sunflower (*Helianthus annuus)* proteins, *J. Agric. Food. Chem.* **50**:1713.

Hamel, J. P., and Hunter, J. B., 1990, Modelling and applications of downstream processing, *ACS Symposium Series*, 419, pp. 1–35.

Grigelmo-Miguel, N., and Martin-Belloso, O., 1999a, Characterization of dietary fiber from orange juice extraction, *Food Res. Int.* **31**:355.

Grigelmo-Miguel, N., and Martin-Belloso, O., 1999b, Comparison of dietary fiber from by-products of processing fruits and greens and from cereals, *Lebensm. Wiss. Technol.* **32**:503.

Herbafood, 2005, Apple and citrus fiber; http://www.herbafood.de.

Kandaswami, C., and Middleton, E., 1997, Flavonoids as antioxidants, in: *Natural Antioxidants, Chemistry, Health Effects and Practical Applications,* F. Shahidi, ed., AOCS Press, Champaign Illinois, pp. 174–194.

Kadam, S. S., Wankier, B. N., and Adsule, R. N., 1991, Processing, in: *Potato: Production, Processing and Products*, D. K. Salunke, S. S. Kadam, and S. J. Jadhav, eds., CRC Press, Boca Raton, pp. 111–154.

Karson, K. J., Collins, J. L., and Penfield, M. P., 1994, Unrefined, dried apple pomace as a potential food ingredient, *J. Food Sci.* **59**:1213.

Katalinic, V., 1999, Grape catechins-natural antioxidants, *J. Wine Res.* **10**:15.

Kew, T. J., and Berry, R. E., 1970, Citrus product color enhancement using extracts of peel of different varieties, *J. Food Sci.* **35**:436.

Kimball, D. A., 1999, *Citrus Processing: A Complete Guide*, 2nd ed., Aspen Inc., Gaithersburg.

Kiokias, S., and Gordon, M. H., 2004, Antioxidant properties of carotenoids in vivo and in vitro, *Food Rev. Int.* **20**:99.

Kosseoglu, S. S., and Rhee, K. C., 1994, New technologies in processing protein products, *Presented at the 85th AOCS Annual Meeting*, Atlanta, GA, USA, May 8–12.

Larrauri, J. A., 1999, New approaches for the preparation of high dietary fibre powders from fruit by-products, *Trends Food Sci. Tech.* **10**:3.

Larrauri, J. A., Ruperez, P., and Saura-Calixto, F., 1997, Effect of drying temperature on the antioxidant activity of red grape pomace peel, *J. Agric. Food. Chem.* **45**:1390.

Laufenberg, G., Kunz, B., and Nystroem, M., 2003, Tranformation of vegetable waste into value added products: (A) the upgrading concept; (B) practical implementations, *Bioresour. Technol.* **87**:167.

Lawhon, J. T., and Lusas, E. W., 1984, New techniques in membrane processing of oilseeds, *Food Technol.* **38**:97.

Lawhon, J. T., Manak, L. J., Rhee, K. C., Rhee, K. S., and Lusas, E. W., 1981, Combining aqueous extraction and membrane isolation techniques to recover protein and oil from soybeans, *J. Food Sci.* **46**:912.

Lehtinen, P., and Laasko, S., 1998, Effect of extraction conditions on the recovery and potency of antioxidants in oat fiber, *J. Agric. Food. Chem.* **46**:4842.

Liadakis, G., 1999, *Utilization of By-Products of Tomato Industrial Processing,* Ph.D thesis, National Technical University of Athens, Athens, Greece.

Liadakis, G. N., Floridis, A., Tzia, C., and Oreopoulou, V., 1993, Protein isolates with reduced gossypol content from screw-pressed cottonseed meal, *J. Agric. Food. Chem.* **41**:918.

Liadakis, G., Tzia, C., Oreopoulou, V., and Thomopoulos, C. D., 1995, Protein isolation from tomato seed meal, extraction optimisation, *J. Food Sci.* **60**:477.

Liadakis, G. N., Tzia, C., Oreopoulou, V., and Thomopoulos, C. D., 1998, Isolation of tomato seed meal proteins with salt solutions, *J. Food Sci.* **63**:450.

Louli, V., Ragoussis, N., and Magoulas, K., 2004, Recovery of phenolic antioxidants from wine industry by-products, *Bioresour. Technol.* **92**:201.

Lu, Y., and Foo, L. Y., 2000, Antioxidant and radical scavenging activities of polyphenols from apple pomace, *Food Chem.* **68**:81.

Lusas, E. W., Lawhon, J. T., and Rhee, K. C., 1982, Producing edible oil and protein from oilseeds by aqueous processing, *Oil Mill Gazetteer*, **86**:28.

Martin-Cabrejas, M. A., Esteban, R. M., Lopez-Andreu, F. J., Waldron, K., and Selvendran, R. R., 1995, Dietary fiber content of pear and kiwi pomaces, *J. Agric. Food. Chem.* **43**:662.

Mazza, G., 1995, Anthocyanins in grapes and grape products, *Crit. Rev. Food Sci.* **35**:341.

Mieth, G., and Kroll, J., 1984, Sonnenblumensamen und deren verarbeitungsprodukte 2. Mitt. Gewinnung proteinangereicherter verarbeitungsprodukte, *Die Nahrung.* **28**:907.

Milligan, E. D., Amlie, J. H., Reyes, J., Garcia, A., and Meyer, B., 1981, Processing for production of edible soya flour, *JAOCS* **58**:331.

Miyake, Y., Yamamoto, K., Morimitsu, Y., and Osawa, T., 1997, Isolation of C- glucosyflavone from lemon peel and antioxidated activity of flavonoid compounds in lemon fruit, *J. Agric. Food. Chem.* **45**:4619.

Moure, A., Cruz, J. M., Franco, D., Dominguez, J. M., Siniero J., Dominguez, H., Nunez, M. J., and Parajo, J. C., 2001, Natural antioxidants from residual sources, *Food Chem.* **72**:145.

Mustakas, G. C., Kirk, L. D., and Griffin, E. L. Jr., 1962, Flash desolventizing defatted soybean meals washed with aqueous alcohols to yield a high-protein product, *JAOCS* **39**:222.

Natu, R. B., Mazza, G., and Jadhav, S. J., 1991, Waste utilization, in: *Potato: Production, Processing and Products*, D. K. Salunke, S. S. Kadam, and S. J. Jadhav, eds., CRC Press, Boca Raton, pp. 175–201.

Onyenebo, S. N., and Hettriarachchy, N. S., 1991, Effect of Navy bean hull extract on the oxidative stability of soy and sunflower oils, *J. Agric. Food. Chem.* **39**:1701.

Oreopoulou, V., 2003, Extraction of natural antioxidants, in: *Extraction Optimization in Food Engineering*, C. Tzia, and G. Liadakis, eds. Marcel Dekker Inc., New York, pp. 329–346.

Pagan, J., Ibarz, A., Llorca, M., and Coll, L., 1999, Quality of industrial pectin extracted from peach pomace at different pH and temperature, *J. Sci. Food Agric.* **79**:1038.

Pagan, J., Ibarz, A., Llorca, M., Pagan, A., and Barbosa-Canovas, G. V., 2001, Extraction and characterization of pectin from stored peach pomace, *Food Res. Int.* **34**:605.

Park, S. H., and Bean, S. R., 2003, Investigation and optimization of the factors influencing sorghum protein extraction, *J. Sci. Food Agric.* **51**:7050.

Pazmino-Duran, E. A., Giusti, M. M., Wrolstad, R. E., and Gloria, M. B. A., 2001, Anthocyanins from banana bracts (*Musa* × *paradisiacal*) as potential food colorant, *Food Chem.* **73**:327.

Rahma, E. H., and Abd El-Aal, M. H., 1988, Chemical characterization of peach kernel oil and protein: functional properties in vitro digestibility and amino acids profile of the flour, *Food Chem.* **28**:31.

Rodrigues de Sotillo, D., Hadley, M., and Holm, E. T., 1994, Potato peel waste: stability and antioxidant activity of a freeze-dried extract, *J. Food Sci.* **59**:1031.

Rodrigues de Sotillo, D., Hadley, M., and Wolf-Hall, C., 1998, Potato peel extract a nonmutagenic antioxidant with potential antimicrobial activity, *J. Food Sci.* **63**:907.

Sahu, S. C., and Green, S., 1997, Food antioxidants: Their dual role in carcinogennesis, in: *Oxidants, Antioxidants and Free Radicals*, S. Baskin and H. Salem, eds., Taylor and Francis, Washington, pp. 329–330.

Saint-Cricq de Gaulejac, N., Glories, Y., and Vivas, N., 1999, Free radical scavenging effects of anthocyanins in red wine, *Food Res. Int.* **32**:27.

Salcedo-Chávez, B., Osuna-Castro, J., Guevara-Lara, F., Domínguez-Domínguez, J., and Paredes-López, O., 2002, Optimization of the isoelectric precipitation method to obtain protein isolates from Amaranth (*Amaranthus cruentus*) seeds, *J. Sci. Food Agric.* **50**:6515.

Saura-Calixto, F., 1998, Antioxidant dietary fiber product: a new concept and a potential food ingredient, *J. Agric. Food. Chem.* **46**:4303.

Schieber, A., Hilt, P., Streker, P., Endreb, H. U., Rentschler, C., and Carle, R., 2003, A new process for the combined recovery of pectin and phenolic compounds from apple pomace, *Innovative Food Science and Emerging Technologies.* **4**:99.

Schieber, A., Stintzing, F. C., and Carle, R., 2001, By-products of plant food processing as a source of functional compounds-recent developments, *Trends Food Sci. Tech.* **12**:401.

Sims, C. A., Balaban, M. O., and Matthews, R. F., 1993. Optimization of carrot juice color and cloud stability, *J. Food Sci.* **58**:1129.

Stoll, T., Schweiggert, U., Schieber, A., and Carle, R., 2003, Process for the recovery of a carotene-rich functional food ingredient from carrot pomace by enzymatic liquefaction, *Innovative Food Science and Emerging Technologies.* **4**:415.

Suh, H. J., Lee, J. M., Cho, J. S., Kim, Y. S., and Chung, S. H., 1999, Radical scavenging compounds in onion skin, *Food Res. Int.* **32**:659.

Taha, F. S., Abbasy, M., El-Nockrashy, A. S., and Shoeb, Z. E., 1981, Countercurrent extraction-Isoelectric precipitation of sunflower seed protein isolates, *J. Sci. Food Agric.* **32**:166.

Tandy, D. C., 1991, Oilseed extraction, in: *Introduction to Fats and Oils Technology*, P. J. Wan, ed., AOCS, Champaign, pp. 59–84.

Tang, Y. C., and Chen, B. H., 2000, Pigment change of freeze-dried carotenoid powder during storage, *Food Chem.* **69**:11.

Tchorbanov, B., Ushanova, G., and Litchev, V., 1986, Lipid-protein concentrations from tomato and paprika seeds, *Nahrung*. **30**:408.

Thebaudin, J. Y., Lefebvre, A. C., Harrington, M., and Bourgeois, C. M., 1997, Dietary fibers: Nutritional and technological interest, *Trends Food Sci. Tech.* **8**:41.

Ting, S. V., and Hendrickson, R., 1969, Natural color enhancers—Orange peel carotenoids for orange juice products, *Food Technol.* **23**:947.

Ting, S. V., and Newhall, W. F., 1965, The occurrence of a natural antioxidant in citrus fruit, *J. Food Sci.* **30**:57.

Tzia, C., 2003, Optimization methods, in: *Extraction Optimization in Food Engineering*, C. Tzia and G. Liadakis, eds., Marcel Dekker Inc., New York, pp. 137–180.

Tzia, C., Oreopoulou, V., Liadakis, G. N., and Bitzeli, E., 1998, Oilseed meal processing for protein isolate production, *Advances in Oils and Fats, Antioxidants, and By-Products,* Vol II, in: *Proceedings of the World Conference and Exhibition on Oilseed and Edible Oils Processing*, 6–10 October, 1996, Istanbul, Turkey, 1998, AOCS Press, pp. 107–111.

Velioglu, Y. S., Mazza, G., Gao, L., and Oomah, B. D., 1998, Antioxidant activity and total phenolics in selected fruits, vegetables, and grain products, *J. Agric. Food Chem.* **46**:4113.

Visser, A., and Thomas, A., 1987, Review: Soya protein products—Their processing, functionality, and application aspects, *Food Rev. Int.* **3**:1.

Waggle, D. H., and Kolar, C. W., 1979, Type of soy protein products, in: *Soy Protein and Human Nutrition*, H. L. Wilcke, D. T. Hopkins, and D. H. Waggle, eds., Academic Press Inc., New York, pp. 19–52.

Wanasundara, P. K. J. P. D., and Shahidi, F., 1996, Optimization of hexametaphosphate assisted extraction of flaxseed proteins using response surface methodology, *J. Food Sci.* **61**:604.

Wank, H. J., and Thomas, R. L. 1989, Direct use of apple pomace in bakery products, *J. Food Sci.* **54**:618.

Wettasinghe, M., and Shahidi, F., 1999, Antioxidant and free radical-scavenging properties of ethanolic extracts of defatted borage (*Borago officinalis* L.) seeds, *Food Chem.* **67**:399.

Wheelwright, S. M., 1991, *Protein Purification. Design and Scale up of Downstream Processing*, Hanser Publishers, New York, pp. 88–91.

Xing, Y., and White, P., 1997, Identification and function of antioxidants from oat groats and hulls, *J. Am. Oil Chem. Soc.* **74**:303.

Yazicioglu, T., Gokcen, J., and Karaali, A., 1981, Studies on extraction of protein isolates from Turkish sunflowerseed and cottonseed extraction cakes and on their functional properties, *Fette Seifen Anstrichm.* **83**:474.

# 12

# Utilization of By-Products in the Fish Industry

Kristberg Kristbergsson and Sigurjon Arason

## 12.1. INTRODUCTION

The global annual harvesting of fish from the wild in oceans and lakes was 90.2 and 93 million tons in 2002 and 2003, respectively, and has been around 90 million tons in the last decade (STATICE, 2005). The Pacific Ocean has been the main harvesting ground with about 50 million tons followed by the Atlantic Ocean at around 20 million tons (Figure 12.1). The number one species has been Peruvian anchovy, ranging from 6.2 to 11.3 million tons in the past few years, but the El Nino storm has caused some major fluctuations in the stock and harvesting. Alaskan Pollock has been in second place at around 3 million tons, with Atlantic herring and Skipjack tuna following at about 2 million tons each 1, about 1 million tons of Capelin, which is a small pelagic schooling fish like the anchovy. Around 1 million tons of Atlantic cod have been harvested in recent years (Figure 12.1). The leading fishing nation in the world has been China with more than 16 million tons, followed by Peru at 6 and the United States at 5 million tons. Russia has ranked number 8 with between 3 and 4 million and Norway and Iceland have been the leading European fishing nations ranking as number 11–15 worldwide with 2.5 and 2 million tons but at the top of the list for the North Atlantic.

Wild fish stocks are a limited resource and overfishing has been a problem in many fishing grounds, although the severity has varied greatly between fish species and regions. Modernization and increased capacity of fishing vessels has increased opportunities and ability to find and harvest fish in the ocean; now we are at a stage where most fish species are being harvested to their limit. Sustainability of the stock and increased or total utilization of the harvested fish therefore is very important from an environmental point of view. The particular species of fish being harvested and utilized is also important with respect to environmental sustainability.

KRISTBERG KRISTBERGSSON AND SIGURJON ARASON • Department of Food Science and Human Nutrition, University of Iceland, Hjardarhaga 2–6, Reykjavík, Iceland. SIGURJON ARASON • Icelandic Fisheries Laboratories, Skulagotu 4, 101 Reykjavík, Iceland.

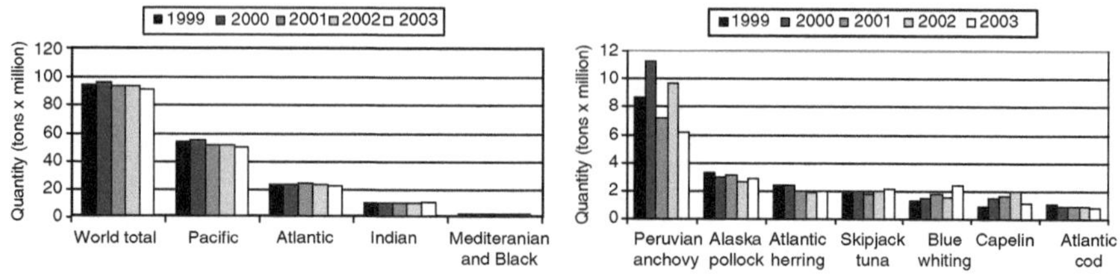


**Figure 12.1.** (a) Total global harvesting of fish indicating main fishing grounds, and (b) global harvesting of some important fish species (1999–2003) (adapted from STATICE, 2005).

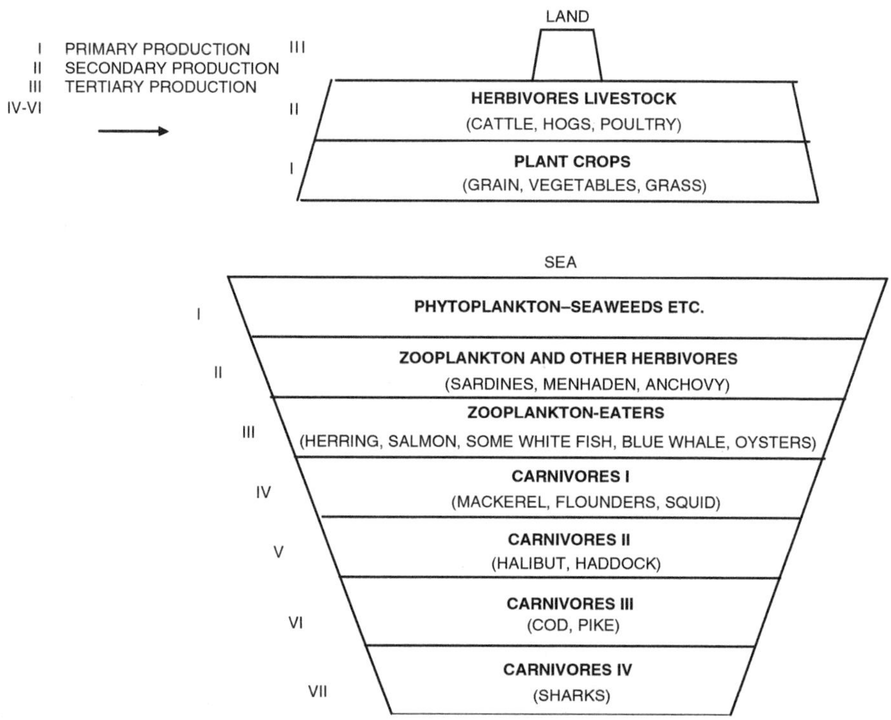


**Figure 12.2.** The production pyramid (reproduced from Borgstrom, 1971).

All life in the oceans is based on phytoplankton like seaweed, but phytoplankton production is gigantic; with respect to organic material it has been estimated to be ~10,000 times the world production of wheat and rice. There are not many fish species in the ocean feeding directly on plankton, like livestock on land feeds on grass; only some sardines, menhaden, and anchovies. The plankton feeders belonging to the second production step according to the ''production pyramid'' put forth by Georg Borgstrom some 50 years ago (Figure 12.2). His pyramid is still valid and it would be beneficiary if we could utilize more of the species in the lower production steps for human consumption.

Going from one production step to the next above may be estimated to be a factor of ten with respect to feed utilization, or in other words to produce 1 kg of anchovies we will require 10 kg of phytoplankton, but to produce 1 kg of cod which belongs to the sixth production step we will need 100,000 kg of phytoplankton. The pyramid is only a very rough estimate; for example, although cod may belong to the sixth step it also can be at a lower step since cod feeds heavily on capelin and by so doing will move one or two steps down the pyramid, reducing the quantity of phytoplankton to 1,000 or 10,000 tons depending on whether capelin is feeding on phytoplankton or zooplankton. Never the less increased human consumption of some of the fatty fish like herring and capelin would be beneficial with respect to utilization of world resources. It also is important to fully utilize all harvested fish, and furthermore to treat the whole fish of the total harvest like the valuable raw material it can be, rather than to discard portions of the fish like the heads or guts, which is still a common practice among many fishermen.

In this chapter we will discuss by-products in the fish industry in general but use the Icelandic fish industry as a model, mostly because of our familiarity with that particular sector of the industry but also because the Icelandic fisheries have been managed in a relatively sustainable manner and the processing industry is reasonably advanced. Fisheries are still the most important industry in Iceland and will likely continue to be so for some time to come. The fish industry has been the foundation of the nation's economic growth, changing the country from being one of the poorest in Europe in the early 1900s to being one of the richest today. In recent years the annual harvesting by Icelandic fishing vessels has been approximately 2 million metric tons (Figure 12.3) and the value of exported products from fisheries has accounted for about 60% of the total value of exported goods.

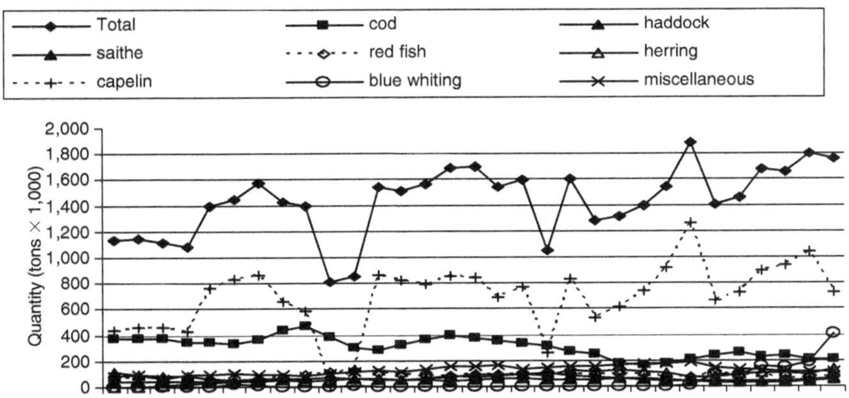


**Figure 12.3.** Total catch and main species of fish harvested in Icelandic fishing grounds from 1973–2003 (adapted from STATICE, 2005).

The most important species are from the gadoids often referred to as round fish, including cod, haddock, saithe (type of pollock), and blue whiting, with cod being by far the most important, from an economical standpoint. Cod is either sold as frozen, salted, or fresh in ice. The frozen fish is mostly sold as fillets or smaller portions suitable for consumption. Salted fish is usually processed as butterfly split fish or as fillets where the fresh fish in ice is either sold as whole gutted fish or as fillets. Although frozen fish only acounts for about 38% of the export volume, it represents, 50% of the value of exported fish products (Figure 12.4).

Other important species include ocean perch, Greenland halibut, and pelagic fish like herring and capelin, crustaceans like shrimp and lobster, and shellfish like scallops. Capelin is a small pelagic schooling fish, mainly found in the North Atlantic and Arctic Oceans.

It is has been harvested by Icelandic fishing vessels since the early 1960s and catches have gradually increased (Vilhjálmsson, 1998). Harvesting today during the winter and summer seasons are at approximately 1 million tons in Icelandic waters, (ICIMF, 2005). Capelin therefore is the single largest species harvested around Iceland representing just under 40% of the total volume but only 8 to 12–13% of the total value of exported fisheries products depending on the price of fish meal and oil which varies greatly with availability of other animal feed. Prices of fish meal also will fluctuate with events like the recent outbreaks, or scare of outbreaks, in life stock like foot-and-mouth disease in cattle or avian flu.

Sustainability of the fish stocks is dependent on the harvesting, both on the percentage harvested from the available stock and also by the harvesting or fishing technique used, whether it be traps, line fishing, nets or trolls, etc., and on the size and technical capability of the vessel involved. Modern fisheries may utilize very powerful vessels with the latest in fish-finding technology including TV monitors guiding the fishing gear, which is a far cry from what was available only a few decades ago. Iceland has been fortunate in its utilization of available stock and overfishing is not as severe a problem as in some other locations. This is most likely due to a legislation on fish quotas that was established for some fish species and enforced within the 200 mile exclusive economic zone (EEZ).

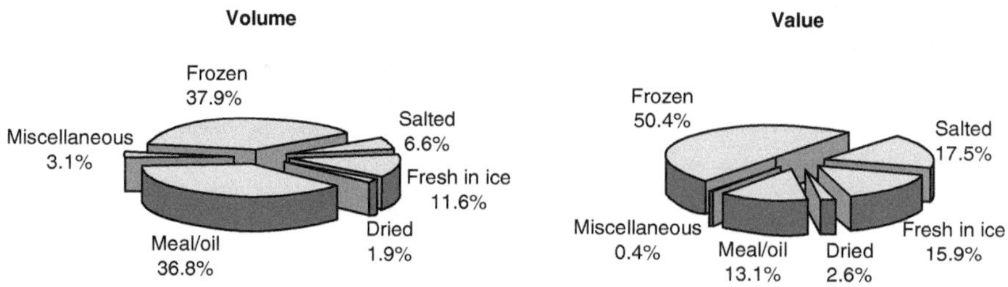


**Figure 12.4.** Seafood exported from Iceland in 2004 classified according to processing with respect to volume and value. The total volume was 827.5 million tons with a total value of 121.746 million Icelandic kronur or ~1740 million US $ (STATICE, 2005).

In 1984, fishing quotas indicating allowable quantities for harvesting were introduced for some of the most important fish species, including cod, in order to control their exploitation (Valdimarsson, 1998). Specific quotas were assigned to each fishing vessel based on prior fishing experience and availability of the particular species as estimated by scientists at the Marine Research Institute in Iceland, but the final decision on annual quantities has been a political decision. The main objective of the legislation on specific quotas was to prevent overfishing and to encourage responsible handling of the harvested fish and to increase exploitation of underutilized seafood resources which could be harvested freely. The quota system has influenced and changed the attitude toward utilization of all harvested fish. Both fishermen and processors have become more interested in making marketable products from raw materials previously used for fish meal or discarded as waste (Arason, 2003).

The utilization of the harvested fish is also much more complete than some years back, partly because of the quota system but also due to new technology and from a revised attitude toward fish as a raw material by fishermen and processors alike. The possibilities for utilization and available processing methods are highly dependent on each particular situation. Fish processing may take place onboard a large oceangoing trawler, which harvest's and processes the freshly caught fish onboard for example, into frozen fillets. Some vessels will keep only the fillets while others will utilize the entire or most of the fish. The utilization of the harvested fish varies between vessels and fish species but generally has increased in recent years. Today onboard meal plants are available in some of the larger vessels. The second type of fish processing is the land-based fish processing, which may vary in size from a small fish-processing outfit with one or two and up to several workers to one or two small to medium fishing vessels supplying raw material. On the other end of the scale we will have large-scale fish-processing plants with hundreds of people working and many boats and trawlers supplying raw material for the operation. The handling of fish is also seasonal where Icelandic regulations require all fish to be gutted during the summertime if the fishing trip will be more than 24 hours. Only the smaller vessels will be bringing ungutted fish ashore and only during the winter seasons. There is no rule of thumb that applies to all of these operations with respect to what type of fish an individual processor will favor or what type of processing is involved. However, many of the smaller operations have specialized in the processing of salted and dried fish or in filleting cod, either by hand or machine, to be packed in polystyrene boxes with ice and airlifted as fresh to the United States or European markets. However, a very large portion of the fish harvested in Icelandic waters is frozen after processing (Figure 12.4) both as IQF fillets or portions and in blocks, later to be portioned into fish fingers or steaks. The freezing is usually performed by the large and modern freezing plants with both plate freezers and IQF freezing techniques available. Many of these modern freezing plants will utilize the fish completely, including bones, intestines, and skin.

## 12.2. MAIN BY-PRODUCTS IN FISH PROCESSING

The definition of by-products in the fish industry varies with fish species as well as both the harvesting and processing methods used. The general understanding of by-products when considering round fish such as cod is that the main body flesh constituting the fillets will be considered to be the main product but the head, backbones, trimmings or cutoffs, skin, and guts (intestines) constitute what is generally thought of as by-products (Table 12.1).

Raw material utilization is not the same in a land-based operation and in offshore processing. The land-based operations may be supplied by small boats, where the fishing trip may last a few hours to a day, or by oceangoing trawlers, where the trip lasts from a few hours to a week or more. In the smaller boat the fish is sometimes brought ashore ungutted, increasing the ease of total utilization of the whole fish compared to the longer trips where the fish is invariably gutted at sea and it depends on the particular crew whether the guts are kept or discarded in ocean. On the oceangoing trawlers with onboard freezing where the trip may take up to a month the utilization varies among vessels and ranges from only keeping fillets to vessels having their own meal plants on board utilizing all of the harvested fish as mentioned previously. Iceland has not practiced the use of large factory vessels staying out at sea for many months and which generally accompanied by a fleet of smaller vessels harvesting fish which is transferred to the factory vessel for processing. These fleets customarily travel long distances and harvest from fishing grounds far from the home port of the fleet.

The possibility for economical utilization of the whole fish varies greatly between these harvesting methods. It also must be kept in mind that parts of the fish other than the fillets may be considered waste on shore but is simply biological material being returned to the ocean which will become feed to other marine life out at sea, although this can be a source of pollution in some closed areas or inland waters; this is generally not a cause of pollution out in the open sea. It also must be

*Table 12.1.* By-products from the land-based fishing industry in Iceland (tons) (STATICE, 2005)

| Product | 2003 | 2004 |
|---|---|---|
| Trimmings | 10,442 | 15,171 |
| Heads | 6,152 | 5,303 |
| Roe | 2,354 | 2,870 |
| Liver | 945 | 1,352 |
| Meal | 625 | 650 |
| Tails | 145 | 168 |
| Tongues | 23 | 22 |
| Oil | 4 | 4 |
| Milt | 1 | |
| Waste | 0 | 0 |

kept in mind that sometimes a product is considered as a by-product or raw material but the same commodity will be regarded as the main product or raw material in other processes. Fish meal and oil are a good example of this. Cod liver oil and fish meal from bones and trimmings generally will be regarded as by-products where meal and oil from pelagic fishes will be regarded as the main product from the harvesting and processing of anchovies or capelin. This is also seasonal in nature and during the spawning season of capelin the females are preferably frozen with the roes and used for human consumption but the male and spent females are processed into fish meal and considered to be a by-product. The feeding summer capelin is almost exclusively harvested for fish meal and oil production, although our research indicated that the feeding capelin is very suitable for fish sauce production, which can be used as human food (Hjalmarsson, 2001). It therefore can be both seasonal and circumstantial what will be considered as the main product or a by-product or raw material. The main objective should be the sustainability of the fish stock and the total utilization of the harvested fish conducted in a manner, which will not pollute or alter the environment in an irrevocable manner.

All portions of the fish should be considered raw material rather than waste or preferably thought of as alternative product streams. With increased scientific understanding of proteins and fish oil chemistry some of these so-called by-products in the future may be turned into far more valuable commodities than the main flesh or fish fillets.

Calculation of yield in the fish processing industry generally is based on gutted fish with head (Table 12.2). Some species, like ocean perch, herring, capelin, and other pelagic, will be delivered to the processing plants ungutted and calculations of processing yield for these species will be based on ungutted fish. Fish meal and oil constitute the bulk of the volume of products from fisheries in Iceland or 63% of total, but their value is far less or only about 14% of the total value of exported seafood products (Figure 12.4).

*Table 12.2*. Processing yields obtained from various processing of cod (Arason, 2003)

| | | Filleting | | | Splitting |
|---|---|---|---|---|---|
| | | Frozen on shore | Frozen at sea | Salting | Salting |
| Gutted fish with head | Product | 48 | 45 | 60 | 72 |
| | Head | 30 | 34 | 23 | 22 |
| | Backbone | 15 | 16 | 16 | 6 |
| | Skin | 3 | 3 | | |
| | Trimmings | 4 | 2 | 1 | |
| | Total | 100% | 100% | 100% | 100% |
| Guts[a] | Intestines | 18% | 18% | 18% | 18% |
| Total by-products: | | 70% | 73% | 58% | 46% |

[a] Guts are calculated as percent of whole fish.

## 12.3. GUTS OR VISCERA

When fish is gutted, the intestinal organs or viscera are removed and often referred to as the guts. This includes both the liver and roes (or milt), which may constitute 10 to 25% of the net weight of round fish depending on maturity and season. Roes and liver have been utilized for human consumption directly or indirectly since ancient times. Other parts of the guts like the pyloric caeca are normally not consumed but may serve as a source of bioactive compounds like enzymes that can be used for various applications. In Iceland, as well as in many other countries, most of the fish intestines are still discarded at sea. However, the guts ought to be regarded as raw material rather than waste. Basic research in the past 25 years on isolation, purification, and characterization of enzymes from fish intestines has led to industrial applications in foods, natural skin care products, cosmetics, and pharmaceuticals. Special attention has been placed on hydrolytic enzymes like cod serine proteases including trypsin (Asgeirsson et al., 1989), chymotrypsins (Asgeirsson and Bjarnason, 1991), elastins (Asgeirsson and Bjarnason, 1993), and serine collagenases (Kristjansson et al., 1995). The research has shown a definite novelty for these cold-active or psychrophilic enzymes since they are frequently more active at low temperatures than their mammalian or bacterial counterparts and have higher catalytic efficiencies, higher sensitivity to heat and low pH, and higher activity toward native proteins. These characteristics may be beneficial in industrial processes, as well as in medical, pharmaceutical, hygienic, and cosmetic applications (Bjarnason, 2001) based on the lower concentrations of enzymes needed due to their high catalytic efficiency. Furthermore, milder conditions are needed to destroy residual enzyme activity due to their temperature sensitivity. They also have been shown to penetrate through the human skin giving rise to a commercial skin care formula called Penzym and marketed by the Zymetech Inc.

Protein hydrolysates generated by cold-active marine enzymes also have proven to be very successful as all natural seafood flavors for soups, surimi, and pet food applications. Various other food processing applications also are being considered, such as chill-proofing of beer, meat tenderization, preparation of minimally treated fruit and vegetable beverages, and hydrolysis of various food proteins, such as gelatin, vegetable proteins, and collagens (Arason, 2003). A mixture of cold-active proteinases (trypsin, chymotrypsin, elastins, and collagenases) is being commercialized under the trade name Cryotin which is claimed to have many potential uses in industry, medicine, and research, especially in food processing applications which require hydrolysis at low temperatures, inactivation under mild conditions, or native collagen digestion (Gudbjarnason, 1999). Cryotin currently is being used in a patented process to prepare high-quality all-natural flavorings for food processing and innovative cooking and marketed as Northtaste® by NorthIce Inc. However, the best-known Icelandic fish by-products are undoubtedly the cod liver oil and cod roes, although enzymes and other substances coming from basic research in biochemistry and molecular biology may very well become as or more important in the near future. Research is being conducted by many groups all over the world on bioactive proteins from fish to be used as nutraceuticals (Alasalvar and Taylor, 2002).

## 12.4. FISH ROE

The availability of fish roes is highly seasonal and roes are only collected during the spawning season of the fish. Quality is highly dependent on the maturity of the roes when harvested (Katsiadaki et al., 1999), which affects the quality of the products made from roes. The flavor profiles of commercially processed ripened roes from Iceland and Norway have been categorized by sensory evaluation and by electronic nose analysis (Jonsdottir et al., 2004). The results may be used as an aide in trade and in product development. Fish roes are used for caviar type products and for various speeds.

Only roes coming from species belonging to the family Acipenseriformes, which includes sturgeons, may be classified as caviar in trade but roes from other fish are often classified as "caviar substitute" (Catarci, 2004). The world trade in sturgeon caviar dropped with the breakup of the Soviet Union but at the same time trade in caviar substitutes and other fish roe expanded (Catarci, 2004) (Figure 12.5).

Lumpfish roes are a very good example of this and are exported from Iceland as "caviar substitute," sometimes referred to as "lumpfish caviar" and priced at a fraction of the cost of real sturgeon caviar. However, lumpfish roes are not considered a by-product since the reason for catching the lumpfish is to collect the roes. The lumpfish roe industry is now well established in Iceland, and most of the roes

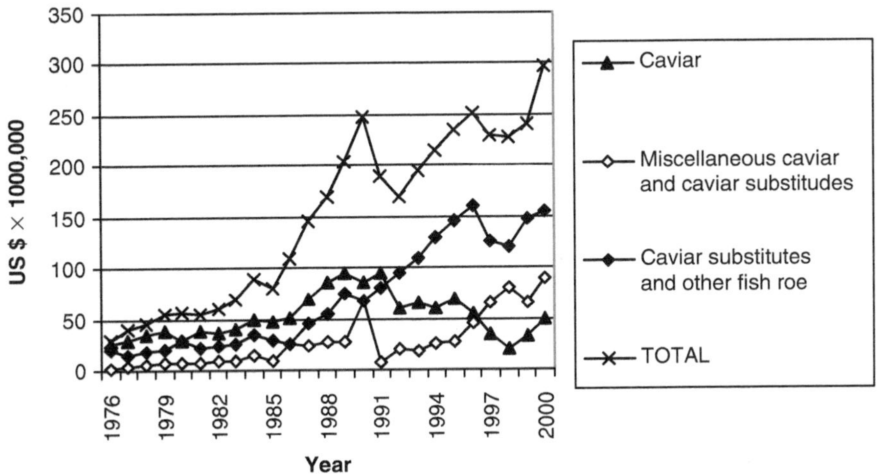


**Figure 12.5.** World trade in caviar, caviar substitutes, and other fish roe, value 1976–2000 (adapted from Catarci, 2004).

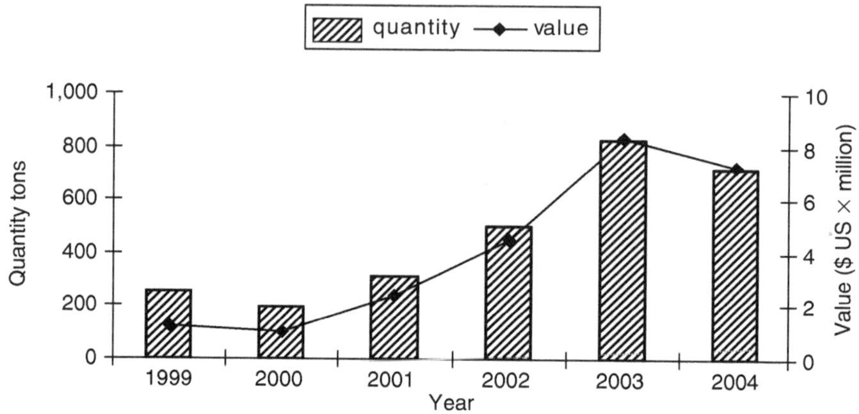


**Figure 12.6.** Export of salted lumpfish roes form Iceland 1999–2004 (STATICE, 2005).

are packed as caviar-type products under various labels specified by individual buyers and sold on the European market. In 2001, the export of lumpfish caviar was 1,184 tons at a value of $10.8 million (Arason, 2003). There also is a large demand for salted unprocessed lumpfish roes exported in larger containers, usually plastic tubs (Figure 12.6).

The eggs in capelin roes are smaller in diameter than the sturgeon and lumpfish roes but also are used for caviar-type products and are especially popular for the traditional Japanese Sushi cuisine dish called Masago where the roes have been cured and processed according to the Japanese tradition. Capelin roes have almost exclusively been sold to Japan, but some producers are now making caviar-type products and spreads for the European market. The quantity of exported capelin roes fluctuates with the annual fluctuation in the harvesting of capelin (Figure 12.3) but was 4,913 tons in 2003 and 9,505 tons in 2004 but only 5% of the capelin is used for human consumption with 1% as frozen roes. The price of capelin roes is approximately 30 times higher than for capelin meal. Special automated processing lines have been developed for capelin roes, for example, by Traust ehf (http://www.traust.is) where the process includes extraction, separation, cleaning, weighing, packing, and preserving of this delicate product.

There is a high demand for frozen and salted cod roes for smoking, canning, and the production of various kinds of spreads which are very popular in the Scandinavian countries. Roes from haddock and saihte also are used similarly. The practice of delivering fresh fish ashore gutted if the individual fishing trip is longer than 24 hours is good for the quality of fish but reduces the availability of roes and other intestinal material for processing, since traditionally the guts have been discarded at sea and only roe from fish has been brought to shore ungutted for salting or freezing have been available for processing. Increase in the price of roes has changed this and today roes from most of the harvested ground fish are being collected and utilized. Most of the cod roes are collected and stored with salt in

*Table 12.3.* Roes from gadoids in Icelandic waters (tons) (STATICE, 2005)

| | 2003 | 2004 |
|---|---|---|
| Frozen on shore | 750 | 797 |
| Frozen at sea | 106 | 191 |
| Salted | 1,051 | 1,494 |
| Canning | 430 | 365 |
| Domestic consumption | 7 | 11 |
| Miscellaneous | 10 | 12 |
| Total | 2,354 | 2,870 |

insulated plastic tubs where they undergo curing during storage (Table 12.3) but they also may be frozen. Further processing includes desalting and membrane removal, mixing of additives such as color and preservatives, packing and pasteurizing in consumer-size containers, generally glass jars, for caviar-type products, or increased mixing and mashing for spread type products, which will be packed in tubs. For some products like Masago special curing techniques and maturing processes are being used.

## 12.5. LIVER AND FISH OIL

The bulk of the fish oil and meal exported from Iceland is processed from pelagic fishes like capelin or herring. In recent years herring mostly has been processed for human consumption by salting or freezing but 95% of the harvested capelin has been used for fish meal and oil production and has been the main source for fish oil. The global production of fish meal and oil is similarly based on pelagic schooling species like menhaden and anchovy. Fish oil from pelagic fish is referred to as fish body oils, since the whole fish is ground during processing.

Generally these fish oils have been used only for the production of animal feed but not directly for human consumption. The value of fish oil has been increasing steadily due to both increased demand and increased quality due to better processing technology (Figure 12.7).

The primary processing of fish oils is generally performed at fish meal plants where the fish is cooked, pressed, and the oil and solid meal are separated with three-phase centrifuges (Figure 12.8). The fish meal plant is perhaps the ultimate factory for by-products in fish processing because that is where raw material ends up if no other means of utilization can be found, for example, bones are generally processed into bone meal in fish meal plants.

The quality of fish meal has increased drastically in the last 20 years. Meal was commonly dried in drum dryers utilizing oil for heat generation, where the meal solids were in direct contact to the heating surface resulting in high degree of denaturation in the heat-labile fish proteins. The requirement for higher quality fish meal with better functionality and bioactivity came with the growth of aquaculture,

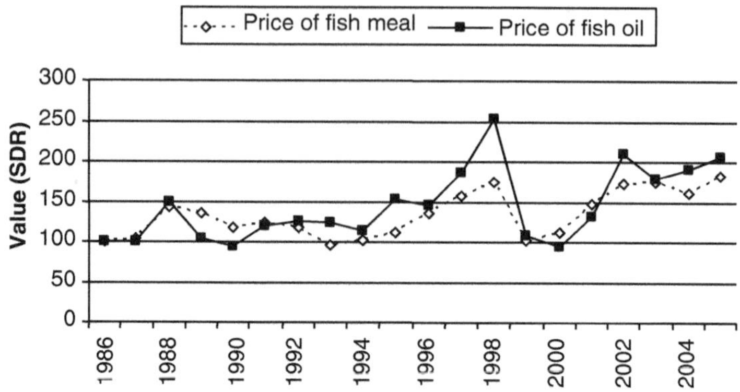


**Figure 12.7.** Value of fish meal and oil from 1986–2005 (STATICE, 2005).

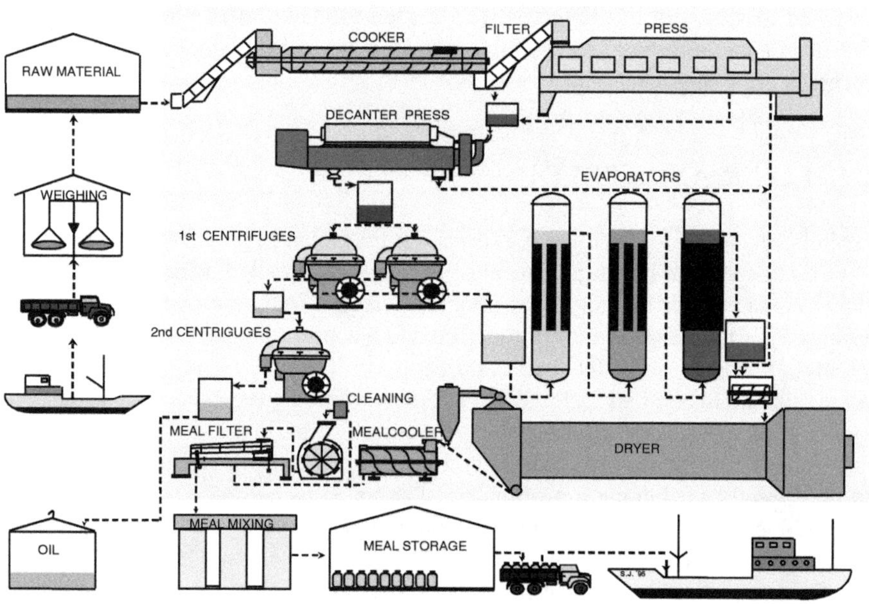


**Figure 12.8.** Layout in a fish meal plant indicating streams for fish meal and oil (curtsey of Mr. Sveinn Jonsson).

but cultured fish is more sensitive to protein quality than domestic animals. Recent developments have included steam dryers and indirect air drying systems like the Dyno-Jet dryers (Atlas-Stord A/S, Denmark)) where the flue gas from the burner is not in direct contact with the air that is used to dry the material. Our studies have shown recently that the functional properties of fish meal processed with these indirect drying methods is superior to the meal obtained with the older drying techniques (Gunnarsson, 2003). After the extraction and separation, the fish oil is purified in oil refineries with the same processing steps that are used for vegetable oils including alkali refining, degumming, bleaching, deodorization, fractionation, and low-molecular-weight distillation for the highly purified suppliments or this may then be followed by steps like hydrogenation and virgule or interesterification depending on the end use of the oil.

Fish oils are highly polyunsaturated with long-chain fatty acids, and therefore are very prone to oxidation. However, some fish oils are less sensitive, which may be due to natural antioxidants in the oil or from the difference in or lack of prooxidants. The keeping quality of fish meal and oil from capelin has been shown to be very good and better than from many other species (Bragadottir et al., 2002, 2004) and research is being directed at increasing the human consumption of capelin oil.

Ever since the Danish researchers Bang and Dyerberger (1981) observed that, regardless of a high-fat diet, heart failure was very low among the Inuit's in Greenland; they subsequently linked this to polyunsaturated fatty acids in their diet and there has been a growing interest in fish oils for human consumption. Since then the beneficial effects from fish oil and omega-3 fatty acids on the cardiovascular system has been well accepted and documented (Kinsella, 1986) and is still being investigated heavily by many researchers. Cod liver is ~10% of the ungutted cod and the bulk of cod liver oil is processed for human consumption and consumed as a dietary supplement in capsules or in liquid form. It has been estimated that an average production of 10,000 kg of cod fillets (gadoid species) will generate by-products with more than 1,000 kg of marine lipids and more than 30% of these lipids are omega-3 fatty acids (Falch et al., 2006). Recent research has been directed toward the production of fish oils and concentrates with increased omega-3 content with special emphasis on both the content and ratio of eicosapentaenoic acid (EPA) to docahexaenoic acid (DHA). Both are all *cis*, long-chain polyunsaturated fatty acids (LCPUFA) also called n-3 or omega-3 fatty acids, indicating that the first double bond is on the third carbon atom from the methyl end of the fatty acid. The DHA contains 22 carbon atoms and 6 double bonds but EPA contains 20 carbons and 5 double bonds. Natural Atlantic cod liver oil contains ~17% EPA and 37% DHA of total fatty acids, but both content and ratio vary among different fish species (Kinsella, 1986). Refined concentrates may have higher concentrations and a 3:1 ratio of EPA to DHA is common in commercial concentrates. Interest in cod liver oil and other fish oils and concentrates recently has been boosted with the linking of infant brain development and function to fatty acids from fish oils (Bourre et al., 1993; Makrides et al., 1995).

Today there is more demand for cod liver than is being supplied from the ungutted fish brought ashore for processing (Ogmundsson, 2005). All the liver brought ashore is being utilized and most of it is used for the production of cod liver oil for dietary supplementations (Table 12.4). However, there are still large quantities of round fish liver discarded at sea. Another popular food item for many is canned cod liver but the availability is low and the annual production in Iceland has only been ~200 tons mostly due to lack of raw material.

*Table 12.4.* Production of different products from liver from gadoids fish in Iceland, 2001 (Arason, 2003)

| | Quantity (tons) | Value (thousands $) |
|---|---|---|
| Purified for supplements | 1,573 | 4,258 |
| Ground fish oil | 2,146 | 2,438 |
| Canned liver | 220 | 1,109 |
| Frozen and fresh liver | 238 | 277 |
| Other liver products | 168 | 149 |
| Total | 4,344 | 8,232 |

In recent years emphasis has been on the health benefits of the LCPUFA in fish oil; however, both whole fish oil and fish liver oil contain other interesting compounds like squalene. Some ether-type fish oils, especially shark liver oil, possess some very unique properties like being fully saturated and therefore not prone to oxidation but still maintain a liquid form at room temperature. The ether oils offer some very unique possibilities for industrial applications. With improved separation techniques and more gentle processing methods, the fish oils might play an even more important role in the pharmaceutical and health food industry in the near future.

## 12.6. MINCED FISH FROM FISH FRAMES, NAPES, AND TRIMMINGS

Minced fish (minced fish meat, minced meat, fish mince, mince) is fish flesh separated from skin and bone in a mechanical bone separator, frequently of the perforated drum and squeezing belt type. Fish or pieces of fish, such as trimmings from fillets, are fed from a hopper to be pressed between a moving, flexible belt and the outside of a counterrotating, perforated, stainless steel drum. The flesh is squeezed through the perforations of the drum and removed from the inside by means of a fixed screw. The coarseness and texture of the mince, the color of the product, and the yield of flesh are affected by the diameter of the drum perforations, the pressure applied to the squeezing belt, and the raw material or trimmings being used. For different applications, the perforations may range from about 1 to 10 mm. Under ideal conditions, the skin and bones remain on the outside of the drum and are removed continuously as waste by a fixed scraping blade to be used for bone meal production (Whittle and Howgate, 2000). After filleting of round fish like cod considerable quantities of flesh may remain on the frames and napes. However, fish mince produced in Iceland is almost all from trimmings produced during the production of boneless fillets (Table 12.1). The mince from the collar bone and the frames is darker in color than the mince from the trimmings or cutoffs and is of poorer quality. The dark to reddish color of frame mince comes mostly from the kidney of the fish located under the spinal cord or the backbone which gets mixed into the

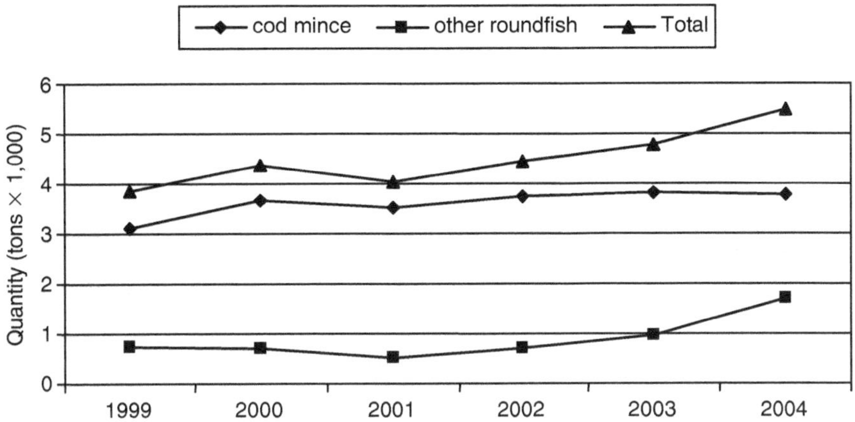


**Figure 12.9.** Export of round fish mince from Iceland (adapted from STATICE, 2005).

flesh with abrasive handling of the frame. Water jet technology has been used to clean the flesh from the backbone of round fish leaving the kidney intact by using high-pressure water jets to wash the meat off the bones. The resulting water jet mince is white in color but has physicochemical properties approaching surimi, which is water-washed fish flesh (Kristbergsson, 1998). Water jet technology gave lower yields than mechanical deboning of pacific whiting frames when tested for surimi production but the quality of the surimi produced from it was high (Wendel et al., 2002). The lower yield may have been from loss of proteins in the water phase, which may be rectified by a better filtration technology for protein recovery or by the relatively new process for surimi production and protein recovery utilizing acid–base reactions (Undeland et al, 2002; Liang and Hultin 2005; Kristinsson et al., 2005) first developed for the production of surimi from fatty fish with dark flesh like herring or capelin. The water jet technology also can be used to remove flesh from bones for further protein recovery for the production of bioactive or functional foods. The production of mince from round fish has been increasing in Iceland, indicating better utilization of the raw material (Figure 12.9).

One ton of gutted cod will give ~100–180 kilos of mince from various parts of the fish. The following is a rough estimate of recoverable mince from round fish.

| | |
|---|---|
| Trimmings | 3–4% |
| Collar bone | 4–5% |
| Head | 3–4% |
| Belly flap | 5–6% |
| Frame | 4–5% |
| Total | 15–18% |

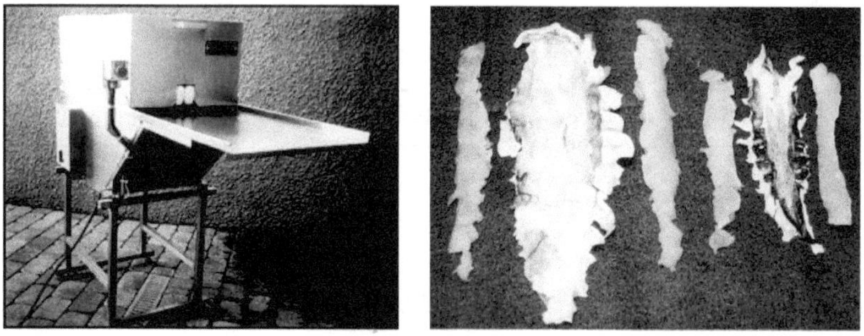

**Figure 12.10.** The MESA 850 cod backbone processor showing recovered meat and swim bladder (courtesy of A.M. Sigurðsson, Hafnarfirði, Iceland).

Machinery for scraping flesh from the backbones and cutting swim bladders and belly flaps from backbones after butterfly splitting or filleting of cod have been developed (Sigurdsson, 1992). The products add to the better utilization of round fish backbones and have been exported and used as value added products on the salted fish markets (Figure 12.10).

Recently, Mei et al., (2003) have developed a method where relatively low-value fish cut-offs and trimmings may be reformed into fillet-like products with textural characteristics resembling intact fish flesh giving increased quality and added value to the products. The method is based on using sodium chloride extracts of comminuted fish trimmings to tie together intact flesh pieces from larger cutoffs and trimmings and mold the product into a fillet shape.

## 12.7. FISH SKIN

The skin on fishes is mostly composed of gelatin and scales. The scales from some species may be utilized as a raw material for pearl essence which is the shimmery substance used in cosmetics like lipstick or eye shadows, and blush (Anon, 2006). Pearl essence is mostly produced form the scales of herring and some US corporations have specialized in this like the Engelhard Corporation of Eastport, Maine. The other main product from skin is gelatin. Gelatin from cold-water fish has lower gelling and melting temperatures than conventional bovine gelatins but similar gel strengths (Gudmundsson, 2002). In Iceland fish skin has been collected by at least one fish processing company and exported to Canada and Spain where it has been used for gelatin production. Fish skin may be stored at refrigerated temperatures or frozen for a short period of time, without negative effect on the functional properties of the gelatin (Gudmundsson and Hafsteinsson, 1997). Interest in fish gelatin has been increasing because it can be used in food formulations where domestic animal products are not considered desirable like in the kosher and halal foods and because some manufacturers have been looking for alternative sources of gelatin due to some recent scare for outbreaks like bovine spongiform encephalopathy (BSE) or mad cow disease. Fish skin also has been tanned and processed into leather by Atlantic Leather Inc. (http://www.atlanticleather.is/). Their products have found applications in locally produced specialty items for the tourist industry; however, the operation is growing into more conventional consumer goods and recently they have signed contracts with Nike for their products to be used in footware and with Dior for some fashion goods.

## 12.8. FISH HEADS: A CASE STUDY

Fish heads contain relatively little meat and often are discarded or used to produce bone meal for animal feed. The different parts of the cod heads are tongue (1–4%), cheeks (5–15%), collar or nape (15–20%), and upper head meat (5–15%) (Figure 12.11).

However, the meat in tongues and cheeks from the fish head is considered to be a delicacy by many consumers due to its unique taste and excellent texture. Until recently the processing of cheeks and tongues could be done only manually, which is labor intensive and thus expensive. Recently mechanical solutions have been developed to automatically cut out fish tongues and cheeks (Sigurdsson, 1993) and for tearing out gills followed by splitting the cod heads (Figure 12.12). The machines can be fitted into automated fish processing lines.

These products are mostly salted and sold on the salt fish markets in Portugal and Spain. About one third of heads from round fish is used for tongue or cheek processing; the other two thirds are used for drying.

Stock fish has been cured by drying in Iceland and northern Norway for more than 1,000 years (Borgstrom, 1971). The fish were hung on drying racks consisting of a series of poles supported by scaffolds. Two fish were tied together at the tails and draped over a pole with one fish hanging on each side (Figure 12.13).

The fish was cured and dried for 40–100 days depending on weather conditions. The resulting stock fish was the main export commodity from Iceland from the 13$^{th}$ century and continued to be so for 400 years when it was replaced by salted

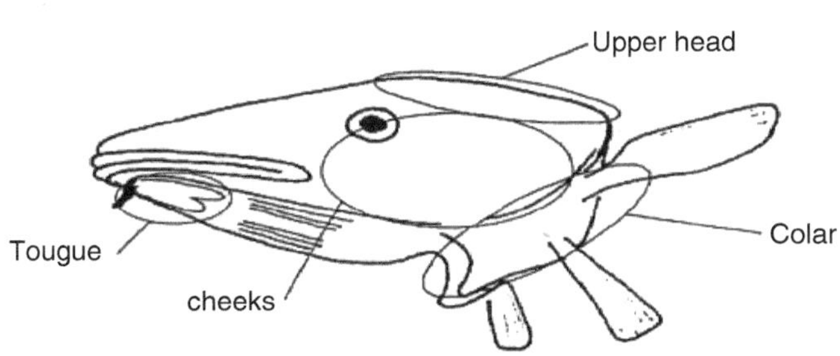


**Figure 12.11.** Diagram of a round fish head like cod indicating location of the fleshy parts (Arason, 2002).

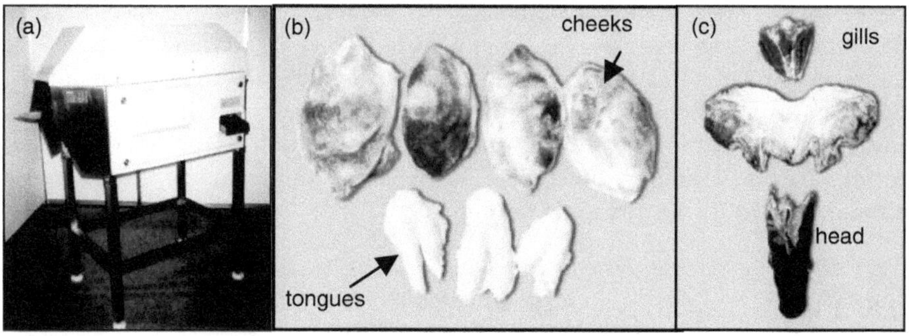


**Figure 12.12.** (a) The Mesa 900 tongue and cheek machine (b) with tongues and cheeks. (c) Also shown are the split heads the Mesa 950 machine (not shown) (courtesy of A.M. Sigurðsson, Hafnarfirði, Iceland).

fish. Stock fish production was reintroduced in the 1930s and again became an important export commodity from Iceland. However, with increased quality of fish brought ashore in ice, other processing methods like freezing and salting produced higher-value products, resulting in a reduction in the production of stock fish from whole round fish.

However, fish heads also were dried by the same methodology and today cod heads have replaced whole fish as the main raw material for dried fish for export. When the export of dried fish is compared to the export of dried fish heads, it will be noted that fish heads represent almost all of the exported fish in this category (Figure 12.14). The export of fish heads has been increasing very quickly in the last decade (Figure 12.15). This is partly due to more uniform drying techniques, but indoor drying techniques for fish have been developed in the past 25 years and the production of dried cod heads has grown from 1,000 tons to about 15,000 tons. Weather conditions in Iceland did hinder a uniform quality of the dried fish and limited the drying of fish to specific seasons. Today modern facilities are used to process the fish heads (Figure 12.16) and recent developments have included computer-controlled air blast cabinet drying processes used to dry fish, fish heads, and backbone fish frames. In Iceland geothermal heat has proven to be the most economical energy source for these indoor fish drying plants (Arason and Arnason, 1992); out of the 20 facilities currently drying cod heads and frames in Iceland 18 are using geothermal heat.

Most of the facilities utilize cabinets for batch processing. For this purpose we used a two-stage drying process. Primary drying where ~25 kg of cod heads per square meter were arranged as a single layer on drying racks. There water content of the heads was reduced from 80% to 50–55% in about 40 hours. This was then followed by secondary drying of the semi-rigid cod heads in drying containers with hot air circulation. In the second drying cabinets the cod heads were dried to about 15% moisture or to a water activity of below 0.6 which was achieved in 3 days. The advantage of using the two-step process is that the quantity of cod heads that can be loaded in the secondary cabinets is much greater than in the primary cabinets

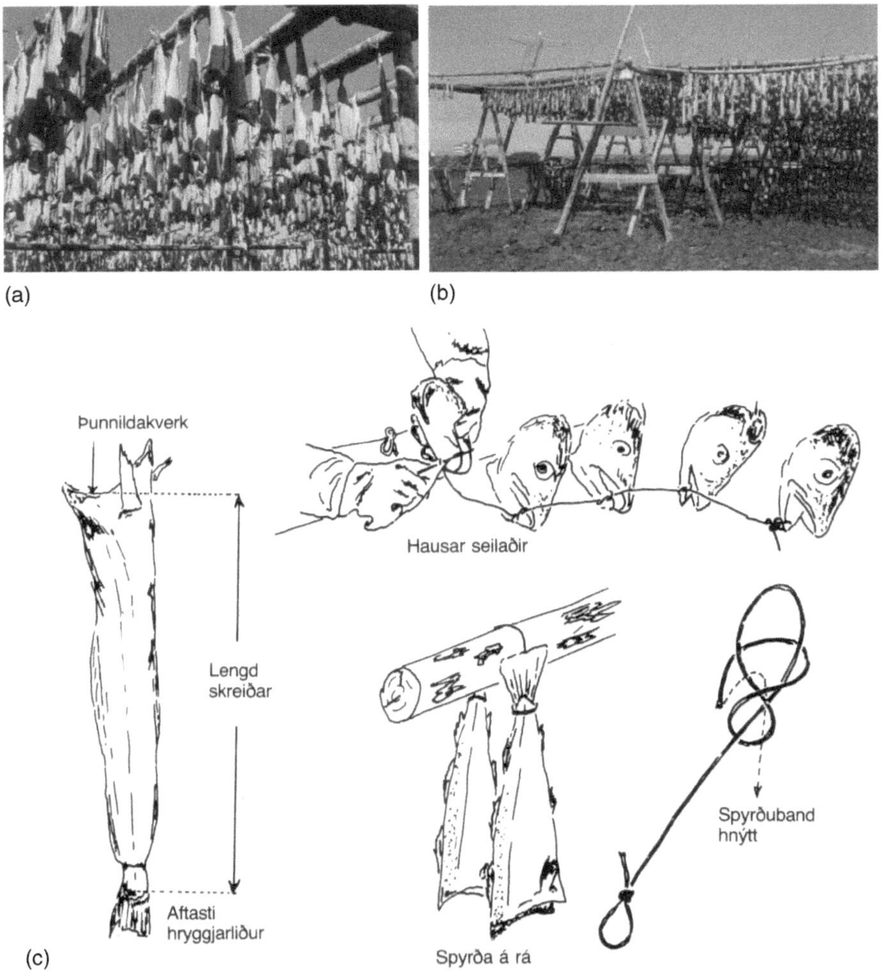


**Figure 12.13.** Drying racks for outdoor drying of fish (above) and schematic of tying of stock fish and heads for drying (from Bjarnason, 1986).

leading to a significant cost reduction; however, the initial and operational cost of the secondary cabinets is much lower than for the primary rack-type cabinets. The method may be used to produce high-quality dried cod heads.

## 12.9. SHRIMP SHELLS: A CASE STUDY

The global production of shrimp from both captured prawns and cultures has increased to over 4.2 million tons in 2002 (FAOSTAT, 2004). The processing of peeled shrimp generates large quantities of waste and has become a major concern

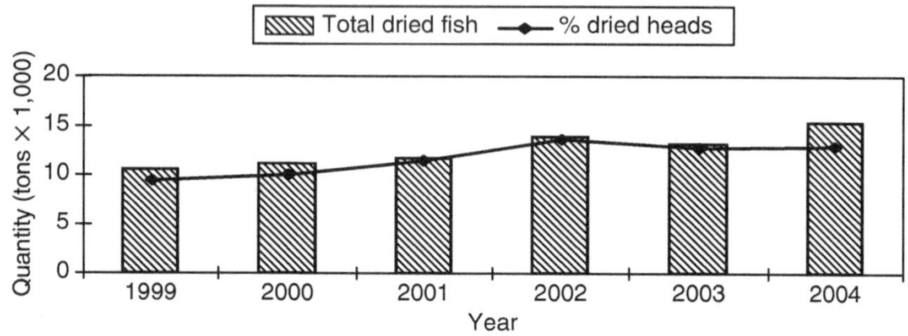


**Figure 12.14.** Export of total dried fish from Iceland 1999–2004 with percentage of dried fish heads (STATICE, 2005)

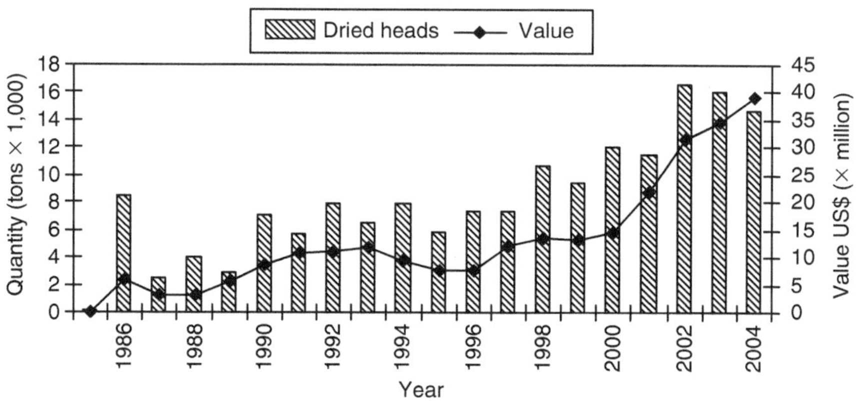


**Figure 12.15.** Export of dried fish heads from Iceland (1985–2004) (adapted from STATICE, 2005).

**Figure 12.16.** The modern computer-controlled drying facility for cod heads, showing the processing of the heads, transport of racks, and racks in the primary drying chamber and control board.

for the shrimp and prawn industry with waste as high as 40% of the original mass (Joensen and Villadsen, 1994). Our investigations indicate that from processing of North Atlantic shrimp 30–40% of peeled shrimp may be expected with 65 to 70% offal, where 20–30% may be soluble material with small particle matter, and another 30 to 40% of solids which is mostly shell material. The slow degradation kinetics of the shell combined with the growth in overall mass of waste from the increased shrimp and prawn industries have led to increased environmental restrictions by various governments (Goosen, 1997). Therefore, interest in alternative use of shrimp and prawn waste has become important. The so-called waste material, however, can just as well be used as raw material for highly profitable processes. The offal contains proteins, chitin, and the coloring agent astaxanthin, which is a necessary ingredient for salmon feeds. The composition of the shell from North Atlantic shrimp was estimated to be ~34% calcium carbonate, 28% chitin, and 38% protein by Veigarsson (1999). The chitin content in crab and shrimp shells can reach as high as 32 and 17% on dry weight basis (Shahidi and Synowiecki, 1991), which is in agreement with our investigations for North Atlantic shrimp (*Pandalus borealis*) where "true" or dry weight basis chitin was 16.55% (Kjartansson et al., 2006b) and protein content was 39.8%. We found less chitin in fresh water prawns (*Macrobrachium rosenbergii*) or 8.28% (Kjartansson et al., 2006a). Cira et al. (2002) reported that waste from white shrimp (*Penaeus spp.)* contained roughly 70.4% water, 14% ash, 45.4% proteins, and 13.1% chitin with small amounts of lipids and pigments. Naturally there is less pigment in both the white and the freshwater species compared to the pinkish to red North Atlantic shrimp. Thus the composition is species dependent but will contain no less than 8 to about 17% of chitin on a dry weight basis.

The astaxanthin extraction has been optimized using vegetable oils (Sachindra and Mahendrakar, 2005) but fish oil also may be used. Astaxanthin is relatively unstable but acid silage of shellfish processing waste has been reported to be a good and economical technique to protect the biomasses from bacterial decomposition (Guillou et al., 1995).

Shrimp shells therefore are an excellent source of chitin, which can be deacetylated to chitosan. Chitin is the second-most abundant biopolymer in nature after cellulose. It is the main structural constituent of the protective shell in insects and crustaceans, like shrimp and lobster. The chitin polymers are embedded in a protein structure which may be calcified with salts forming a hard-shell structure. The biopolymer is composed of poly-*N*-acetyl-*D*-glucosamine (GlcNAc), which is biodegradable but not watersoluble. The polymer has an acetyl group connected to an amine group in the C2 position on the glucan ring, which may be removed by enzymatic or chemical hydrolysis in caustic soda at elevated temperatures, producing a deacetylated form exposing the free amino groups at some of the C2 positions. The amino group in chitosan has a pKa value of 6.3 and it behaves as a polycationic electrolyte in acidic aqueous media where it is soluble. The charge density depends on pH and the percent deacetylation. When the deacetylation reaches 55% or more, the polymer is specified as chitosan.

The properties of chitin and chitosan are highly dependent on the physicochemical properties of the polymers, but many of the reactions and interest in chitosan are due to its cationic nature, since other biopolymers are anionic under similar conditions. Currently, applications have been reported in agriculture, foods, and cosmetics and its properties are being investigated for both medical and pharmaceutical applications. The oligomers generated from the chitosan polymers by hydrolysis are of special interest in medical applications and have been reported to have wound-healing properties (Muzzarelli and Muzzarelli, 2005) and to induce bone regeneration. It is widely used as a flocculent for water purification. Approximately half of the chitosan produced is being sold as a dietary supplement due to suggested lipid binding and hypocholesterolemic properties. A number of studies have demonstrated the ability of ingested chitosan to decrease fat absorption or lipase activity in human or animal digestive tracts (Tsujikawa et al., 2003), while other studies have shown little or no effect (Guerciolini et al., 2001). It is claimed to form a chemical complex with fat in the digestive system or to hinder the absorption of lipids by forming a complex with the bile salts. Our research has indicated that these reactions are both highly dependent on the physicochemical nature of chitosan and environmental conditions involved (Helgason et al., 2006) and on the type of lipid involved.

The processing of chitin from shrimp shells involves removing the mineral and protein matrix in which the chitin is embedded. Generally this has been achieved by washing the dried shells, after grinding, in concentrated acids and alkali, but the conditions used vary greatly with individual processors (Table 12.5); the order of the processing steps also is variable. The astaxanthin may be salvaged with organic solvents after the drying and grinding of the shells, but is not performed in all instances. High-quality chitin may be obtained by performing the demineralization with 1.82% hydrochloric acid for 3 to 4 hours followed by deproteinization with 1% sodium hydroxide for 24 hours. This powder then should be washed with water to give chitin flakes. The flakes then can be dried and sold as chitin or the wet flakes may be deacetylated to chitosan with 50% sodium hydroxide for another 24 hours (Figure 12.17).

This process will yield high-quality chitin and chitosan flakes with various degrees of deacetylation depending on the duration of the final cooking in sodium hydroxide. Recently, researchers at the University of Tennessee and University of Massachusetts in the United States, in cooperation with Primex Inc. (which is the main producer of chitin and chitosan in Europe) and our laboratories, have conducted work for the optimization of chitin and chitosan processing with ultrasound.

*Table 12.5.* Processing parameters for the extraction of chitin from shrimp shells and formation of chitosan with deacetylation

| Step | Reagent | Time (hr) | Temperature |
|---|---|---|---|
| Deproteinization | 0.5–15%NaOH | 0.5–80 | 25–100°C |
| Demineralization | 2.5–8% HCl | 0.5–50 | 15–30°C |
| Decoloration | Various organic solvents like $H_2O_2$, NaOCl | up to 1 | 20–30°C |
| Deacetylation | 40–60% NaOH | 0.5–150 | 60–150°C |

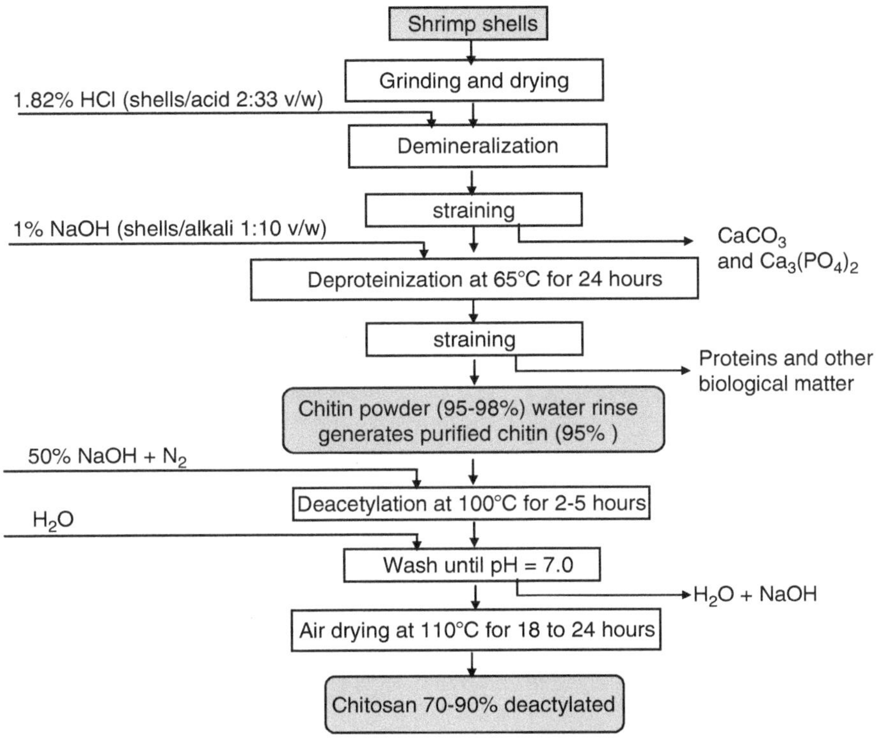


**Figure 12.17.** Processing steps for the production of chitin powder and chitosan.

It was shown that the removal of proteins may be enhanced with the aid of ultrasound in the processing (Kjartansson et al., 2006a, b) and the conversion of chitin to chitosan may be enhanced with the aid of high-intensity ultrasound with the possible reduction in solvent strength and processing time. The properties of chitosan are greatly influenced by the deacetylation process (Berth and Dautzenberg, 1998). Both the degree of deacetylation and the relative location of the resulting cationic amino groups on the polymer chain are very important with respect to functionality. But the removal of the acetyl groups can be random or in blocks. Recently methods to analyze and control the deacetylation process have been put forth (Bahrke et al., 2002). Chitosan therefore can be hydrolyzed into oligomers with specific properties which will greatly enhance how the polymer can be utilized in various applications. This will lead to very highly priced products for medical applications in the area of tissue engineering (Murugan and Ramakrishna, 2006) and bone repair (Samdancioglu et al., 2006). Chitin and chitosan are a very good examples of how potential waste can be turned into highly profitable raw materials or products.

## 12.10. REFERENCES

Alasalvar, C., and Taylor, T., 2002, Seafoods: Technology, Quality and Nutraceutical Applications. Springer-Verlag New York, LLC.

Anonymous, 2006, Cosmetics. http://en.wikipedia.org/wiki/Cosmetics

Arason, S. 2003. Utilization of Fish Byproducts in Iceland. in Advances in Seafood Byproducts 2002 Conference Proceedings. Ed. Bechtel PJ. Alaska Sea Grant College Program, PJ University of Alaska Fairbanks, Fairbanks, 43–62.

Arason, S., and Arnason, H., 1992, Utilization of geothermal energy for drying fish products, *Geothermics* **21**:(5/6):745–75.

Asgeirsson, B, Bjarnason JB. 1991. Structural and kinetic-properties of chymotrypsin from Atlantic cod (*Gadus-morhua*) - comparison with bovine chymotrypsin. Comp. Biochem. Physiol. **99**(2): 327–335.

Asgeirsson, B., Bjarnason, J.B., 1993, Properties of elastase from Atlantic cod, a cold-adapted proteinase. *Bioc. Biophys. Acta* **1164**(1):91–100.

Asgeirsson, B; Fox, JW; Bjarnason, JB. 1989. Purification and characterization of trypsin from the poikilotherm *Gadus-morhua*. Eur J Biochem. **180**(1):85–94.

Bahrke, S.; Einarsson, J. M.; Gislason, J.; Haebel, S.; Letzel, M. C.; Peter-Katalinic, J.; Peter, M. G. 2002, Sequence analysis of chitooligosaccharides by matrix-assisted laser desorption ionization postsource decay mass spectrometry. Biomacromolecules **3**:696–704.

Bang H.O. and Dyerberg J. 1981. The lipid metabolism in Greenlanders. Meddelelser om Gronland, Man & Society, 2. 18 pp.

Bannerman A.M. 1964. Processing Cod Roes. Torry Advisory Note No. 18. (FAO in partnership with Support unit for International Fisheries and Aquatic Research, SIFAR, 2001) http://www.fao.org/wairdocs/tan/x5899E/x5899e00.htm#Contents

Berth, G.; Dautzenberg, H.; G.P., M. 1998. Physico-chemical characterization of chitosan in dilute solution. Advan. Chitin Sci. **2**:429–436.

Bjarnason J. 1986. Handbók fisk vinnslunnar; Skrieðarvinnsla. Icelandic Fisheries Laboratories.

Bjarnason, JB. 2001. ''Biotechnological applications of fish offal in Iceland''. Conference proceedings ''Verdiskaping av marine biprodukter etter år 2000''. Trondheim, Jan. 2001 (http://www.rubin.no/files/projects/attach/jonbragibjarnason.doc).

Borgstrom G. 1971. Principles of Food Science Vol 1. Macmillan Co., New York, New York. pp. 254–255.

Bourre, JM; Bonneil, M; Clement, M; Dumont, O; Durand, G; Lafont, H; Nalbone, G; Piciotti, M. 1993. Function of Dietary Polyunsaturated Fatty Acids in the Nervous System. Prostaglandins, Leukotrienes and Essential Fatty Acids **48**:5–15.

Bragadóttir, M., Pálmadóttir H. and Kristbergsson K. 2002. Seasonal changes in chemical composition and quality parameters of capelin (*Mallotus villosus*). J. Aquat. Food Prod. Technol. **11**(3/4):87–103.

Bragadóttir M, Pálmadóttir H and Kristbergsson K. 2004. Composition and chemical changes during storage of fish meal from capelin (*Mallotus villosus*). J. Ag. Food Chem. **52**(6):1572–1580.

Catarci C. 2004. World markets and industry of selected commercially-exploited aquatic species with an international conservation profile. FAO Fisheries Circulars-C990 (http://www.fao.org/documents/show_cdr.asp?url_file = /DOCREP/006/Y5261E/y5261e06.htm)

Chen, H-M. and Meyers, S.P., 1983. Ensilage treatment of crawfish waste for improvement of astaxanthin pigment extraction. J. Food Sci., **48**:1516–1520 and 1555.

Cira, L. A., Huerta, S., Hall, G. M., Shirai, K. 2002. ''Pilot scale lactic acid fermentation of shrimp waste for chitin recovery.'' Process Biochemistry **37**:1359–1366.

Falch E, Rustad T, Aursand M. 2006. By-products from gadiform species as raw material for production of marine lipids as ingredients in food or feed. Process Biochem. **41**(3):666–674.

FAOSTAT. 2004. Total Amount of shrimp and prawn harvested 2002. FAO Statistical Databases, Food and Agricultural Organization of United Nations, 2002. F. a. A. O. o. U. N. FAO Statistical

Databases, 2002, FAO Statistical Databases, Food and Agricultural Organization of United Nations, 2002.

Goosen, M. F. A. 1997. Application of Chitin and Chitosan. Lancaster, Technomic Publishing Company, Inc.

Gudbjarnasson S. 1999. Bioactive marine natural products. Rit Fiskideildar 16 107–110 (http://www.hafro.is/Bokasafn/Greinar/Rit-fisk/rit_fisk_16–107.pdf).

Gudmundsson, M., and Hafsteinsson, H., 1997, Gelatin from cod skins as affected by chemical treatments, *J. Food Sci.* **62**(1):37–39 & 47.

Gudmundsson, M. 2002. Rheological properties of fish gelatins. J. Food Sci. **67**(6):2172–2176.

Guerciolini, R.; Radu-Radulescu, L.; Boldrin, M.; Dallas, J.; Moore, R. 2001. Comparative Evaluation of Fecal Fat Excretion Induced by Orlistat and Chitosan Obes. Res. **9**(6):364–367.

Guillou A, Khalil M, Adambounoub L. 1995. Effects of silage preservation on astaxanthin forms and fatty acid profiles of processed shrimp (*Pandalus borealis*) waste. Aquaculture **130**:351–360.

Gunnarsson, J.R. 2003. The functional properties of fishmeal; The effects of dryers. [MSc thesis] University of Iceland. University Library, Reykjavik, Iceland.

Helgason T, McClements DJ, Kristbergsson K, Weiss J. 2006. Relationship between molecular characteristics of chitosan and itsability to bind fat droplets: An *in vitro* digestion model study. *In Press, Corrected Proof.*

Hjalmarsson, G.H. 2001. Fish sauce from capelin (Mallotur villosus) as affected by harvest season. [M.Sc. thesis]. University of Iceland. University Library, Reykjavík, Iceland.

ICIMF - Information Center Icelandic Ministry of Fisheries. Available from:http://www.fisheries.is/stocks/capelin.htm (Accessed May 10, 2005).

Joensen, O. Villadsen, A. 1994. Ecological sustainable production of chitin and chitosan. Chitin World. Z. S. Karnicki, Brzeski, M. M., Bykowski, P. J.,Wojtasz-Pajak, A. Bremerhaven, Wirtschaftsverlag NW:38–47.

Jonsdottir, R; Olafsdottir, G; Martinsdottir, E; Stefansson, G. 2004. Flavor characterization of ripened cod roe by gas chromatography, sensory analysis, and electronic nose. J. Ag. Food Chem. **52**(20): 6250–6256.

Katsiadaki LG, Taylor KDA, Smith G. 1999. Assessment of quality of cod roes and relationship between quality and maturity stage J. Sci. Fd. and Agric., **79**(10):1249–1259.

Kinsella JE. 1986. Food components with potential therapeutic benefits - the n-3-polyunsaturated fatty-acids of fish oils Food Tech. **40**(2):89.

Kjartansson G, Zivanovic S, Kristbergsson K, Weiss J. 2006a, Sonication assisted extraction of chitin from shells of fresh water prawns (*Macrobrachium rosenbergii*) J. Agr. Food Chem. **54**(9): 3317–3323.

Kjartansson G, Zivanovic S, Kristbergsson K, Weiss J. 2006b, Sonication assisted extraction of chitin from North Atlantic shrimps (*Pandalus borealis*). *In Press, Corrected Proof.*

Kristbergsson, K. 1998. Water Jet Deboning of Meat and Fish Bones. - NordFood Project P93185. Final report. Nordic Industry Fund, Oslo, Norway.

Kristinsson, HG; Theodore, AE; Demir, N; Ingadottir, B. 2005. A comparative study between acid- and alkali-aided processing and surimi processing for the recovery of proteins from channel catfish muscle. J. Food Sci, **70**(4):C298-C306.

Kristjansson, M.M; Gudmundsdottir, S; Fox, JW; Bjarnason, J.B. 1995. Characterization of a collagenolytic serine proteinase from the Atlantic cod (*Gadus-morhua*). Comparative Comp. Biochem. Physiol. **110**(4):707–717.

Liang, Y; Hultin, HO. 2005. Effect of pH on sedimentation of cod (Gadus morhua) muscle membranes. J. Food Sci. **70**(2):C164-C172 MAR 2005.

Makrides, M; Neumann, M; Simmer, K; Pater, J; Gibson, R. 1995. Are Long-Chain Polyunsaturated Fatty-Acids Essential Nutrients In Infancy. Lancet, **345**(8963):1463–1468.

Mei M, Gudmundsdottir G, Thorkelsson G, Arason S and Kristbergsson K. 2003. Reforming fish cut-offs into fillets with texture resembling intact fish flesh. In: IFT Annual Meeting Book of Abstracts; 2003 July 12–16; Chicago, Il. Institute of Food Technologists. 40–9.

Murugan, R; Ramakrishna, S. 2006. Nano-featured scaffolds for tissue engineering: A review of spinning methodologies. Tissue Engineering, **12**(3):435–447.

Muzzarelli RAA, Muzzarelli C. 2005. Chitosan chemistry: Relevance to the biomedical sciences. Adv. Polymer Sci. **186**:151–209.

Ogmundsson J. 2005. Personal communication with the director of product development and quality control at Lysi hf the main oil refinery in Iceland.

Sachindra, NM; Mahendrakar, NS 2005. Process optimization for extraction of carotenoids from shrimp waste with vegetable oils. Bioresource Technology, **96**(10):1195–1200.

Samdancioglu, S; Calis, S; Sumnu, M; Hincal, AA. 2006. Formulation and in vitro evaluation of bisphosphonate loaded microspheres for implantation in osteolysis. Drug Development and Industrial Pharmacy, **32**(4):473–481.

Shahidi F, Synowiecki J. 1991. Isolation and characterization of nutrients and value-added products from snow crab (chinoecetes-opilio) and shrimp (pandalus-borealis) processing discards. J. Ag. Food Chem. **39**(8):1527–1532.

Sigurdsson, A. M. (1992). Apparatus for scraping fish flesh off the backbone after splitting. (In Icelandic). Icelandic patent no. 1491.

Sigurdsson, A. M. (1993). Machine for cutting tongues, cheeks and belly flaps from fish heads. US patent no. 5,226,848.

STATICE. 2005. Statistical Series Fisheries **90**:61 (2005:6) 1-16ISSN 0019-1078 (www.hagstofa.is/hagtidindi www.statice.is/series)

Tsujikawa, T; Kanauchi, O; Andoh, A; Saotome, T; Sasaki, M; Fujiyama, Y; Bamba, T. 2003. Supplement of a chitosan and ascorbic acid mixture for Crohn's disease: A pilot study Nutr. **19**(2): 137–139.

Undeland, I; Kelleher, SD; Hultin, HO. 2002. Recovery of functional proteins from herring (Clupea harengus) light muscle by an acid or alkaline solubilization process. J. Ag. Food Chem. **50**(25): 7371–7379.

Valdimarsson, G. 1998. Developments in Fish Food Technology – Implications for Capture Fisheries. J. Northw. Atl. Fish. Sci. **23**:233–249.

Veigarsson H. 1999. Water Management in Northern Shrimp Muscle (*Pandalus borealis*) with Emphasis on Processing. [M.Sc. thesis]. University of Iceland. University Library, Reykjavík, Iceland.

Vilhjálmsson, H. 1998. Loðna. Námsgagnastofnun-Hafrannsóknastofnunin: **12**, (Capelin. Educational publications og the Marine Research Institute of Iceland: 12), Reykjavík, Ísland.

Wendel A, Park, JW, Kristbergsson K. 2002. Recovered meat from Pacific whiting frame. J. Aquat. Food Prod. Tech. **11**(1):5–18.

Whittle KJ and Howgate P. 2000. Glossary of Fish Technology Terms. Food and Agriculture Organization of the United Nations.

# 13

# Examples of Special Case Studies in Different Branches

Winfried Russ

## 13.1. FOOD WASTE USED AS RAW MATERIAL FOR NONFOOD PRODUCTS AS A VIABLE ALTERNATIVE TO DISPOSAL

Waste from the food industry can be classified by regarding the different attributes of the matrix, which includes constituents, storage stability, or physical properties. It is coherent to later utilization to base an exemplary classification system on the important components of product-specific food waste (see Figure 13.1).

The traditional methods for utilizing food waste have been to use it as animal feed or to spread it over agricultural land. If these primary methods were not feasible, for example, because of toxic residues, only possibility was to dump or incinerate the waste. In recent years more problems are appearing in disposal or utilization of food waste with conventional methods (Figure 13.2A). Further processing of food waste is not economically feasible and therefore rare, because in most cases no added net value can be gained by doing so. Examples include (see Section 2, this chapter) drying used yeast (Wysocki and Glöckner, 1971; Macher, 1968), spent grains (Prestel, 1994; Hug, 1992; Beckhoff, 1986; Meyer-Pittroff, 1994), or whey (Franzen et al., 1999), as well as cleansing waste oils (Jurisch and Meyer-Pittroff, 1995).

Recently, new methods for utilizing food waste have been developed which are designed to create added net value or to exploit one or more of the components of the food waste (Figure 13.2B). Agricultural methods of disposal are less important with these new methods. The methods of waste utilization can be divided into two fundamentally distinct groups:

1. Fully utilizable food waste.
2. Food waste that can be partly utilized.

WINFRIED RUSS • Technische Universität München, Chair for Energy and Environmental Technologies of the Food Industry, Weihenstephaner Steig 22, 85350 Freising-Weihenstephan, Germany.

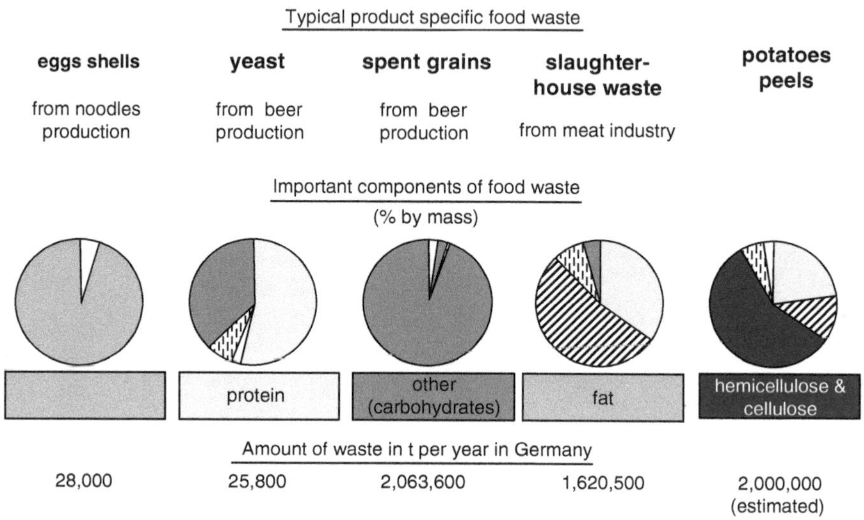


**Figure 13.1.** Typical product-specific food waste, their main components, and amounts produced annually.

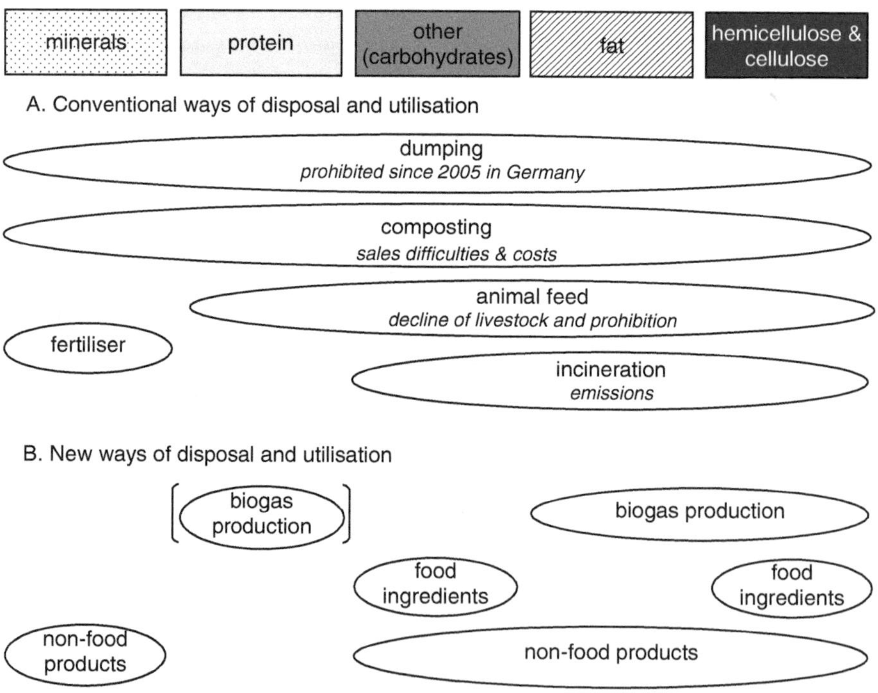


**Figure 13.2.** Conventional ways, problems (in italic), and new ways of disposal and utilization of different kinds of food waste.

The key factor for food waste from the first group (which can be fully utilized) is the existence of one or more substances, which have not been removed from the waste and which also possess valuable properties. Valuable, in this sense, refers to the fact that these substances have something positive to contribute to the utilization of this waste. Food waste can be processed for further usage in such a way that it is able to be stored without danger of spoilage, or as a result of processing, the valuable substances can be made more accessible, for example, pressing spent grains to reduce the water content (Zanker and Kepplinger, 2002) or the cleansing and esterification of waste oil so that it may later serve as combustible fuel (Russ et al., 1997, 2004).

In the second group, the desired substances are removed from the waste and can either be directly utilized or further processed before utilization. For instance, the lactose contained in whey can be used as a substrate for bacteria to produce fats or surfactants (Daniel et al., 1999). An additional process involves the extraction of lactoferrin from whey (Plate et al., 2002). The production of grape seed oil from pomance or the utilization of carrot, citrus, and apple processing waste as a source of fiber or for use as thickening agents have proven to be economical. With these methods, a portion of the waste remains. Its disposal cost is offset by the monetary gain from the utilization of a portion of this waste.

In these cases, the waste produced by the food industry is used in the same way as renewable raw materials and should be viewed as such. However, it must be acknowledged that the previous extraction of valuable substances can result in a concentration of noxious substances in the remainder of the waste. The methods for the utilization of this waste should offer a possibility for eliminating these dangerous substances or facilitate their conversion to inert substances. For example, utilization of waste from harvesting grapes can result in a concentration of the pesticides present in the waste. Spreading this waste over vineyards or other agricultural areas results in the wide dispersal of the problematic substances contained in the waste.

It is crucial that these new methods are based on the fact that waste under consideration for further processing contains one or more substances of potential value. An exemplary method was developed to utilize food waste characterized by rapid spoilage due to protein decomposition. It must be noted, however, that the negative effects of other substances present in the waste also must be minimized. In addition to protein, hot and cool trub from the brewery also contain many bittering compounds derived from hops. The bitter taste of these substances prevents the hot and cool trub from being fed directly to animals. One solution is to dry the material in order to preserve it, which has to bring with it the added benefit of reducing the bitterness through the heating process.

Fibers are added to many materials for the purpose of strengthening the matrix of the materials. The most important example of this application is fiberglass. Many types of food waste contain cellulose fibers. By way of example, it will be shown how these fibers must be processed and prepared so that they can be used in the matrix to increase the strength of fired clay products.

The addition of spent grains to increase porosity in brick is a typical example of how food waste can be utilized as a filler on the basis its size and form. Further possibilities for the use of spent grains as a filler and structural material in fiber board and as an insulating material already has been investigated and described by a team of researchers (Höhn et al., 1999; Mayer, 1998; Blümelhuber, 2002).

In kieselguhr sludge from the brewing industry the mineral component, kieselguhr, represents the substance of value. It will be demonstrated how the kieselguhr can be recycled. A significant advantage is that the quality of the kieselguhr as a filter aid can be achieved and even surpassed as a result of the recycling process. The utilization of kieselguhr in the production of calcium silicate bricks fulfils several fundamental goals of a modern utilization concept. On the one hand, the silicon dioxide of the kieselguhr can be used as an easily accessible source of silicon. Conversely, a method must be developed to avoid possible negative environmental influences (the evolution of odors during storage) with no detrimental effects on the utilization.

In order to utilize waste oils as fuel, the existing technology and machinery currently in use for the combustion of rapeseed oil must be modified for use with waste oils. In this case, the feasibility of this method already has been established; however, one must consider the potential environmental damage through the combustion caused by possible impurities contained in this fuel.

## 13.2. USE OF PHENOLIC PROTEIN PRECIPITATES (TRUB) FROM BEER PRODUCTION IN ANIMAL FEED

During beer production, proteins and polyphenols coagulate, forming complexes which then fall out of solution. This precipitate is separated from the wort in the form of hot trub at the end of wort production and after cooling but before fermentation in the form of cool trub. The amount of hot trub, cool trub, and yeast accumulated in the brewing process varies widely with the equipment and technology used throughout the brewing industry. The specific ratios of trub and yeast measured in a study with 12 breweries by Russ and Meyer-Pittroff (2003) are given in Table 13.1.

The current disposal of the hot trub with the spent grains and of the cool trub in the wastewater is neither economical nor practical due to their nutritional value.

*Table 13.1.* Accumulated amounts (kg/1000 liter beer) of hot trub, cool trub, and yeast in breweries (Russ and Meyer-Pittroff, 2003)

| | Minimum | Average | Maximum |
|---|---|---|---|
| Hot trub | 5.2 | 19.2 | 50.5 |
| Cool trub | 8.7 | 56.2 | 182.0 |
| Spent yeast | 18.8 | 174.4 | 467.2 |

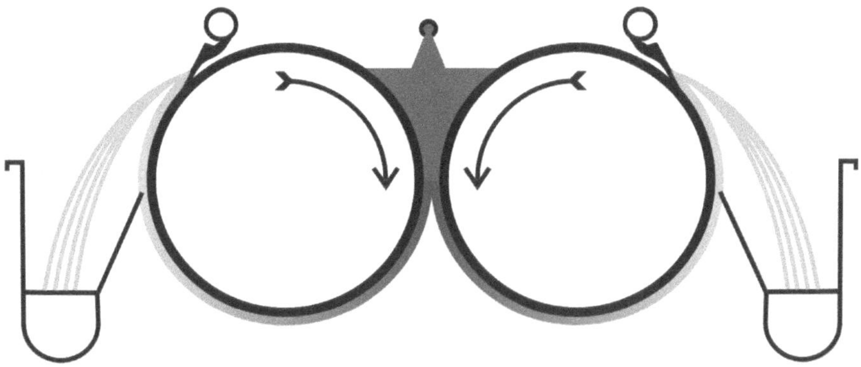

**Figure 13.3.** Schematic drawing of a double drum dryer with a nip feed.

The primary objective is to treat the protein–polyphenol precipitate in order to create a nonperishable animal feed which despite the bitter taste will still be accepted by animals.

Experiments show that all types of trub as well as a mixture of trub and spent yeast can be dried using a drum dryer (Figure 13.3) to create a usable livestock feed additive. Short exposure to high temperatures noticeably reduced the bitterness of hot and cool trub. The composition of the trub–yeast mixture is comparable to that of dried spent yeast. Dried trub consists mainly of nonnitrogenous extract, which is made up mostly of sugars.

The feeding experiments using these dried products show that a combination of trub and yeast can be used in dried protein feed preparations. Depending on the ratio between trub and yeast a rate up to 40% in feed composition is possible. This contradicts findings often reported in the literature, which suggest that livestock will not accept trub in their feed.

## 13.3. KIESELGUHR SLUDGE FROM THE DEEP BED FILTRATION OF BEVERAGES AS A SOURCE FOR SILICON IN THE PRODUCTION OF CALCIUM SILICATE BRICKS

Kieselguhr consists primarily of $SiO_2$, part of which occurs in the amorphous form. Kieselguhr sludge is generated during deep bed filtration of beverages. Currently, the most frequent disposal method is to spread the used kieselguhr over agricultural land. This is not a satisfactory solution to the problem of disposal. In experiments it was established whether kieselguhr sludge can function as a source for silicon during the formation of CSH (calcium silicon hydrate) phases in calcium silicate material.

Generally, the utilization of kieselguhr sludge in the production of calcium silicate material yielded positive results, which can be attributed to the small particle size, the large specific surface area, and its amorphous structure. As a result of these characteristics, kieselguhr behaves in the same way as a fine silicate during the hardening process. It dissolves quickly and completely and is converted into binding medium. The organic components in the used kieselguhr decompose during hardening.

Increases in strength and density were achieved through the utilization of kieselguhr sludge. No reduction in flexural strength was observed through decreasing the lime content by up to 40% by mass. The compressive strength of samples after hardening increases with increasing quantities of lime and kieselguhr (Figure 13.4). Compared to the control samples, the samples containing kieselguhr sludge displayed not only a 50% increase in strength but also a 50% reduction in the required amount of lime (Russ et al., 2006).

A kieselguhr sludge–lime mixture can be integrated successfully into the production of calcium silicate material. While still meeting the requirements for raw density and compressive strength, less quicklime or sand is needed for production. This is most evident in the improved particle size distribution of the raw mixture.

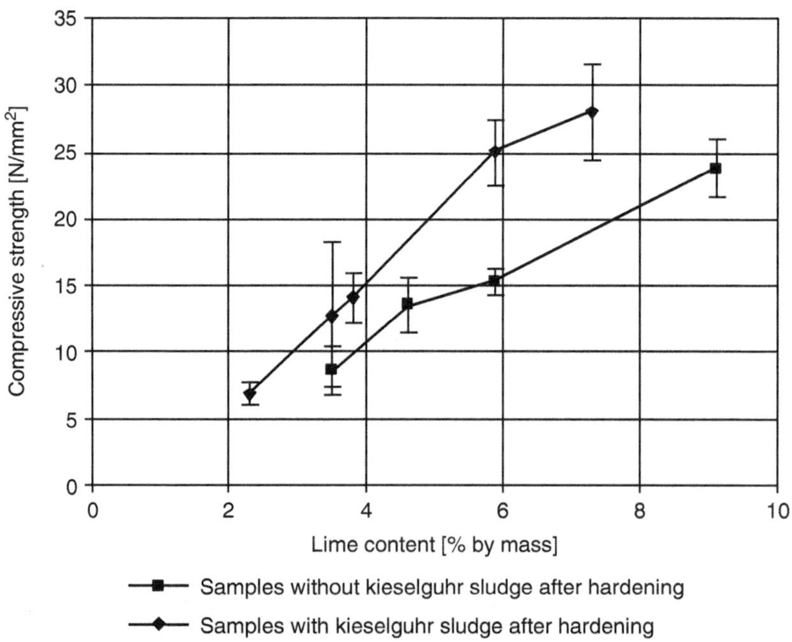


**Figure 13.4.** Compressive strength of the test samples with and without kieselguhr sludge after hardening as a function of the calcium hydroxide content (Russ et al., 2006).

The control samples showed little loss in strength, because the contribution of the CSH phases to their strength was minimal. Additionally, the calcite formation progressed without affecting the structure.

Not all preservatives designed to reduce odors and suppress microbial growth in used kieselguhr are suitable for the production of calcium silicate material. Alkali and sulfate ions in preserved kieselguhr sludge reduce the quality and strength of the finished calcium silicate products. Other influences resulting from the addition of preservatives could not be established.

No problems were encountered in the large-scale experiment. An unpleasant odor of ammonia was detected upon opening the autoclave, which can be attributed to the hydrolytic disproportion of the proteins. Visually, the samples appeared to be flawless, and the edges showed no signs of instability. The raw densities of the calcium silicate products containing kieselguhr sludge were the same as the standard products made without kieselguhr sludge. There were no significant changes in the compressive strength of the calcium silicate material.

## 13.4. APPLICATION OF SPENT GRAINS TO INCREASE POROSITY IN BRICK

Spent grains are a form of waste, which accumulates during the brewing process. This waste is a result of the mashing process, in which desired substances in the malt (sprouted and kilned grain) are extracted using water. Spent grains are routinely used as animal feed; however, they can be fed only to ruminants, because of their high cellulose (23–25% by mass) and hemicellulose (30–35% by mass) content. The utilization of spent grains as animal feed has been decreasing steadily due to the declining number of cattle being raised in Germany. Furthermore, spent grains spoil within 3 days due to their high protein (19–23% by mass) content, and therefore cannot be stored and automatically fed.

One possibility for the disposal of spent grains is to use them to increase the porosity in brick (Knirsch et al., 1998; Russ et al., 2005). Laboratory experiments have shown that the addition of sawdust, commonly used in the brick-making process to increase porosity, can be substituted with spent grains.

Values of the plasticity of the samples show that through the substitution of sawdust with spent grains, practically no adjustment to the water content was necessary. The small variations can be attributed to the different amounts of sewage sludge added to the green clay. The flexural strength of fired samples did not vary with the addition of the spent grains and measured approximately 9.0 MPa. Air-dried samples exhibited a small increase in flexural strength, which contributed to improvement in handling properties of the green clay bricks. The results of shrinkage and true density were almost identical, while the raw density, the water absorption capacity, and the open porosity showed minor differences, which still fall within the normal margin of error. A lower raw density and either a higher level of open porosity or a higher water absorption capacity can result

*Table 13.2*. Technical analysis results for test bricks containing spent grains compared to the regular factory mixture (Russ et al., 2005)

| | Factory mixture | Spent grains (fine) | Spent grains (coarse) |
|---|---|---|---|
| Moisture content [% by mass] | 13.5 | 19.3 | 19.4 |
| Dry shrinkage $S_T$ [%] | 5.76 ± 0.11 | 5.93 ± 0.18 | 6.71 ± 0.10 |
| Flexural strength $\sigma_{air\text{-}dried}$ [MPa] | 6.84 ± 0.54 | 6.59 ± 0.52 | 6.62 ± 0.36 |
| Firing shrinkage $S_B$ [%] | 0.42 ± 0.12 | 0.48 ± 0.17 | 0.35 ± 0.20 |
| Total shrinkage $S_G$ [%] | 6.18 ± 0.12 | 6.41 ± 0.24 | 7.06 ± 0.21 |
| Density $\rho[g/cm^3]$ | 1.77 ± 0.07 | 1.77 ± 0.09 | 1.70 ± 0.01 |
| Density $\rho_{theor.\ [g/cm^3]}$ | 1.71 ± 0.01 | 1.72 ± 0.27 | 1.70 ± 0.22 |
| True density $\rho[g/cm^3]$ | 2.59 ± 0.19 | 2.69 ± 0.02 | 2.71 ± 0.02 |
| Water absorption WA [%] | 18.93 ± 0.47 | 20.38 ± 0.52 | 20.28 ± 0.08 |
| Flexural strength $\sigma_{fired}$ [MPa] | 8.68 ± 0.40 | 8.42 ± 0.84 | 8.20 ± 0.45 |

in an increase of the insulating properties of the bricks. Additionally, the important criterion of color was not influenced by the substitution of the sawdust with the spent grains.

In a large-scale experiment, fine and coarse spent grains were used in place of sawdust to increase porosity. No problems were observed during the mixing, pressing, and cutting stages of production. The bricks produced from clay with spent grains possessed on average a higher compressive strength. As shown in Table 13.2 the bricks produced with spent grains possessed a comparable strength and a higher porosity (higher water absorption capacity). The sintering processes began at lower temperatures, thus resulting in a fired clay product that was more strongly sintered, exhibiting both greater strength and higher porosity.

## 13.5. INVESTIGATING THE USE OF FIBERS TO IMPROVE THE STRENGTH OF GREEN CLAY BRICKS

The use of fibers to increase green strength in brick is comparable to their use in bonded fiber materials. Different investigations showed the feasibility of incorporating processed fibrous organic waste from the food industry into brick production, in order to improve the green strength of brick (Knirsch et al., 1998; Russ et al., 1994).

In the food industry, labels for reusable containers are made of paper that contains a large amount of long cellulose fibers. This paper mixture of the labels is designed to remain whole even when wet. The fibers for these investigations were removed from returnable bottles in a bottle washing machine. Upon their return, reusable containers are cleaned in a caustic soda solution, whereby the labels are removed and transported out of the washing machine. Furthermore, the characteristics of nonlabel paper containing long fibers, which had been laminated on both sides with plastic (such as the double-sided plastic laminate used in Tetra

Pak packaging), were also tested. Regardless of the source, all fibrous waste of biological origin is composed of cellulose, and therefore the results from these experiments are applicable to any form of this waste.

Special attention must be given to particle size distribution of the fibers; that means to shred the material finely enough that the majority of the particles were measured 2 mm in length and that the largest particles in the distribution were no greater than 4 mm. The shredding was done using cutting mills, hammer mills, shredders, pelletizers, and high-pressure water jets. Cutting mills can consistently create a product of a suitable particle size distribution, given that the water content remains below 35% by mass. Shredding with a high-pressure water jet is very effective to fulfil the required parameters for shredding the labels; however, the resulting energy costs are too high to make this method economical. Processing label or Tetra Pak waste with a die pelletizer creates a shredding effect through compression and transport. Shredding and pelletization are particularly effective with a screw extruder.

A laboratory extruder formed the clay into samples of approximately 200 mm in length and 16 mm in diameter. The samples were air-dried for a period of 3 days and subsequently placed in a drying oven for an additional 3 days at approximately 130°C.

The degree of fiber exposure and the type of shredding has a considerable influence on the dry strength of the brick. Exposure of the fibers, combined with the fact that a portion of the paper mass remains intact, results in a noticeable increase in strength. The fibers in the material must first be exposed for them to be utilized in brick production. Since a single fiber is unable to prevent cracking and is too short to span the total width of a crack, it is necessary to preserve the fibers structurally as a unit. Individual fibers do not increase brick strength as well as fibers which retain their textural unity. Analogous to the use of fibers in composites such as asbestos cement or fiberglass, the fibers strengthen the bricks structurally. For this reason, the correct preparation, especially the shredding, of the label waste is critical.

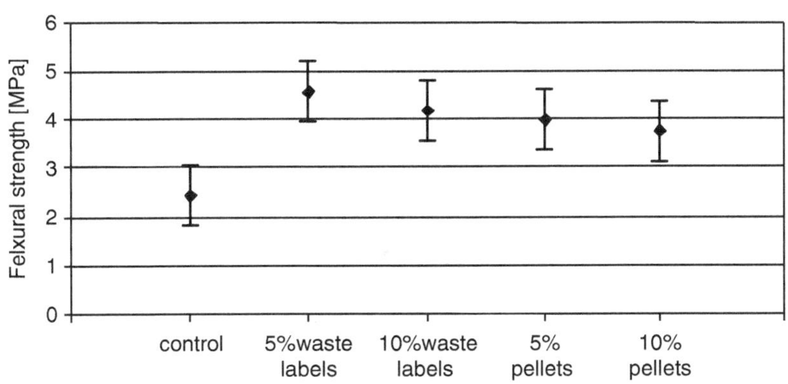


**Figure 13.5.** Increase in flexural strength in green clay brick with the addition of different waste fibers.

Figure 13.5 shows the change in flexural strength of dried bricks when using waste labels and pelletized Tetra Pak waste as porosity-increasing materials. Label waste and pellets were added in the amount of 5% and 10% by mass to the Keuper clay samples.

The flexural strength can be increased up to 50% by adding waste labels as compared with the control samples. The addition of pelletized Tetra Pak also increased the strength of the bricks. However, the fibers of the laminated paper were less exposed than those of the paper made from labels. Additionally, shrinkage cracks due to drying are visibly absent in the dried brick.

## 13.6. HEAT TREATMENT OF KIESELGUHR IN ORDER TO IMPROVE ITS FILTRATION PROPERTIES AND FOR ITS RECYCLING

Several processes were developed for recycling of kieselguhr, but the resulting quality is not sufficient for many recycling cycles. Mostly not more than five cycles are possible. That means the dumping problem (concerning odor and legislation) is not solved, only reduced.

A new recycling process on a preindustrial scale has been developed in cooperation with an industry partner for recycling used kieselguhr from deep bed filtration. In addition to this process, the filtration qualities of the kieselguhr have been notably improved as compared with the original kieselguhr (Russ et al., 2003). The filtration properties of the kieselguhr were improved by heat treatment at temperatures no higher than 500°C for an optimum of 120 minutes in the presence of water vapor.

A parameter for the maximum duration of filtration, the cumulative increase in pressure across the filter bed of the pilot filter, rose more gradually over time using the heat-treated kieselguhr than with the untreated kieselguhr. The total amount of filtration time possible was increased by using heat-treated kieselguhr, therefore reducing filter downtime. Given that the heat treatment took place under the optimal conditions described above, the turbidity, as a measure of filtration precision, does not increase in comparison with untreated kieselguhr. The extent to which the filtration properties of heat-treated kieselguhr were able to be improved was also dependent upon the origin and type of kieselguhr.

With heat treatment the water value and the permeability increased. The particle size distribution indicates that structural changes were uncommon as a result of heat treatment. At temperatures below 550°C sintering processes also were negligible. The differential scanning calorimetrie (DSC) results showed no changes of kieselguhr as a result of heat treatment. Neither exothermic nor endothermic reactions took place, nor did a crystalline phase change occur to any extent worth mentioning. The specific surface area of the kieselguhr heated for 120 minutes at 500°C was reduced by 29% and the pore surface area reduced by 70%, as compared to an untreated sample. The pore volume of the heat-treated kieselguhr was 29%

less than that of the same untreated kieselguhr. Additionally, the average pore diameter was almost doubled, from 16 μm to 30 μm.

The improved filtration properties of the heat-treated kieselguhr ultimately could explained by the simultaneous decrease in the specific surface area, increase in the pore diameter, and reduction in the pore surface area. The higher values for turbidity together with the greater decrease in filtration pressure both resulted from a change in pore size and a reduction in the adsorptive characteristics of the particles due to the heating process. The adsorptive nature of kieselguhr is directly related to its amount of accessible surface area.

A presumable result of the heat treatment was a reduction in the smaller pores, and therefore a lessened possibility for interaction between the kieselguhr and the suspended particles. Thus, more suspended particles were able to pass through the filter bed causing the values for turbidity to increase. Only the larger pores remained open, which lowered the resistance of the filter bed, increasing the flow rate through it. As a consequence, the filter bed compressed more slowly, which resulted in a more gradual pressure increase during filtration. The implications of these results are not only that kieselguhr can be successfully recycled, but that its filtration properties can be improved at the same time.

## 13.7. EMISSIONS REDUCTION THROUGH THE USE OF FOOD INDUSTRY WASTE OILS AS FUEL

Unlike fossil fuels, the combustion of vegetable oil is regarded as $CO_2$ neutral and environmentally friendly. In recent years, the application of vegetable oils as a source of fuel has been extensively researched. The limiting factor hindering their more widespread implementation is the amount of land available for the cultivation of these crops.

Beside engines like Elsbett, most common for combustion of vegetable-based fuels are diesel engines. There are basically two kinds: for prechamber (IDI) diesel engines straight vegetable oils can be used after minimal processing, while for direct injection (DI) diesel engines the oils must undergo esterification. In order to test the feasibility of using waste oils from the food industry as a fuel source, and additionally to test emissions generated by the combustion of these fuels, two cogeneration units were each fitted with a prechamber (IDI) diesel engine. Diesel was selected as the standard fuel and rapeseed oil and rapeseed oil methyl ester (RME) as the fuels of agricultural origin, because of their widespread availability. Mustard oil could be used in the nonester form because of its purity. Waste oil methyl ester (WOME) was obtained from a supplier. The animal fat methyl ester (AFME) for the experiment was produced in the laboratory.

These esterified waste oils and animal fats as well as mustard oil were tested and compared to the more or less "common" fuels: diesel, rapeseed oil, and rapeseed methyl ester. The results show that in principle each of these fuels is suitable for use in a prechamber diesel engine. To keep constant electrical power at constant revolution, which means a engine performance at constant level, fuel supply was regulated by the controller. Thus a change in temperature is an indication for a different performance of the fuel. The animal fat methyl ester did not burn as well in the diesel engine, as was evidenced by the increased temperature of the exhaust.

For further measurements a catalytic converter was installed on one of the cogeneration units where also urea could to be injected for denitrification of the exhaust. The carbon monoxide emissions from waste oils are not significantly different from those of ''common'' fuels and can be reduced the same way as those of hydrocarbon emissions, through utilization of a catalytic converter. The rate of carbon monoxide reduction by catalytic conversion was 84 to 86% (Figure 13.6).

A lower hydrocarbon concentration was associated with fuels of agricultural origin, with the catalytic converter a reduction of 29 to 42% was achieved. The nitrogen oxides emissions without catalytic conversion were comparable. A significant reduction in $NO_x$ was achieved through the injection of urea. Combining a urea injection with the SCR catalytic converter reduced $NO_x$ emissions between 53 and 67% (Figure 13.7).

Each prechamber diesel engine exhibited its own characteristic exhaust, which was independent of fuel type. The selective catalytic reduction of the exhaust emissions can be realized without restriction using fuels of agricultural origin.

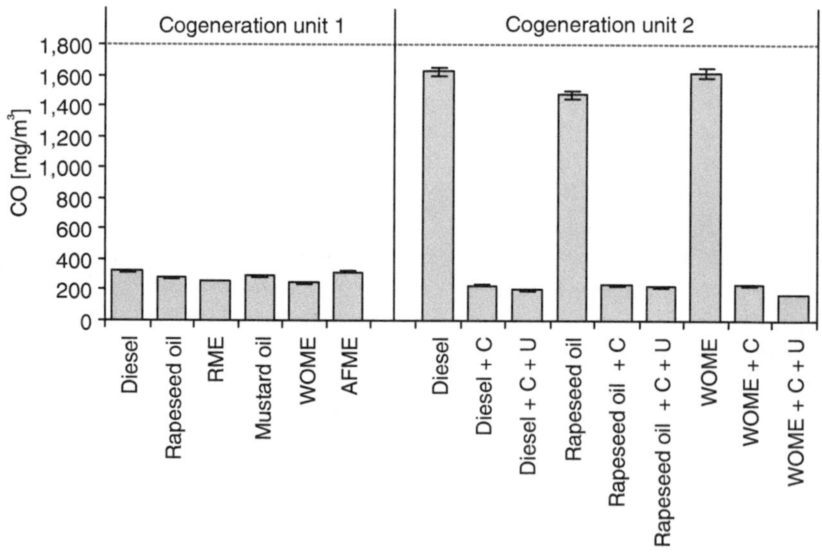


**Figure 13.6.** Carbon monoxide emissions (CO) based on a residual oxygen content of 5% by volume in the exhaust emissions (C, converter; U, urea) (Russ et al., 2004).

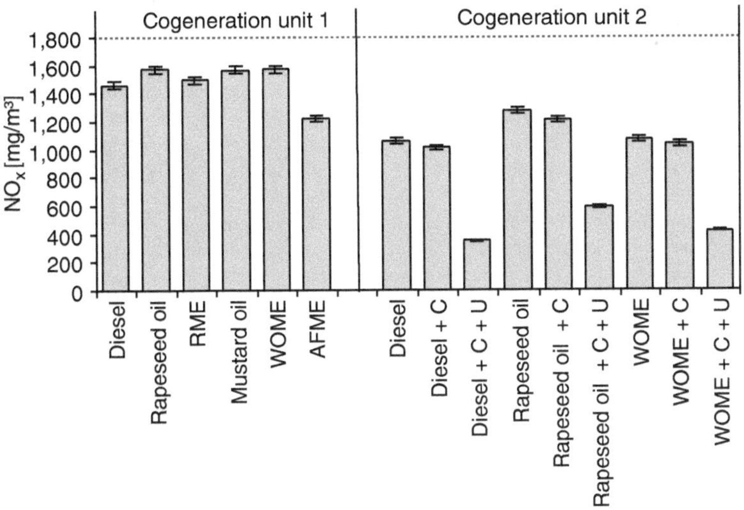


**Figure 13.7.** Nitrogen oxides emissions based on a residual oxygen content of 5% by volume in exhaust emissions (C: converter; U: urea) (Russ et al., 2004).

## 13.8. REFERENCES


Beckhoff, J., 1986, Nährstoffgehalt und Qualität von Biertreber-Silagen. *Brauwelt* **126**(13):498–501.

Blümelhuber, G., 2002, Verbundwerkstoffe aus Fasern und Pflanzenölprodukten, dissertation, Technische Universität München, Freising.

Daniel, H. J., Otto, R. T., Binder, M., Reuss, M., and Syldatk, C., 1999, Production of sophorolipids from whey - development of a two-stage process with cryptococcus curvatus ATCC 20509 and candida bombicola ATCC 22214 using deproteinized whey concentrates as substrates, *Appl Microbiol Biot* **51**.

Franzen, J., Erich, E., Haep, S., Knoch, A., and Steinkamp, H., 1999, Grundlegende Untersuchungen zur Kreislaufwirtschaft in Sennereien und Molkereien durch Aufbereitung von Molke mit Hilfe der Vakuumdestillation, final report of the AiF research project no. 11452.

Höhn, G., Mayer, W. A., Meyer-Pittroff, R., and Russ, W., 1999, Verwertung von faserhaltigen Abfall- und Reststoffen der Lebensmittelindustrie als biologische Isolier- und Dämmstoffe, final report, research project F 97, BayFORREST.

Hug, H., 1992, Waste utilization in the brewhouse, in: *EBC-Monograph XIX - Waste reduction in brewery operation*, Hans Carl, Nürnberg.

Jurisch, C., and Meyer-Pittroff, R., 1995, Verwertung von pflanzlichem Altfett als biogener Kraftstoff, in: *Proceedings of the 3rd symposium Im Kreislauf der Natur*, CARMEN e.V., Wüzburg, pp. 109–126.

Knirsch, M., Penschke, A., Krebs, S., Russ, W., Mörtel, H., Mayer, W. A., and Meyer-Pittroff, R., 1998, Application of brewery wastes in the production of bricks, *Tile & Brick Int* **14**(2).

Macher, L., 1968, Fütterungstechnische Verwertung eiweißreicher Brauereiabfälle, *Brauwelt* **108**(23): 408–411.

Mayer, W. A., 1998, Verwertung von Trebern zur Herstellung von Isolier- und Dämmstoffen, dissertation, Technische Universität München, Freising.

Meyer-Pittroff, R., 1994, Möglichkeiten der alternativen Treberverwertung, *Brauwelt* **134**(1):8–14.

Plate, K., lber, R., Scheper, T., Demmer, W., and Buchholz, H., 2002, Wertschöpfende Aufarbeitung von Molke – Gewinnung von Lactoferrin über Membranadsorber, in: *Stoffliche Nutzung von nachwachsenden Rohstoffen*, W. Naundorf et al., ed., TU Bergakademie, Freiberg.

Prestel, W., 1994, Verfütterung von Biertrebern, *Brauwelt* **134**(22):1029–1030.

Russ, W., and Meyer-Pittroff, R., 2003, The use of phenolic protein precipitates (trub) from beer production in animal feed, *Monatsschr Brauwiss* **56**(5/6):84–88.

Russ, W., Meyer-Pittroff, R., Feix, R., and Kirnbauer, P., 1994, Recycling of brewery residuals on the construction materials industry, *Brauwelt Int* **II**(4):370–375.

Russ, W., Dobiasch, A., Herdegen, V., Meyer-Pittroff, R., Schöberl, H., and Vogel, R. F., 1997, Vom Bioabfall zum Kraft- und Futterstoff, in: *Bioabfall-Management '97*, H. Herbers, ed., Köln, pp. 203–232.

Russ, W., Schmid, N., and Meyer-Pittroff, R., 2003, Heat treatment of kieselguhr in order to improve filtration properties and for recycling, *Monatsschr Brauwiss* **56**(7/8):134–140.

Russ, W., Rauber, M., Ionel, I., and Meyer-Pittroff, R., 2004, Emissions generated by the use of waste oils from the food industry as alternative fuel for diesel engines, in: *Proceedings of 5$^{th}$ International World Energy System Conference*, Oradea, ISBN 1198–0729.

Russ, W., Mörtel, H., and Meyer-Pittroff, R., 2005, Application of spent grains to increase porosity in bricks, *Constr Build Mater* **19**:117–126.

Russ, W., Mörtel, H., Meyer-Pittroff, R., and Babeck, A., 2006, Kieselguhr sludge from the deep bed filtration of beverages as a source for silicon in the production of calcium silicate bricks, *J Eur Ceram Soc* **26**(13):2547–2559.

Wysocki, G., and Glöckner, H., 1971, Zur Abwasserbehandlung von Brauereiabwasser – Verwertung von Brauereiablaufhefe, *Brauwelt* **111**(75):1691–1694.

Zanker, G., and Kepplinger, W. L., 2002, Die Verwertung von Biertrebern im Brauereiverbund; *Brauwelt* **142**(46/47).

# 14

# Incineration of Solid Food Waste: A Project About Spent Grain

Gerald Zanker, Werner Kepplinger, and Christian Pecher

## 14.1. INTRODUCTION

Over the next few years, a considerable change will take place in the brewing sector by way of closing natural loops. As a result of various waste disposal regulations coming into force, dumping or open ground tipping of organic waste with a total organic carbon content (TOC) exceeding 5% will be prohibited. As a result, industry now must look for ways in which waste substances from breweries can be efficiently used. An effort should be made in any case to have a process generating no waste.

Spent grain are a by-product in the brewing process. The spent grain is either discharged by a lauter tun or a mash filter and holds a moisture content of approximately 75–80%, i.e., dry matter portion is only 20–25%. With this moisture content the spent grain is not biologically stable. It has to be disposed or further treated to prevent biological degradation. Approximately 200 kg arise per 1 $m^3$ of beer. A utilization route will be described, i.e., how these spent grains can be processed practically residue-free in an existing brewery with a capacity of 100,000 $m^3$ per annum. The capacity of this plant is about 20,000 tons of wet spent grain per year.

Spent grains are usually fed to cattle as a fodder supplement. As spent grain are not biologically stable in a wet state, they have to be used within 2–3 days. In order to obtain stable spent grain and make them suitable for longer storage, it is necessary to dry them. Heretofore, this has been done mainly with so-called contact dryers as these have low emission values. This method is relatively cost-intensive and raises the sales price considerably. As costs of thermal drying are significantly higher than mechanical removal of water, the mechanical drying route should be

GERALD ZANKER • Brau Union AG, A-8055 Graz, Austria. WERNER KEPPLINGER • University of Leoben, A-8700 Leoben, Austria. CHRISTIAN PECHER • Flottweg GmbH & Co. KgaA, Industriestrasse 6–8, D-84137 Vilsbiburg, Germany.

favored. At a certain moisture content the thermal drying again becomes more economic and shall be used.

Another problem relates to future development of cattle herds. In Styria, a state of Austria, numbers have dropped roughly 50% since the 1970s. This trend is evident throughout Central Europe. In the future, this problem could become more acute as a result of various animal epidemics. In addition, there is the problem of very expensive dumping. As a result of new regulations, all waste having an organic carbon content above 5% has to be disposed of in special waste disposal sites (Kepplinger and Zanker, 2001).

Various utilization possibilities for spent grain have been proposed and investigated in recent years:

- Additive in brick manufacture in order to produce lightweight bricks.
- Production of pressboard for the construction industry.
- Additive in foodstuff.
- Additive in other animal feed.
- Combustion as ancillary fuel in power stations, though the very low calorific value is problematic here.

If the spent grain is not utilized, the brewery would have to pay for every disposed ton. Under this precondition the possibility of using the spent grain as source for thermal energy was investigated.

Thermal energy is a basic production factor of the brewing process. The energy prices have risen in the past constantly, and they will keep on rising also in the future. Therefore, the energy supply of a brewery is a growing cost factor. A promising alternative for the utilization of spent grain has been found.

Using the spent grain as energy source was an interesting alternative that had to be further developed. A process had to be found in order to make the energy accessible. The focus was set on high energy efficiency, a treatment without residues, and continuous utilization of the spent grain (Zanker and Kepplinger, 2002; Kepplinger and Zanker, 2003).

The problems and specifications for process development can be summed up as follows:

- High water content (about 80%) leads to very low calorific value.
- Mechanical drying is necessary as thermal drying is too expensive.
- Water pressed out has a very high organic compound content.
- Separate thermal drying should be avoided.
- Energy recovery should be via combustion, even though the $NO_x$ formed in the combustion is more than 20 times above the limit.
- As dumping is very costly, this should be avoided by a secure process.
- Utilization of the resulting ash; about 150 tons of ash would have to be otherwise dumped each year.
- Complete resource cycles within the brewery.

## 14.2. PROCESS DESCRIPTION

The concept of burning the spent grain to produce thermal energy is summarized as ''cycle of energetic treatment of spent grain'' (see Figure 14.1). The spent grain is dewatered mechanically in the first stage. Afterwards the material is burned in a biomass vessel. The hot gases from the combustion are used to generate saturated steam or hot water. The generated thermal energy can cover up to 80% of the breweries' heat demand. Arising by-products such as presswater and ashes are also recycled.

The core components for the energetic treatment of spent grain are the press and the combustion. In order to find the right pieces of machinery several trials were performed. The key factors for a suitable technology were the performance figures realized in the process and the costs of the machinery. For the mechanical dewatering stage several continuous pressing devices such as screw presses and belt presses were tested. It became evident that with belt presses the highest possible dry matter content in the first stage could be generated. For the combustion it turned out that standard biomass vessels for wood were not suitable for the special burning characteristics of the spent grain. Both, the belt press and furnace had to be further developed before installed in the spent grain combustion process that is shown in Figure 14.2.

The wet spent gain is pumped with a monopump out of the spent grain silo into the belt press. There the water content is reduced from 80% below 58%. With this moisture content the material can be stored between 2 and 3 days. The spent grain can be buffered to cope with production stops, for example, on the weekends where no thermal energy is needed. In the next stage the spent grain is fed to a moving

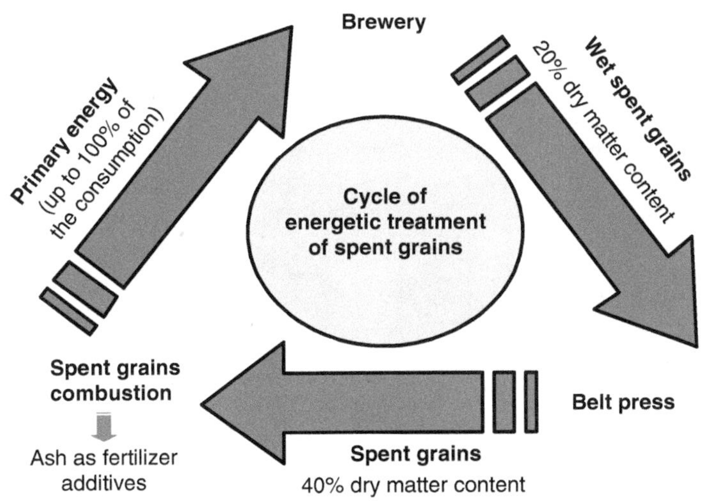


**Figure 14.1.** Energetic treatment cycle.

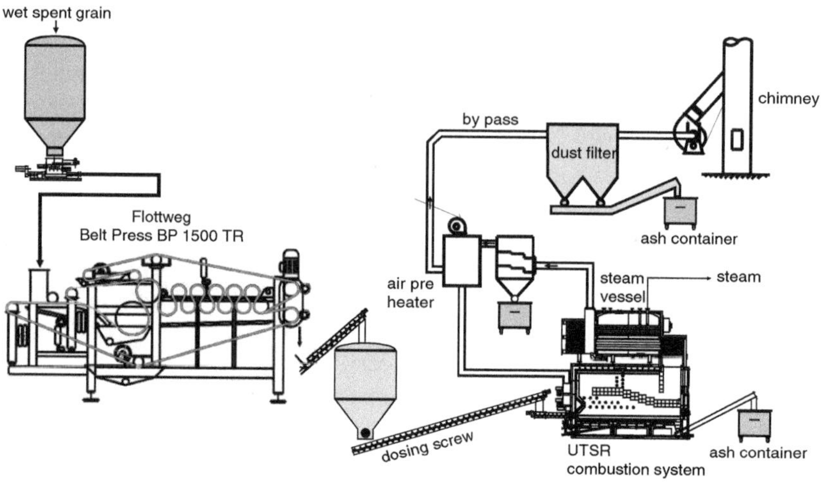


**Figure 14.2.** Spent grain combustion system.

grate furnace with a screw conveyor. The spent grain is burned and saturated steam or hot water is produced by the hot gases. The resulting gases are filtered in a cyclone and an e-filter to comply with existing emission laws. In order to use the rest energy after steam production the incoming air is preheated.

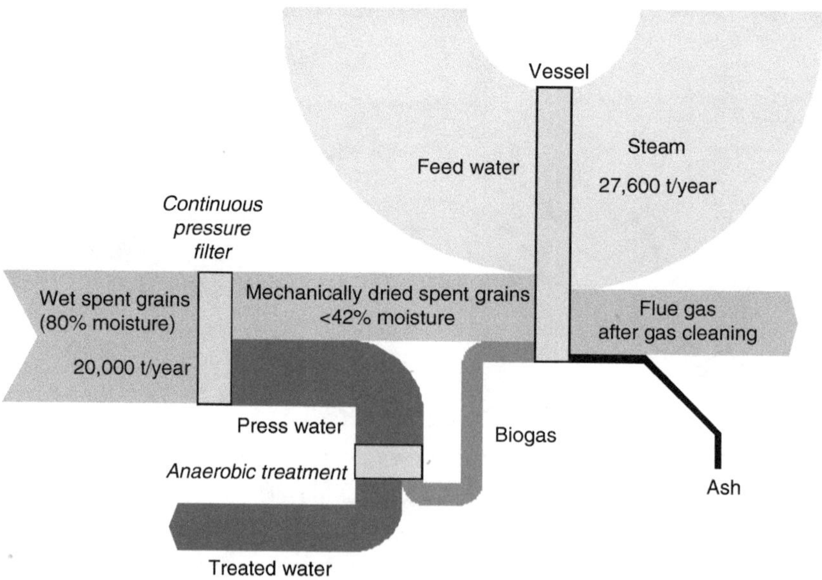


**Figure 14.3.** Sankey diagram of dewatering and combustion process.

On the Sankey diagram in Figure 14.3 all product flows in the combustion process are illustrated. The wet spent grain is dewatered. This results in approximately 50% pre-dried material and 50% water. This presswater is added to an anaerobic treatment where biogas is produced. The combustion results in generated steam, flue gases, and ash that are used as fertilizer ingredients (Groß et al., 1998).

## 14.3. PLANT DESCRIPTION

The spent grain combustion works fully automatically. It is a flexible system that can be adjusted to changing energy demands. The used belt press is a continuous pressing device (Figure 14.4). The spent grain is pressed between an upper and lower belt. A number of rollers with decreasing diameters results in steadily increasing shear and compressive forces. The formation of boundary layers is prevented. The belt press was further developed to cope with the special characteristic of spent grain. High belt pressures and a special cleaning system are needed to process up to 7 t/h at a dry matter content of 42%.

An air-cooled moving grate furnace is used for the combustion. Approximately 1,000 kWh of thermal energy can be obtained by processing 1 ton of wet spent grain. The furnace is designed to the characteristics of burning wet biomass. The thermal rating of these boiler types is between 2,000 and more than 10,000 kW/hour. It is possible to run the boilers at ± 25% of nominal rating, i.e., a boiler with a nominal rating of 4,000 kW/h can be run at between a minimum 3,000 and maximum 5,000 kW/h. This flexibility facilitates a response to changes in the amount of energy required during the brew cycle. The refractory lining in the boiler is designed such that radiation heat is available for additional preheating and drying of spent grain. This enables a very uniform combustion, and fluctuations in the moisture content of the spent grain can be accommodated.

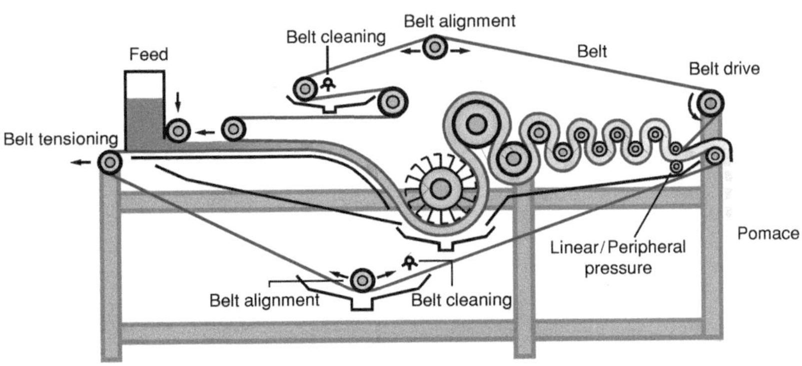


**Figure 14.4.** Flottweg belt press for dewatering the spent grains.

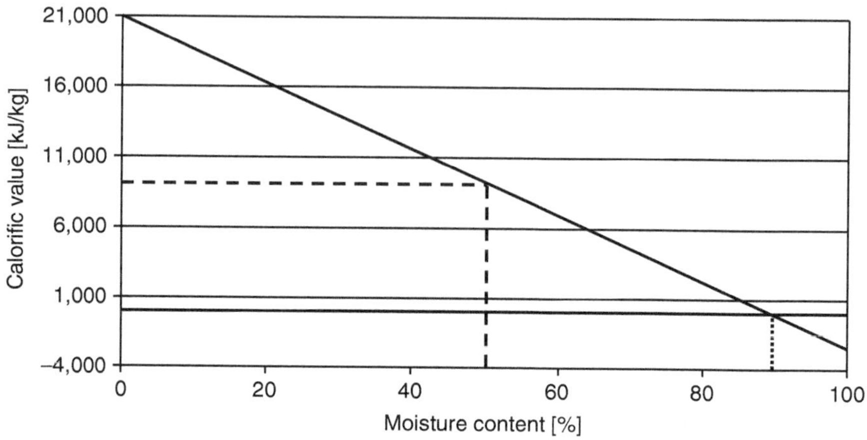


**Figure 14.5.** Calorific value of spent grain correlated with the moisture content.

Figure 14.5 shows that the calorific value of the spent grain is directly connected to the moisture content. The calorific value of the dry substance is similar to lignite coal at approximately 21 MJ/kg. So, the dryer the spent grain is, the more heat can be obtained. This relationship is summarized in a linear function that is idealized:

$$H = (1 - w) \cdot 21{,}000 - w \cdot 2{,}250$$

where H = calorific value and w = water content.

In order to achieve the best possible efficiency, water content of spent grain has to be below 60%. In several trials it became evident that a dry matter content above 40% before entering the furnace improves the combustion properties significantly. At a water content of about 80%, wet spent grain has practically no calorific value because all the energy goes toward evaporating water. Wet spent grain will not burn; therefore predrying is necessary. The most economical way of dewatering can be achieved mechanically with a press.

The curve in Figure 14.5 is idealized. Every percentage point of the dry matter content over 40% increases the energetic efficiency factor exponentially rather than linearly. The mechanical dewatering stage is crucial for the entire process and determines the overall energetic efficiency (Kepplinger, et al., 1998).

## 14.4. USE AND PROPERTIES OF BY-PRODUCTS

Emissions to the atmosphere are of major importance. The limit values set out in TA Luft (German technical recommendation for emissions to atmosphere) for biomass plants were used as a basis. The emitted gases are cleaned in a cyclone

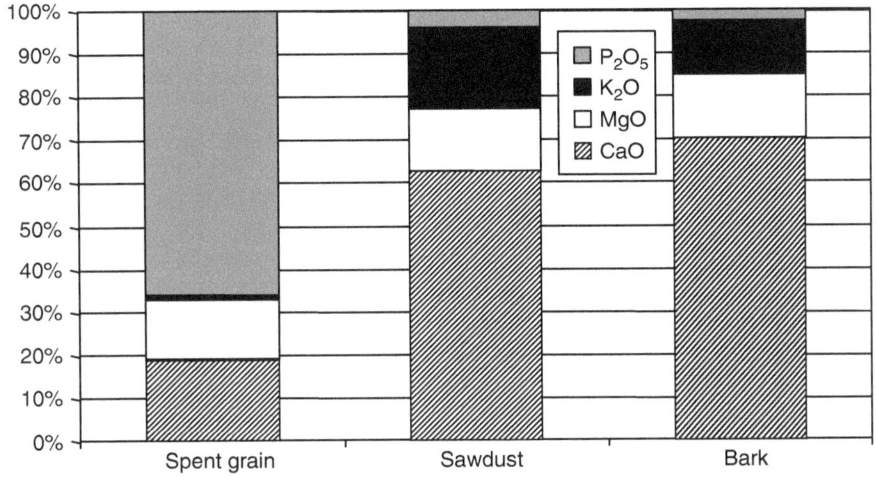


**Figure 14.6.** Composition of biomass ashs.

and in an electric filter in order to comply with the existing emission standards. The content of $NO_x$, sulfur dioxide, hydrocarbon, carbon monoxide, and dust do not exceed 50% of the Austrian emission standard.

It can be seen that ash from spent grain has a totally different composition to that of ash from wood (Figure 14.6). Per 1 ton of wet spent grain approximately 9 kg ash are generated. This ash has a high $P_2O_5$ content above 60%. The high $P_2O_5$ level is of major consequence as this represents a very important component in fertilizers. As a result, this ash can be added to the class of NPK fertilizers (N = nitrogen, P = phosphorus, K = potassium). As a function of ash content, with 20,000 tons of wet spent grain 180 tons of ash per annum can be generated. This amount is sufficient to produce 920 tons of NPK fertilizer in the present case. From a commercial standpoint, this value of $P_2O_5$ makes a contribution toward operational expenses.

Two different methods to recycle the presswater are applicable: It can be either treated in an anaerobic wastewater plant or be mixed with brew water.

The mechanical dewatering from 20% to 42% dry matter results in 520 liter press water per 1 ton of wet material. This water has a chemical oxygen demand between 10,000 and 20,000 mg $O_2$ per liter. In an anaerobic fermentation appr. 6–8 $m^3N$ biogas per ton of water can be generated. That means with 20,000 tons wet spent grain 80,000 $m^3$ of biogas can be produced per year (Kepplinger, 2004).

By mixing press water with brew water for mashing at a ratio of 10:90 the brewery gains additional extract. One percent more yield in the brew house is possible. The benefits are water savings, energy savings, and malt savings. An important issue is the effect on beer quality that was researched in a prestudy. It was found that in total the chemical and physical stability of the beer is not adversely effected. An increased content of amino acids and unsaturated higher fatty acids were detected. These can be classified as yeast nutriment to enhance fermentation. There were no negative findings regarding ions, foam stability, or by-products of the fermentation. In addition no accumulation of heavy metals could be verified. Only hexan-octan acids were increased. An HPCL analysis of the low-molecular polyphenols resulted in no noteworthy variations. The tasting of the beer was rated as positive.

The existing installations show that the major financial benefit from spent grain combustion is the savings from the reduction of external energy consumption such as natural gas or fuel oils. Additional financial benefits could be obtained by the reuse of press water, combustion of old labels, and the sale of ash and $CO_2$ certificates.

The data in Figure 14.7 represent a 250,000 $m^3$ brewery that pays 0.3 € per $m^3$ natural gas and can still achieve 10.5 € per ton of wet spent grain. The figures are based on a green field plant where a gas-fired boiler system is compared with spent grain combustion. The overall costs for thermal energy with a spent grain combustion average at 10% compared to a gas boiler system. Installation of spent grain combustion means 1.8 Mio € additional investment, but 810,000 € additional savings. Hence, the amortization time of a spent grain combustion is 2.36 years. Considering an energy price increase of 7.5% and spent grain revenue decrease of 5.5% the cost savings will rise by 79% in 5 years.

Technical feasibility and commercial application of the process described have been proven in large-scale plant tests. All applicable environmental regulations could be complied with. Spent grain represents a fuel with no effect on $CO_2$ levels. Using this process, a complete integration into the brewery complex up to a waste-free brewery can be achieved. The process can be economically operated. The hitherto existing dependence on external users (fodder, fillers, etc.) disappears, as do the costs for dumping the waste substances.

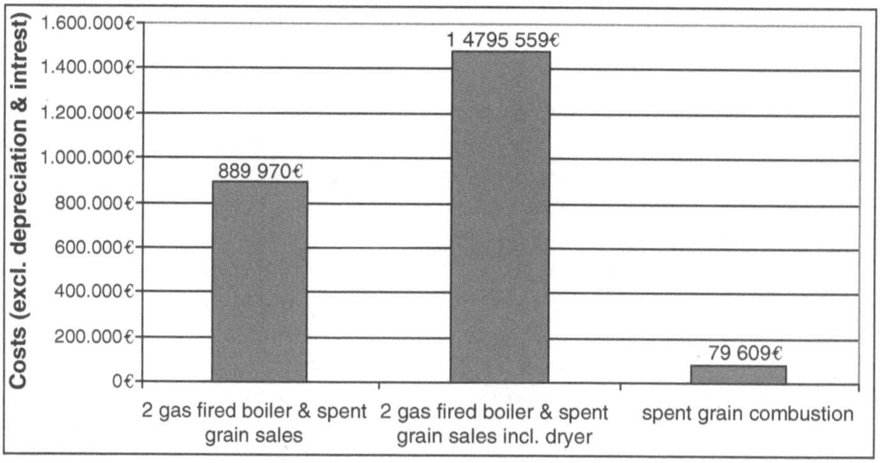


**Figure 14.7.** Costs for thermal energy supply.

## 14.5. REFERENCES

Groß, B., Kepplinger, W., Zanker, G., and Pelz, D., 1998, Energie- und Reststoffmanagement von Biertrebern im Brauereiverbund, *BHM – Berg- und Hüttenmännische Monatshefte* **143**(10).

Kepplinger, W., 2004, Die Brauerei der Zukunft, in: *Proceedings of the 8th Hannoversche Industreiwasser-Tagung*.

Kepplinger, W., and Zanker, G., 2001, Use of spent grains, in: *Proceedings of the 28th EBC Congress Budapest*, European Brewery Convention.

Kepplinger, W., and Zanker, G., 2003, Utilisation of brewers' grains within a brewery complex, *Brauwelt Int* **21**(3):162–165.

Kepplinger, W., Pelz, D., and Zanker, G., 1998, *Verfahren zur thermischen Verwertung von Biertrebern*, European Patent No. 1007 884.

Zanker, G., and Kepplinger, W., 2002, Die Verwertung von Biertrebern im Brauereiverbund, *Brauwelt* **02**(46–47):1742–1747.

# 15

# Composting of Food and Agricultural Wastes

Stefan Shilev, Mladen Naydenov, Ventsislava Vancheva, and Anna Aladjadjiyan

## 15.1. INTRODUCTION

The rapid increase of world population in the last years accompanied by the intensification of human activities brought serious environmental problems such as the pollution of soil, water, and air, forest destruction, etc. In the future these negative impacts may cause global climatic changes (greenhouse effect) and might be a menace for the existence of the human race. Immediate measures to avoid the negative influence of human activities are necessary. Many industrial processes result in a large amount of wastes. Food and agriculture industry are among the oldest of human practices, but as a source of wastes it does not make any exception from other industrial activities. In the near future the management of food and agricultural wastes will play an important role in the conservation of the natural resources in many countries, including the Balkan region.

The term ''composting'' is used here to define the process of controlled biological maturity under aerobic conditions, where organic matter of animal or vegetal origin is decomposed to materials with shorter molecular chains, more stable, hygienic, humus rich, and finally beneficial for the agricultural crops and for recycling of soil organic matter (Sequi, 1996). The process is mediated by different microorganisms actuating in aerobic environment: bacteria, fungi, actinomycetes, algae, and protozoa, which participate naturally in the organic biomass or are added artificially (Tuomela et al., 2000). The process can be described by the follow equation:

$$\text{organic matter} + O_2 \rightarrow \text{compost} + CO_2 + H_2O + NO_3^- + SO_4^{2-} + \text{heat}$$

Now the interest regarding the composting process is related to the following points:

STEFAN SHILEV, MLADEN NAYDENOV, AND VENTSISLAVA VANCHEVA • Department of Microbiology and Environmental Biotechnologies, Agricultural University–Plovdiv, 12 Mendeleev Str., 4000-Plovdiv, Bulgaria. ANNA ALADJADJIYAN • Department of Mathematics and Physics, Agricultural University of Plovdiv 12, Mendeleev Str., 4000-Plovdiv, Bulgaria.

1. Environmental point of view, because during this process the biomasses are transformed to material rich in nutritional substances that can improve the structural characteristics of the soil (Sommer and Dahl, 1999).
2. Hygienic point of view, because during the process the organic matter is disinfected by the influence of the high temperatures (Dumontet et al., 1999).
3. Energy management point of view, because during the process energy is released through the degradation of large organic molecules (Schaik et al., 2000; Sonesson et al., 2000).

The term ''composting'' is usually used (although it is not correct) for the description of aerobic stabilization of the organic matter (solid wastes), obtained without separation of different fractions (Sequi, 1996; Tuomela, 2002). Compost offers many benefits to the landscape and garden. For example, compost (1) improves soil tilth condition and structure; (2) increases the soil's ability to hold water and nutrients; (3) supports living soil organisms; (4) helps to dissolve mineral forms of nutrients; (5) buffers soil from chemical imbalances; (6) may provide biological control of certain soil pests; and (7) helps to return organic materials to the soil and keeps them out of landfills and waterways. Compost can be used as a mulch, a liquid ''fertilizer,'' or incorporated into the soil or potting mixes.

## 15.2. COMPOSTING AGENTS

The relationships in a compost pile are very complicated and could be represented by one pyramid of primary, secondary, and tertiary level consuments (Figure 15.1).

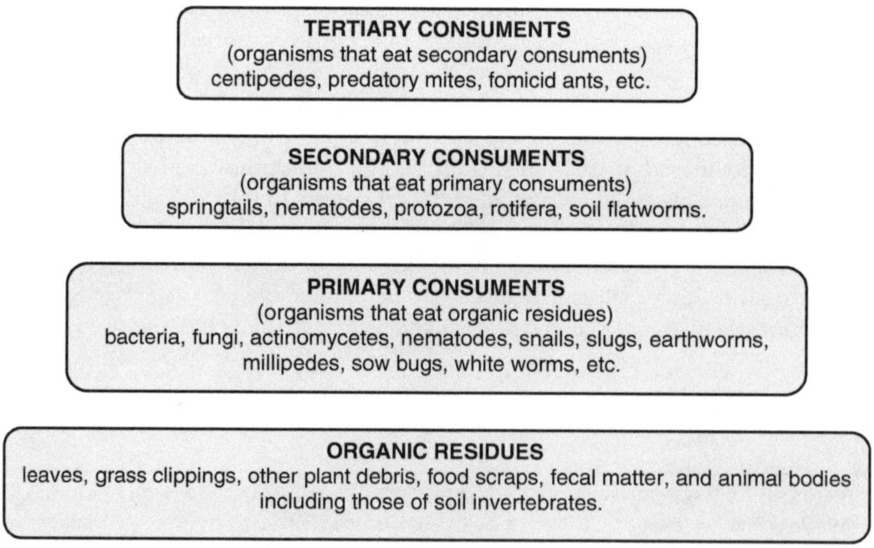


**Figure 15.1.** Food pyramid in the compost.

The base of the pyramid is made up of organic matter including plant and animal residues.

The organic residues, such as leaves or other plant materials, are eaten by some types of invertebrates, like millipedes, sow bugs, snails, and slugs. These invertebrates shred the plant materials, creating more surface area for action of fungi, bacteria, and actinomycetes, which are in turn eaten by organisms such as mites and springtails. Many kinds of worms, including earthworms, nematodes, red worms, and pot worms, eat decaying vegetation and microbes and excrete organic compounds that enrich compost. There tunnels aerate the compost and their feeding increases the surface of organic matter for microbes to act upon. As each decomposer dies or excretes, more food is added to the system for other decomposers.

## 15.3. SOURCES OF FOOD AND AGRICULTURAL WASTES

### 15.3.1. Food Wastes

Food-processing wastes are those end products of various food-processing industries that have not been recycled or used for other purposes. Food industry produces large volumes of wastes, both solids and liquid, resulting from the production, preparation, and consumption of food. These wastes pose increasing disposal and potentially severe pollution problems and represent a loss of valuable biomass and nutrients. In general, wastes from the food-processing industry have the following characteristics (Litchfield, 1987):

1. Large amounts of organic materials such as proteins, carbohydrates, and lipids
2. Varying amounts of suspended solids depending on the source
3. High biochemical oxygen demand or chemical oxygen demand

Fruits, vegetables, dairy products, grains, bread, unbleached paper napkins, coffee filters, eggshells, meats, and newspaper can be composted. Food waste has unique properties as a raw compost agent. Because it has a high moisture content and low physical structure, it is important to mix fresh food waste with a bulking agent that will absorb some of the excess moisture as well as add structure to the mix. Bulking agents with a high C:N ratio, such as sawdust and yard waste, are good choices. Composting provides a way in which solid wastes, water quality, and agricultural concerns can be joined. Benefits of compost to the food industry:

1. Reduces solid waste disposal fees.
2. Ends wasting large quantities of recyclable raw ingredients.
3. Educates consumers on the benefits of food waste composting.
4. Markets your establishment as environmentally conscious.
5. Markets your establishment as one that assists local farmers and the community.

*Table 15.1.* Agricultural residues from different crops in Bulgaria (n.d., no data)

| Indicator | Utilized residues (t ha$^{-1}$) | Waste (t ha$^{-1}$) | Area sown in 1999 (thousand ha) |
|---|---|---|---|
| Wheat | 1.5 | 0.55 | 966.3 |
| Rye | 2.0 | 0.55 | n.d. |
| Barley | 1.5 | 0.55 | 254.7 |
| Oats | 1.5 | 0.55 | n.d. |
| Rape | n.d. | 3–4 | n.d. |
| Maize | n.d. | 3–4 | 455 |
| Soybeans | n.d. | 0.5–0.6 | n.d. |

6. Helps close the food waste loop by returning it back to agriculture.
7. Reduces the need for more landfill space.

### 15.3.2. Agricultural Wastes

Cereals are a major source of agricultural waste in many countries. For example, in Bulgaria about 60% of arable lands in (approximately 4,000,000 ha) are sown by cereals (both bread and fodder crops, as well as oil crops). Part of plant residues is utilized. Non-utilized residues are considered as agricultural wastes. The wastes from plant production can be estimated on the basis of data given in Table 15.1, describing agricultural residues from different crops. Both utilized and unutilized plant residues are shown.

The data presented in Table 15.1 lead to the conclusion that a considerable part of crops residues in Bulgaria is not utilized. An amount of agricultural residues from vineyards counting for 15.12 tons ha$^{-1}$ year$^{-1}$ must be added to the upper one. The total area occupied by vineyards in Bulgaria is about 100,000 hectares. Therefore, the total amount of nonutilized agricultural residues per year can be estimated at 5,000,000 tons. Keeping in mind that the calorific value of plant residues is between 15 and 20 MJ kg$^{-1}$ (Aladjadjiyan, 1992) nonutilized agricultural residues can be evaluated at 80.8 PJ, which is equal to 1.9 Mt of oil equivalent. It means that agricultural residues provide a significant resource for biomass-to-energy processing.

Livestock is other important part of agriculture. The industrial livestock complex is an artificial ecological system. The amount of wastes from livestock and especially of liquid manure is quite large. Liquid manure contains different microorganisms that are dangerous for people as well as for animals. On the other hand, the manure has high energy potential and it is a significant source of renewable energy. The total amount of liquid manure produced in Bulgaria during 2002 can be estimated at 11.2 million tons.

## 15.4. PHASES OF COMPOSTING

Generally, during composting the organic matter is submitted to changes, which could be separated to two phases: (1) degradation, and (2) maturation.

The first phase of the composting process starts with the degradation of the most easily degrading organics (sugars, organic acids, amino acids). It is mediated by aerobic microorganisms with consumption of oxygen and release of carbon dioxide and energy. This is a thermophilic phase, which proceeds with high velocity; its duration is between several weeks and a few months, depending of the substrate characteristics. The vigorous aeration or mixing of the composting components is an obligatory condition to ensure the cooling of the substrate, but also to support the oxygen supply to biomass. Phytotoxins are produced during this phase of the process, which proceeds by the decomposition of the organics (Beck-Friis et al., 2000). The high temperatures, pH conditions, and humidity are the reasons that the most active microorganisms during this phase are the bacteria. In the end of the first phase fresh compost is obtained.

When the easily degradable compounds that are metabolized during the first stage are exhausted, the composting process continues with the more complex organic molecules; that is why the time for degradation is longer. The beginning of this stage is followed by the death of a huge part of microbial population, because of lack of available food. Additionally, the fast decrease of temperature is followed by changes in the population of active microorganisms from thermophilic to mesophilic. It is true that during this stage the temperature reaches 40–45°C to decrease until room temperature. According to several authors (Maynard, 2000; Sequi, 1996; Tuomela, 2002; Wu et al., 2000), this stage could last a few months.

During the mesophilic phase actinomycetes appear. They degrade actively the starch, cellulose, and lignin compounds necessary for the synthesis of humus components. They play also a basic role in the humus formation and produce aromatic compounds, which give to the end product a specific smell of wood soil. The maturation stage is characterized also with disintegration of the compost material: formation of small particles by a lot of invertebrates (earthworms, ticks, and centipede).

The humus formation is strongly related to chemical and biological transformation of animal and vegetable wastes and with microbial synthesis. The humufication is attributed to oxidative polymerization of phenolics obtained by the catabolism of lignin, tannins, and polyphenols, or to new microbial biosynthesis.

## 15.5. COMPOSTING PARAMETERS

### 15.5.1. Temperature

During the composting the microorganisms break down organic matter and produce carbon dioxide, water, heat, and humus, the relatively stabile organic end product. Under optimal conditions the composting temperature changes during the process, and three phases can be distinguished (Figure 15.2): (1) the mesophilic phase (moderate temperature phase), with a duration of a couple of

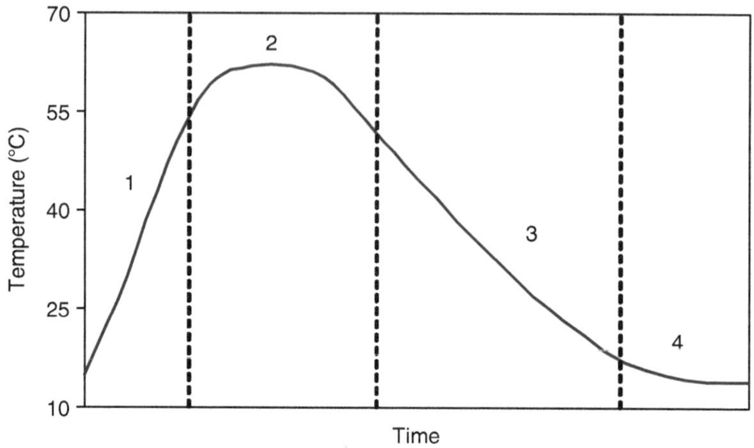


**Figure 15.2.** Temperature variation during the composting and phases of microbial activity (1, mesophilic, 2, thermophilic, 3, cooling, 4, maturing).

days; (2) the thermophilic phase (high temperature phase), with duration from some days until several months; and finally (3) a several-months cooling and maturation phase.

Different microbial communities predominate during the composting phases. The initial decomposition is carried out by mesophilic microorganisms, which rapidly break down the biodegradable compounds. The heat produced by them causes the fast increase of compost temperature. When the temperature reaches 40°C, the mesophilic microorganisms become less competitive and are replaced by the thermophilic ones. At temperature above 55°C, many human and plant pathogens are destroyed, while seeds of weeds and parasites are destroyed when it reaches 60°C. Temperature must not exceed 65°C, because many useful microbes will be killed, resulting in a decrease of the rate of decomposition. For that purpose, the compost managers should use aeration and mixing to maintain the temperature below this point (Figure 15.3).

During the thermophilic phase, high temperatures accelerate the breakdown of proteins, fats, and complex carbohydrates like cellulose and hemicellulose, the major structural molecules in plants. When these compounds are exhausted, the compost temperature gradually decreases and mesophilic microorganisms once again take part for the final phase of ''curing'' or maturation of the remaining organic matter.

The transformation of organic matter is possible due to the enzymes, which are catalysts for all biochemical reactions. They are produced by active microbes and could be classified as intracellular and extracellular enzymes. The extracellular ones are useful for the degradation of big molecules in the environment. When the huge molecules are destroyed out of the cell, small particles enter the microbial cell via

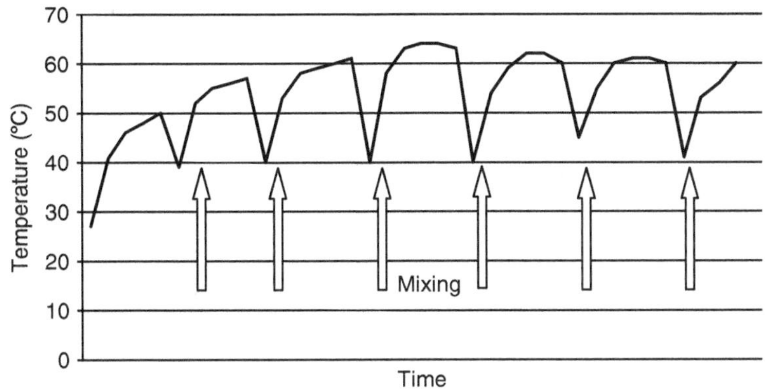


**Figure 15.3.** Changes in compost temperature after mixing (Chiumenti and Chiumenti, 2002).

*Table 15.2.* C/N ratio of some organic materials

| Organic Material | C/N |
|---|---|
| Leaves | 34–85 |
| Wood material (juniper, poplar) | 80–145 |
| Straw | 50 |
| Sawdust | 150 |
| Bark | 110 |
| Paper | 100 |
| Vegetable scrap | 20 |
| Sewerage sludge | 6 |
| Organic fraction of solid municipal wastes | 12 |
| Manure | 5–25 |

different mechanisms and after treatment by intracellular enzymes are included in microbial metabolism.

### 15.5.2. C/N Ratio

The most important elements required for microbial decomposition are carbon and nitrogen. Carbon provides an energy source and the building material representing 50% of the microbial cell biomass. Nitrogen is a critical component of the proteins, nucleic acids, enzymes, and coenzymes necessary for cell growth and function. To set the optimal amounts of these two elements is necessary to consider the carbon-to-nitrogen (C/N) ratio of each compost ingredient. The ideal C/N ratio for composting is generally considered to be around 30:1. To obtain this optimum ratio it is necessary to know the C/N ratio of the organic materials that will be used as compost ingredients and to mix them at appropriate quantities. In Table 15.2 the C/N ratios of some organic materials are presented.

During the composting, the C/N ratio gradually decreases from 30:1 to 10–15:1 at the final product, because two thirds of the carbon of the organic compounds consumed by the microorganisms is converted to carbon dioxide. The rest is incorporated together with the nitrogen into the microbial cell.

Another problem related to C/N ratio is the bioavailability of these elements. Some of the carbon may be bound up in compounds highly resistant to biological degradation. Corn stalks and straw, for example, break down slowly because they are made up of a strong form of cellulose. Although these materials can be composted, they have a relatively low rate of decomposition, which means that not all of the carbon is available for the microorganisms.

Other elements, like phosphorus, calcium, and sulfur, are also important for the compost microbial activity. Correlations between some of them also have been established, e.g., the optimal one for C/P is known between 100 and 200, while for C/S is 100–300.

### 15.5.3. Humidity

Water is a fundamental life factor of active compost microorganisms, because:

- It is necessary for the nutrients exchange via cell membranes.
- It is a transport environment for the extracellular enzymes and soluble substrates.
- Is the environment where the chemical reactions occur.

Optimal humidity of the compost is 50–60%. When it increases higher than 65%, anaerobic conditions can be established, while values below 40% reduce the biological activity of the compost (Keener et al., 2000). On the other hand, composting of high-humidity matter (80%) is also possible through the addition of matters with a low-humidity content to stabilize the compost (Table 15.3). During the composting process the humidity decreases naturally and reaches dangerous levels for the biological activity. When this occurs, it is necessary to add water to increase it.

*Table 15.3.* Humidity of different organic matters

| Organic matter | Humidity (%) |
|---|---|
| Organic fraction of municipal solid wastes | 70–80 |
| Sludge | 70–80 |
| Beef manure | 75–80 |
| Birds farm litter | 40–60 |
| Juniper's wood | 70–75 |
| Wheat straw | 15–20 |
| Maize's steams | 25–35 |

### 15.5.4. Oxygen and pH

The oxygen is also an essential ingredient for successful composting because it is an aerobic process. Although the atmosphere contains 21% oxygen, the aerobic microbes can survive at concentrations as low as 5%. In this sense a content of 10% of oxygen is considered optimal for the composting.

On the other hand, a pH between 5.5 and 8.5 is optimal for compost microorganisms. When the bacteria and fungi digest organic matter, they release organic acids. As the pH decreases, the growth of fungi increases, followed by decomposition of lignin and cellulose. If the system becomes anaerobic, the acid accumulation can lower the pH to 4.5, thus limiting the microbial activity. In such cases, the introduction of air is usually sufficient to return the compost pH to acceptable ranges.

### 15.5.5. Particle Size

*Total free space* is defined here as the relation between the volume of free space and the volume of biomass, expressed in percentage. On the other hand, *free air space* (FAS) is the volume occupied by the air as a part of the total free space.

$$\text{Total free space} = V_v/V_t;\ \text{FAS} = (V_v - V_a)/V_t$$

where $V_v$ is volume of free space, $V_a$ is volume of water, and $V_t$ is total volume of biomass.

FAS is a very important parameter, because it influences the compost capacity to retain oxygen. In fact, the composting is realized when water, air, and organic matter are present in the system. In normal conditions, the value of total free space is between 35% and 50% (Keener et al., 2000).

The particle size depends on the composting material, humidity content, and total volume of composted organics. The finer and wetter are the composted particles, the greater is the compost compactness. Optimal particle size is 25–75 mm. When the composted organic matter has smaller size (leaves, grass, animal excrements, etc.), the addition of more inert materials (bark, sawdust) is needed.

### 15.5.6. Stability Parameters

The biological processes of degradation of organic matter are negatively influenced by the presence of heavy metals in high concentrations. Heavy metals lead to toxic effects, which result in strong suppression of microbial activity or to death of nontolerant strains.

On the other hand, some parameters exist that allow an estimation of a biological stability of composted material: germination, nitrogen mineralization, respiration, and humification (Chiumenti and Chiumenti, 2002). The term *biological stability* describes the degradation of organic matter, where the processes are retarded because of the lack of optimal conditions for microbial growth (Keener et al., 2000; Scaglia et al., 2000).

## 15.6. COMPOSTING METHODS

There are many ways to compost organic materials. Some of the most popular will be reviewed.

### 15.6.1. Cold (Slow) Composting

Cold (slow) composting is appropriate with carbon-rich rather than nitrogen-rich material and if there is no concern about a slow composting rate, a desire for weed seed destruction, or a need for plant disease suppression. The advantages of cold/slow composting include ease of implementation and a lower level of management required. The disadvantages of cold/slow composting include a slow rate of decomposition and potential for pests to excavate buried wastes. Additionally, if the raw materials contain weed seed or plant pathogens, these will not be destroyed in the composting process.

This type of composting could include sheet, trend, cold bin, and heap composting. This method could be used to build up organic matter throughout the yard. Cold piles can be built wherever compost is needed, under trees, in washed out areas, in the space that will be next year's garden, etc. Over a year or two, the material will decompose, adding valuable organic matter to the soil without the need for a formalized bin or composting activity.

### 15.6.2. Hot (Fast) Composting

Hot (fast) composting will yield the fastest rate of composting and best control of weed seed and plant pathogens. Hot composting is also the most intensive method and requires several elements to succeed, including a minimum of 1 $m^3$ of material to start the pile; a blend of greens and browns (to obtain a good C/N ratio); proper moisture content; frequent turning to provide aeration; and particle size of less than 2–3 cm.

### 15.6.3. Using Earthworms

Many people know the value of worms in their garden. Worms are great decomposers, especially red wigglers and African night crawlers. They often are used if kitchen and table scraps are among the composting materials.

### 15.6.4. Mixing Method

The mixing method is appropriate if ''no-turn'' method is used or if one wants to speed up the composting process. Materials with different C/N ratio and moisture are simply mixed up and added to the compost system. This prevents formation of compact layers that may restrict the flow of water and oxygen through the pile. The addition of the mixture to a compost system is done in batches. Each batch has to be watered so that the moisture is evenly distributed. It is really difficult to get water into the entire pile after the pile has been built, so water should be added as one builds the pile.

It is also possible to add fresh materials to an actively (or passively) composting pile. One way to add materials to an existing pile is to add them during the mixing or turning on the pile. Burying new materials in the pile also works well. Eventually, new fresh materials will be used to start a second batch of compost. This will give the first batch of compost time to stabilize and mature.

The addition of fresh materials will supplement the existing food base. If more materials with low C/N ratio are used, the effect will be adding nitrogen and potentially speeding up the composting process, increasing moisture, and/or heating up the pile. If more materials rich in carbon are used, the effect will be slowing the composting process, drying out the pile, and/or reducing the pile temperature.

## 15.7. PRODUCT QUALITY

Complete compost analysis includes: Total N, P, K, Ca, Mg, Zn, Fe, Cu, and Mn; C/N ratio; pH and EC.

### 15.7.1. Particle Size

Different end uses of compost have different particle size requirements. Composts used in greenhouse potting mixes need a specific size to maintain correct porosity and water-holding capacity. Particle size is less critical when compost is applied to farmland, although large particles can affect the spreadability of the compost. Many composters screen their product to remove large sticks and twigs. Laboratories will do particle size analysis, but for land application, visual inspection is sufficient.

### 15.7.2. Ratio of Carbon to Nitrogen–C/N Ratio

Enhancing the organic matter content of soil by adding compost will promote the root growth of plants, make a clay soil lighter and easier to work, make a sandy soil richer, and make soil better able to hold water and nutrients.

The C/N ratio of stable compost at the end of the process will be between 10:1 and 20:1, with the most stable composts falling in the lower end of this range. Compost with high C/N ratio will reduce N availability to plants due to nitrogen immobilization caused by available microflora. A final C:N above 20:1 may indicate a compost that will not readily release nitrogen, while a final C/N ratio above 30:1 may indicate a compost that will inhibit nitrogen mineralization and tie up nitrogen from the soil. This affects compost use for farmland application and additional nitrogen supplementation is necessary for better plant growth.

### 15.7.3. Compost Maturity

Maturity describes a compost's fitness for a specific use. Stability describes a compost's resistance to further biological breakdown. Immature composts can contain phytotoxic organic acids. Stability and maturity are critical for compost used in greenhouse potting mixes and bagged products, but less critical for application to farmland, especially when several weeks elapse between application and planting.

The C/N ratio is one of the indicators of the compost maturity. According to the California Compost Quality Council (CCQC) to qualify as ''mature'' or ''very mature,'' a compost must have a C/N ratio of less than or equal to 25 and pass two additional tests performed concurrently on the same sample, one test from ''group A'' and one from ''group B''. Group A tests, which indirectly measure the degree of organic matter decomposition, include carbon dioxide release or respiration, oxygen demand, and Dewar self-heating test. Group B tests, which measure chemical characteristics of the product (some of which can be toxic to plants) include ammonium nitrate ratio, ammonia concentration, volatile organic acids concentration, and plant bioassays.

### 15.7.4. Electrical Conductivity (EC): A Measure of Soluble Salts

High concentration of salts may injure plants. EC is critical for greenhouse potting mixes and less critical for farmland application especially in humid areas. The best EC is not higher than 4 dS/m.

### 15.7.5. Compost Acidity–pH

The pH of the growing medium plays a major role in the availability of plant nutrients. The best value of pH is between 6 and 7, which is common for most of the plants.

### 15.7.6. Ammonium Nitrogen ($NH_4^+$) and Nitrate Nitrogen ($NO_3^-$)

Ammonium and nitrate nitrogen is a plant-available N form, but high values of ammonium nitrogen can injure plants and its concentration in compost should be less than 500 mg/kg dry weight. In aerobic conditions ammonium nitrogen is easily converted into nitrates. The concentration of nitrate nitrogen should be between 200 and 500 mg/kg dry weight of compost. Low values indicate lack of plant-available nitrogen.

### 15.7.7. Moisture

Compost with more than 60% moisture may be hard to spread. High moisture also means more water and less organic matter. Low moisture materials (less than 40%) may be dusty.

### 15.7.8 Organic Matter

Organic matter is defined as the present percentage of dry amendment. Low values (less than 30%) usually indicate that organic matter has been mixed with sand or soil. High values (more than 60%) indicate fresh, uncomposted material.

## 15.8. APPLICATION OF COMPOSTING PROCESSES

### 15.8.1. Bioremediation

The composting process and the use of compost provide a solution for managing industrial waste and for remediation of soil contaminated with toxic organic compounds and toxic metals. During the composting contaminated soil is excavated and mixed with bulking agents such as wood chips, straw, hay, corn cobs, manure, and vegetable wastes. The types of amendments used depend on the type of the soil and the balance of carbon and nitrogen needed to promote microbial activity and make it easier to deliver the optimum levels of air and water to the microorganisms. What remains after the remediation should be a humus-rich soil with no toxic intermediates that is a value-added soil additive. Bulk density has to be low enough (less than 650 $kg/m^3$) to allow for good aeration. When dense materials such as manure and sludge are used, it is necessary to add bulking agents, such as wood chips, corn cobs, and straw.

Three common designs are: *static biopile composting* (compost is formed into piles and aerated with blowers or vacuum pumps); *mechanically agitated in-vessel composting* (compost is placed in a treatment vessel where it is mixed and aerated); and *windrow composting* (compost is placed in long piles known as windrows and periodically mixed by mobile equipment). Windrow composting is usually considered to be the most cost-effective composting alternative.

The composting process may be applied to soils and lagoon sediments contaminated with biodegradable organic compounds. Pilot and full-scale projects have demonstrated that aerobic, thermophilic composting is able to reduce the concentration of a large number of common industrial and agricultural contaminants, such as hydrocarbons, pesticides, mineral oil, explosives, volatile organic solvents, and various aromatic compounds including chlorophenols and polyaromatic hydrocarbons (PAH).

Significant changes in chemical composition occur during the composting process. The plant-derived residues consist of polysaccharides (cellulose and hemicellulose), lignin, and tannin. The end product has a low polysaccharide content, most of which is microbial cell wall and extracellular substances, with about 25% of the initial carbon content present in the form of highly stabilized humic substances. Organic matter content ranges from 30% to 50% of dry weight, with the remainder being minerals. The combination of high organic content and a variety of minerals makes compost an excellent adsorbent for both organic and inorganic chemicals.

#### 15.8.1.1. Compost Composition

Different compositions of compost were investigated, but most of the researchers found that the best results during remediation are provided by a mixture of 30% contaminated soil with 70% initial compost feedstock. (Brinton et al., 1996). When 40% of contaminated soil is included to a composting mix, the process could not reach the necessary thermophilic conditions and degradation of contaminants is suppressed.

Most of the chemical reactions of contaminants degradation in soil, where temperature is only 15°C to 30°C, are relatively slow. The high temperature achieved during composting accelerates the degradation. Typical temperatures of composting are 50°C or higher as a result of thc heat produced by microorganisms during the degradation of the organic material. In most cases, this is achieved by the use of indigenous microorganisms.

In cleanup processes of soil contaminated with oil good results are achieved by using ratio of 75% contaminated soil, 20% compost, and 5% turkey manure (United States EPA, 1997) or 25% finished compost and 75% contaminated soil, but that varies depending on the degree of contamination.

Combining feedstocks such as raw manure with contaminated soil is effective as a bioremediation method, but lengthens degradation time and costs. Using fresh manure requires additional structuring agents, leading to higher processing costs. When finished, compost plus compost tea as an inoculant are used, the microbial activity increases as much as 100 times that of raw manure, and in return yields a much more cost- and time-effective solution. When compost prepared from contaminated soils is used as a feedstock for compost tea, the degradation rate speeds up due to the adaptation of available microorganisms to the contaminants.

#### 15.8.1.2. Benefits

Traditional remediation can cost several times more than bioremediation by composting technology, as addition of compost to contaminated soils accelerates plant and microbial degradation of organic contaminants and improves plant growth in toxic soils as well. Using the composting process or adding compost to a biopile-type remediation process may decrease remediation time, and compost bioremediation could be completed in weeks instead of months. The extended time period increases cost, since the site must be monitored and operated for a long-term period.

Compost has a high microbial diversity with microbial populations much higher than fertile soils and many times higher than in highly disturbed or contaminated soils. Therefore, compost bioremediation takes far less time than land farming. Since the microbes are the primary agents for degradation of organic contaminants in soil, increasing microbial density can accelerate degradation of the contaminants.

15.8.1.3. Limitations

The following factors may limit the applicability and effectiveness of the bioremediation process:

- Substantial space is required for composting.
- Excavation of contaminated soils is required and may cause the uncontrolled release of volatile organic compounds (VOC). High levels of fugitive emissions also are observed during windrow composting. If VOC contaminants are present in soils, off-gas control may be required.
- Composting results in a volumetric increase in material because of the addition of amendment material. The final, decontaminated mix has about twice the volume of contaminated soil, because the volume loss of a composted mix is typically about 50% of initial. This could increase a volume of a contaminated soil when the bioremediation process has not achieved the expected level.

### 15.8.2. Composting of High-Metal-Containing Plant Biomass

Phytoextraction is a promising and cost-effective method for remediating soils contaminated with toxic metals. At present two basic strategies of phytoextraction are being developed: induced phytoextraction and continuous phytoextraction (Salt et al., 1998). The former relies on the growth of high biomass crops, which are induced to accumulate high concentrations of heavy metals by application or not of chemical amendments to the soil (Blaylock et al., 1997; Salt et al., 1998, Shilev et al., 2003). Such plants are compatible with routine agricultural practices and allow repeated planting and harvesting of the metal-rich tissues. The second approach depends on the natural ability of some plants to accumulate, translocate, and resist high amounts of heavy metals over the complete growth cycle (McGrath et al., 2002). Such plants are known as hyperaccumulators (Baker and Brooks, 1989). The plants that have these capabilities produce relatively low biomass and are not compatible with routine agricultural practices (Blaylock et al., 1997).

Composting has been proposed as a postharvest biomass treatment by some authors (Kumar et al., 1995; Salt et al., 1998; Raskin et al., 1997). Hetland et al. (2001) carried out laboratory experiments with lead-contaminated plant material (small sunflowers, grasses) obtained after induced phytoextraction. The disintegrated biomass (particles less than 0.16 cm in diameter) was composted in 125-ml borosilicate bottles with constant aeration for 2 months. Total dry weight loss was about 25%. Leaching tests of the composted material showed, however, that the composting process formed soluble organic compounds that enhanced lead solubility. These results documented that composting can significantly reduce the volume of harvested biomass; however, lead-contaminated plant biomass would still require treatment prior to disposal. Plant biomass harvested after induced phytoextraction can contain very mobile and leachable metal–chelate complexes. Moreover, Zhao et al. (2000) also showed that most of zinc within the leaves of hyperaccumulators is present also in water-soluble forms. This means that composting process should be conducted carefully in order to avoid nondesirable leachates regardless of the type of phytoextraction used. It is necessary to emphasize that the purpose of composting is to reduce the volume and weight of plant material, with no consideration to the agricultural properties of the final product. Total dry weight loss of contaminated plant biomass is an advantage of composting as pretreatment step. It will lower costs of transportation to a hazardous waste disposal facility and costs of deposition or costs of transportation to other facilities, where final crop disposal will take place. However, composting lasts from 2 to 3 months, extending time from harvesting to a final disposal. Furthermore, contaminated decomposed biomass should be treated as hazardous material. Further investigations are needed to assess the effect of the presence of chelating agents in harvested plant material, in conjunction with metals, on composting process. Furthermore, there is no information on the stability or transformation of metal–EDTA complexes during the process.

### 15.8.3. Plant Disease Suppression

Compost possesses an ability to suppress plant diseases when used as a soil amendment. The astonishing variety of microbes, many of which may be beneficial in controlling pathogens, is the main reason for disease suppression. Beneficial microbes help to control plant pathogens through either specific or general suppression. General suppression occurs when a beneficial microbe fills an ecological niche that would otherwise be exploited by a pathogen. For example, a beneficial organism may out-compete a pathogen for energy, nutrients, or "living space," thereby decreasing the survival of the pathogen. Specific suppression occurs when some of the metabolites secreted by beneficial organisms are toxic to a pathogen or when the pathogen is parasitized by specific organisms. Many plant pathogens contain cellulose or chitin (commonly found in insects and fungi) and all contain sugar polymers (commonly found in all life). Certain compost microorganisms, such as *Gliocladium*, *Pseudomonas*, *Trichoderma*, and *Streptomycetes*, produce enzymes capable of breaking these compounds down and killing the pathogens in the process.

These specific disease-fighting microorganisms are naturally available in compost or they are added after the thermophilic stage of composting process. Exposure to heat during this stage of composting is often responsible for killing not only pathogens but also those beneficial microorganisms that cannot tolerate the high temperature. Thus for compost to serve as a means for minimizing plant pathogens in the field, it must be re-colonized by beneficial microorganisms. When composts are not amended with beneficial microorganisms, reinoculation occurs naturally. However, some studies suggest that controlled inoculation of compost with known biocontrol agents (fungi and bacteria) is necessary for consistent levels of pathogen suppression in the field after application. This amended compost then can be applied to crops infected by known diseases. Research has shown that compost significantly reduced or replaced the application of pesticides, fungicides, and nematicides, which could adversely affect water resources, food safety, and worker safety.

Application of compost has shown successful control to different plant diseases and pests. The damping off of peas and cotton caused by *Pythium ultimum* and *Rhizoctonia solani* suppressive effect was sustained over 4-year period even though compost was added only during the first 2 years (Lewis et al., 1992). Compost addition increased disease suppressiveness of *Fusarium oxysporum* f. sp. *lini* in soil/compost mixtures. This is caused by both soil and compost microflora and mainly acted through nutrient and space competition toward the population of the pathogen. Compost showed a high ability to suppress *Fusarium* wilt caused by *Fusarium oxysporum* f. sp. *lycopersici* and *Fusarium oxysporum* f.sp. *basilici* as well (Serra-Wittling et al., 1996; Cotxarrera et al., 2002; Reuveni et al., 2002). Compost also is used in nematode management. Addition of compost reduced Columbia lance nematode (*Hoplolaimus Columbus*) and root-knot nematodes (*Meloidogyne javanica*) (Marull et al.,1997; Khalilian et al., 2002).

Addition of compost also has been reported to reduce wheat take-all (*Gaeumannomyces graminis*), root-rot of peas (*Phoma medicaginis*), red-core of strawberries (*Phytophthora fragariae*), club root of brassicas (*Plasmodiophora brassicae*), root infection of *Phytophthora nicotianae* in citrus seedlings, snow molds of turf grass caused by *Microdochium nivale* and *Typhula ishikariensis*, and *Rhizoctonia* large-patch disease in mascarene grass.

High resistance to different airborn diseases and pests induced by compost in substrate cultivation of strawberry was observed in a recent investigation (Bobev et al., 2004). Gray mold incidence (*Botrytis cinerea*) on mature strawberry and inhibition of fungal sporulation was strongly repressed when compost was used. The index of leaf damage by powdery mildew (*Sphaerothecia maculans*) was 30 times higher in the current than in the organic culture. Similar suppression of spider mites and aphids also was observed. Spider mites (*Tetranychus urticae*) and aphids (*Chaetosiphon fragefolii*), which become epidemic in the common culture, were controlled in the neighboring organic culture. A 15 and 18 time difference in population densities on the leaves were scored with both pests, respectively.

### 15.8.4. Impacts of Using Composted Products in the Food and Agricultural Context

Many growers think of compost primarily as a source of nutrients to add to the soil. However, its contribution of a diverse set of microorganisms combined with its high levels of organic matter may offer even more significant benefits. The application to soil of mature composted products may result in:

- Increased organic matter in soils, improved drainage in clay soils, and controlled soil erosion.
- Improved soil properties and associated plant growth; it builds sound root structure, reduces plant stress from drought and frosts, improves nutritional content of food grown in compost-rich soils.

- Reduced environmental impacts due to reduced soil erosion, waterlogging, nutrient loss, surface crusting, eutrophication of waterways, siltation of waterways, etc.
- Boosts soil microbial population.
- Improved water retention in sandy soils and reduces water demands of plants.
- Attracts and feeds earthworms.
- Balances soil pH (acidity/alkalinity).
- Reduces waste.
- Replaces application of chemical fertilizers, herbicides, and pesticides. This is very important for production of a quality agricultural product for food industry.

## 15.9. REFERENCES

Aladjadjiyan, A., 1992, Lessons from Denmark and Austria on the Energy Valorization of Biomass. Contract No: JOU2-CT92-0212, coordinator for Bulgaria.

Baker, A. J. M., and Brooks, R. R., 1989, Terrestrial higher plants which can hyperaccumulate metallic elements—a review of their distribution, ecology and phytochemistry, *Biorecovery* **1**:81.

Beck-Friis, B., Pell, M., Sonesson, U., Jonsson, H., and Kirchmann, H., 2000, Formation and emission of $N_2O$ and $CH_4$, from compost heaps of organic household waste, *Environ Monitoring Assessment* **62**:317.

Blaylock, M. J., Salt, D. E., Dushenkov, S., Zakharova, O., Gussman, C., Kapulnik, Y., Ensley, B. D., and Raskin, I., 1997, Enhanced accumulation of Pb in Indian mustard by soil-applied chelating agents, *Environ. Sci. Technol.* **31**:860.

Bobev S., Willekens, K., Goeminne, G., Desmet, T., Van Driessche, G., Claeyssens, M., Van Beeumen, J., Messens, E., and Maes, M., 2004, Resistance to air-born diseases and pests induced by compost in substrate cultivation of strawberry. 56$^{th}$ *International symposium of crop protection*

Brinton, W. F., Trankner, A., and Droffner, M., 1996, Investigations Into Liquid Compost Extracts, *BioCycle* **37**:68.

Chiumenti, R., and Chiumenti A., 2002, *Technology of composting. Agencia regionale per la prevencione e protezione ambientale del Veneto*, ISBN 954–8853–59–0, Sofia. (BG).

Cotxarrera, L., Trillas-Gay, M. I., Steinberg, C., and Alabouvette, C., 2002, Use of sewage sludge compost and Trichoderma asperellum isolates to suppress Fusarium wilt of tomato, *Soil Biol. Biochem.* **34**:467.

Dumontet, S., Dinel, H., and Baloda, S. B., 1999, Pathogen reduction in sewage sludge by composting and other biological treatments: a review, *Biol. Agric. & Hortic.* **16**:409.

Hetland, M. D., Gallagher, J. R., Daly, D. J., Hassett, D. J., and Heebink, L. V., 2001, Processing of plants used to phytoremediate lead-contaminated sites, in: *Phytoremediation, Wetlands, and Sediments, The Sixth International in situ and on-site Bioremediation Symposium* A. Leeson, E.A. Foote, M.K. Banks, V.S. Magar, eds., Battelle Press, Columbus, Richland, pp. 129–136.

Keener, H. M., Dick, W. A., and Hoitink, H. A. J., 2000, Composting and beneficial utilization of composted by-product materials, in: *Land application of agricultural and municipal by products*, J. F. Power, W. A. Dick, R. M. Kashmanian, J. T. Sims, R. J. Wright, M. D. Dawson, and D. Bezdicek, eds., Soil Science Society of America Book Series pp. 315–341.

Khalilian, A., Williamson, R. E., Sulivan, M. J., Mueller, J. D., and Wolak, F. J., 2002, Injected and broadcasted application of composted municipal solid waste in cotton, *Applied Engineering in Agriculture* **18**(1):17.

Kumar, N.P.B.A., Dushenkov, V., Motto, H., and Raskin I., 1995, Phytoextraction: the use of plants to remove heavy metals from soils, *Environ. Sci. Technol.* **29**:1232.

Lewis, J.A., Lumsden, R.D., Millner, P. D., and Keinath, A. P., 1992, Suppression of damping-off of peas and cotton in the field with composted sewage sludge, *Crop Protection* **11**:260.

Litchfield, J.H., 1987, Food Biotechnology, **1**(1):29.

Marull, J., Pinochet, J., and Rodríguez-Kábana, R., 1997, Agricultural and municipal compost residues for control of root-knot nematodes in tomato and pepper, *Comp. Sci. Util.* **5**(1):6.

Maynard, A. A., 2000, *Compost: the process and research*, The Connecticut agricultural experiment station Bulletin no 966, Luglio 2000, 14 pp.

McGrath, S.P., Zhao, F. J., and Lombi, E., 2002, Phytoremediation of metals, metalloids, and radionuclides, *Advances in Agronomy* **75**:1.

Raskin, I., Smith, R. D., and Salt, D. E., 1997, Phytoremediation of metals: using plants to remove pollutants from the environment, *Curr. Opin. Biotechnol.* **8**:221.

Reuveni R., Raviv, M., Krasnovsky, A., Freiman, L., Medina, S., Bar, A., and Orion, D., 2002, Compost induces protection against *Fusarium oxysporum* in sweet basil, *Crop Protection* **21**:583.

Salt, D.E., Smith, R. D., Raskin, I., 1998, Phytoremediation, *Annu. Rev. Plant Physiol. Plant Mol. Biol.* **49**:643.

Scaglia, B., Tambone, F., Genevini, P. L., Adani, F., 2000, Respiration index determination: dynamics and static approaches, *Compost Sci. & Util.* **8**:90.

Schaik, C., van, Murray, H., Lamb, J., and Di-Giacomo, J., 2000, Composting reduces fuel and labour costs on family farms, *Biocycle* **41**:72.

Sequi, P., 1996, The role of composting in sustainable agriculture, in: *The Science of Composting*, M. Bertoldi, P. Sequi, B. Lemmens, and T. Papi, eds., Blackie Academic & Professional, London, pp. 23–29.

Serra-Wittling C., Houot, S., and Alabouvette, C., 1996, Increased soil suppressiveness to Fusarium wilt of flax after addition of municipal solid waste compost, *Soil Biol. Biochem.* **28**(9):1207.

Shilev, S., Benlloch, M., Sancho, E., 2003, Utilization of rhizobacteria *Pseudomonas fluorescens* in phytoremediation strategies, pp 39. in: Vanek, T. and J.–P. Schwitzguebel (eds.), *Phytoremediation Inventory COST Action 837 view*, Published by VOCHB AVČR, Czech Republic.

Sommer, S.G., and Dahl, P., 1999, Nutrient and carbon balance during the composting of deep litter, *J. Agric. Eng. Res.* 7**4**:145.

Sonesson, U., Bjorklund, A., Carlsson, M., and Dalemo, M., 2000, Environmental and economic analysis of management systems for biodegradable wastes, *Resources, Conservation and Recycling* **28**:29.

Tuomela M., Vikman, M., Hatakka, A., and Itavaara, M., 2000, Biodegradation of lignin in a compost environment: a review, *Bioresource Technology* **72**:169.

Tuomela, M., 2002, *Degradation of Lignin and Other* $^{14}$*C-Labelled Compounds in Compost and Soil with an Emphasis on White-Rot Fungi*, Dissertation thesis, Dpt. Applied Chemistry and Microbiology, University of Helsinki, ISBN: 952–10–0311–1.

United States Environmental Protection Agency, EPA530-F-97-042, October 1997.

Wu L., Ma, L. Q., and Martinez, G. A., 2000, Comparison of methods for evaluating stability and maturity of biosolids compost, *J. Environ. Quality* **29**:424.

Zhao, F.J., Lombi, E., Breedon, T., and McGrath, S. P., 2000, Zinc hyperaccumulation and cellular distribution in Arabidopsis halleri, *Plant Cell Environ.* **23**:507.

# INDEX

**B**

SPRINGER NATURE

# GPSR Compliance

*The European Union's (EU) General Product Safety Regulation (GPSR) is a set of rules that requires consumer products to be safe and our obligations to ensure this.*

*If you have any concerns about our products, you can contact us on ProductSafety@springernature.com*

In case Publisher is established outside the EU, the EU authorized representative is:

Springer Nature Customer Service Center GmbH
Europaplatz 3
69115 Heidelberg, Germany

FSC
www.fsc.org
MIX
Papier | Fördert gute Waldnutzung
FSC® C083411

Kolibri 360 GmbH
Ferdinand-Jühlke-Straße 7
99095 Erfurt, Deutschland
produktsicherheit@kolibri360.de